IET TELECOMMUNICATIONS SERIES 89

Information and Communication Technologies for Humanitarian Services

Other volumes in this series:

Volume 9	**Phase Noise in Signal Sources** W.P. Robins
Volume 12	**Spread Spectrum in Communications** R. Skaug and J.F. Hjelmstad
Volume 13	**Advanced Signal Processing** D.J. Creasey (Editor)
Volume 19	**Telecommunications Traffic, Tariffs and Costs** R.E. Farr
Volume 20	**An Introduction to Satellite Communications** D.I. Dalgleish
Volume 26	**Common-Channel Signalling** R.J. Manterfield
Volume 28	**Very Small Aperture Terminals (VSATs)** J.L. Everett (Editor)
Volume 29	**ATM: The broadband telecommunications solution** L.G. Cuthbert and J.C. Sapanel
Volume 31	**Data Communications and Networks, 3rd Edition** R.L. Brewster (Editor)
Volume 32	**Analogue Optical Fibre Communications** B. Wilson, Z. Ghassemlooy and I.Z. Darwazeh (Editors)
Volume 33	**Modern Personal Radio Systems** R.C.V. Macario (Editor)
Volume 34	**Digital Broadcasting** P. Dambacher
Volume 35	**Principles of Performance Engineering for Telecommunication and Information Systems** M. Ghanbari, C.J. Hughes, M.C. Sinclair and J.P. Eade
Volume 36	**Telecommunication Networks, 2nd Edition** J.E. Flood (Editor)
Volume 37	**Optical Communication Receiver Design** S.B. Alexander
Volume 38	**Satellite Communication Systems, 3rd Edition** B.G. Evans (Editor)
Volume 40	**Spread Spectrum in Mobile Communication** O. Berg, T. Berg, J.F. Hjelmstad, S. Haavik and R. Skaug
Volume 41	**World Telecommunications Economics** J.J. Wheatley
Volume 43	**Telecommunications Signalling** R.J. Manterfield
Volume 44	**Digital Signal Filtering, Analysis and Restoration** J. Jan
Volume 45	**Radio Spectrum Management, 2nd Edition** D.J. Withers
Volume 46	**Intelligent Networks: Principles and applications** J.R. Anderson
Volume 47	**Local Access Network Technologies** P. France
Volume 48	**Telecommunications Quality of Service Management** A.P. Oodan (Editor)
Volume 49	**Standard Codecs: Image compression to advanced video coding** M. Ghanbari
Volume 50	**Telecommunications Regulation** J. Buckley
Volume 51	**Security for Mobility** C. Mitchell (Editor)
Volume 52	**Understanding Telecommunications Networks** A. Valdar
Volume 53	**Video Compression Systems: From first principles to concatenated codecs** A. Bock
Volume 54	**Standard Codecs: Image compression to advanced video coding, 3rd Edition** M. Ghanbari
Volume 59	**Dynamic Ad Hoc Networks** H. Rashvand and H. Chao (Editors)
Volume 60	**Understanding Telecommunications Business** A. Valdar and I. Morfett
Volume 65	**Advances in Body-Centric Wireless Communication: Applications and state-of-the- art** Q.H. Abbasi, M.U. Rehman, K. Qaraqe and A. Alomainy (Editors)
Volume 67	**Managing the Internet of Things: Architectures, theories and applications** J. Huang and K. Hua (Editors)
Volume 68	**Advanced Relay Technologies in Next Generation Wireless Communications** I. Krikidis and G. Zheng
Volume 69	**5G Wireless Technologies** A. Alexiou (Editor)
Volume 70	**Cloud and Fog Computing in 5G Mobile Networks** E. Markakis, G. Mastorakis, C.X. Mavromoustakis and E. Pallis (Editors)
Volume 71	**Understanding Telecommunications Networks, 2nd Edition** A. Valdar
Volume 72	**Introduction to Digital Wireless Communications** H.-C. Yang
Volume 73	**Network as a Service for Next Generation Internet** Q. Duan and S. Wang (Editors)
Volume 74	**Access, Fronthaul and Backhaul Networks for 5G and Beyond** M.A. Imran, S.A.R. Zaidi and M.Z. Shakir (Editors)
Volume 76	**Trusted Communications with Physical Layer Security for 5G and Beyond** T.Q. Duong, X. Zhou and H.V. Poor (Editors)

Volume 77 **Network Design, Modelling and Performance Evaluation** Q. Vien
Volume 78 **Principles and Applications of Free Space Optical Communications** A.K. Majumdar, Z. Ghassemlooy and A.A.B. Raj (Editors)
Volume 79 **Satellite Communications in the 5G Era** S.K. Sharma, S. Chatzinotas and D. Arapoglou
Volume 80 **Transceiver and System Design for Digital Communications, 5th Edition** S.R. Bullock
Volume 81 **Applications of Machine Learning in Wireless Communications** R. He and Z. Ding (Editors)
Volume 83 **Microstrip and Printed Antenna Design, 3rd Edition** R. Bancroft
Volume 84 **Low Electromagnetic Emission Wireless Network Technologies: 5G and beyond** M.A. Imran, F. Héliot and Y.A. Sambo (Editors)
Volume 86 **Advances in Communications Satellite Systems Proceedings of the 36th International Communications Satellite Systems Conference (ICSSC-2018)** I. Otung, T. Butash and P. Garland (Editors)
Volume 95 **ISDN Applications in Education and Training** R. Mason and P.D. Bacsich

Information and Communication Technologies for Humanitarian Services

Edited by
Muhammad Nazrul Islam

The Institution of Engineering and Technology

Published by The Institution of Engineering and Technology, London, United Kingdom

The Institution of Engineering and Technology is registered as a Charity in England & Wales (no. 211014) and Scotland (no. SC038698).

First published 2020

The Institution of Engineering and Technology
Michael Faraday House
Six Hills Way, Stevenage
Herts, SG1 2AY, United Kingdom

www.theiet.org

British Library Cataloguing in Publication Data
A catalogue record for this product is available from the British Library

ISBN 978-1-78561-996-0 (hardback)
ISBN 978-1-78561-997-7 (PDF)

Typeset in India by MPS Limited
Printed in the UK by CPI Group (UK) Ltd, Croydon

Contents

About the editor

Muhammad Nazrul Islam is an associate professor at the Department of Computer Science and Engineering in the Military Institute of Science and Technology (MIST), Dhaka, Bangladesh, and also serving in Bangladesh Army in the corps of signals in the rank of lieutenant colonel. He was awarded a Ph.D. in Information Systems from Åbo Akademi University (Finland) in 2014, an M.Sc. in Computer Engineering from Politecnico di Milano (Italy) in 2007, and a B.Sc. in Computer Science and Information Technology from Islamic University of Technology (Bangladesh) in 2002. Before joining MIST, he was working as a visiting teaching fellow at Uppsala University, Sweden, and as a post-doctoral research fellow at Åbo Akademi University, Finland. He was also a lecturer and assistant professor at the Department of Computer Science and Engineering at the Khulna University of Engineering and Technology (KUET), Bangladesh, during 2003–2012. Dr. Islam served as an IT and Communication officer of the United Nations Multidimensional Integrated Stabilization Mission in Mali (MINUSMA) for one year during 2018–2019. His research areas include, but not limited to, human–computer interaction, humanitarian technology, health informatics, military information systems, information systems usability, and computer semiotics. Dr. Islam is a frequent reviewer of journals, book chapters, and international conferences. He is the associate editor of BMC Research Notes and MIST International Journal of Science and Technology. He has published over 100 peer-reviewed articles in reputed international journals and conferences, besides two books, and was the Principal Investigator of five research projects from the Government and the Industry. He has received a number of prestigious scholarships, grants, and awards in recognition of his research contribution. Dr. Islam has also received several *best paper* awards from international conferences, and the *best faculty in research work* award in 2016, 2017, and 2019 at MIST. He is a member of the IEEE and the IEB (Engineering Institute of Bangladesh) and a TPC member of over 15 international conferences.

Foreword

1.1 About the book

This book provides new developments, innovations, and research outcomes; case studies and lessons learned; and other considerations for the creation and deployment of effective ICTs to provide humanitarian services for the resource-constrained and vulnerable populations in the world in order to improve their lives. It contains contributions from researchers, professionals in humanitarian assistance, postgraduate students, and from academia and industry, with varied backgrounds working in the area of ICTs and humanitarian services. The chapters are particularly designed as a series of independent modules, whereas each chapter explores some aspect of humanitarian services through ICTs. Topics of this book include connectivity and communications technologies for humanitarian services; ICTs in disaster mitigation, relief, and recovery; humanized and inclusive education; technologies for the women, disabled, and aged populations; IoT, big data, and blockchain for humanitarian engineering; adopting and adapting ICTs in humanitarian sectors; and other technological advancements for humanitarian assistance.

1.2 Intended audience

This book is specially designed for undergraduate and postgraduate students with different disciplinary backgrounds such as computer science, ICT, software engineering, social science, and many others; and for the engineers, scientists, professionals, and researchers involved in the design, development, testing, deployment, operation, and feedback assessment of humanitarian ICT systems with the aim of drawing special attention to the societal expectation on humanitarian technologies' abilities to provide practical and lasting solutions. The book is also suitable for the policymakers, entrepreneurs, professionals, and researchers from across government, non-government, and industry organizations with the aim of drawing special attention to ICTs for humanitarian services. The book could be a good resource for some specific professional associations and societies, including the IEEE Humanitarian Activities Committee (HAC), IEEE Special Interest Group on Humanitarian Technology (SIGHT), Information Technology for Humanitarian Assistance, Cooperation and Action (ITHACA), The International Humanitarian Studies Association (IHSA), Association of Information Systems (AIS), United Nations and The United Nations Office for the Coordination of Humanitarian Affairs (OCHA), UNICEF, Cooperative for Assistance and Relief Everywhere (CARE), International Federation of Red Cross and Red Crescent Societies (IFRC), Action Against Hunger (AAH), and The Institution of Engineering and Technology (IET).

Chapter 1
Introduction

Muhammad Nazrul Islam[1]

1.1 Motivation for this book

Humanitarian services are generally concerned with or seeking to promote human welfare. Their primary purpose is to save lives, maintain human dignity, alleviate sufferings, prevent and strengthen preparedness, and provide material or logistic assistance in response to humanitarian crises including man-made crises and disasters caused by natural hazards. The humanitarian service is thus different from the development aid, which mainly seeks to address the underlying socio-economic factors and provides support for the social, economic, and political development of developing nations over the long term. Humanitarian services influence the population around the world and, in turn, lead to making a better world.

Nowadays, information and communication technologies (ICTs) are becoming the backbone to provide quality and efficient services, and are playing an increasingly important and more sophisticated role in humanitarian-service activities involving education, logistics, disaster management, organizational learning, health systems, and the like [1]. The ICT solutions are not only limited to providing services or aid during the crisis, rather also focused to capacity building, empowerment, and assessment; to establish socio-technical strategies; and to increase emphasis on demonstrating effectiveness, improving efficiency, and collaborating with other organizations [1]. ICT applications are developed to access by the humanitarian practitioners for the people in need, or by people in need of essential services or goods [2].

ICTs are changing not only the world but also the humanitarian world by providing direct aid or services to support the work of humanitarian organizations; development of electronic and mobile solutions to provide health services; connectivity and communications technologies for humanitarian service; disaster mitigation, relief, and recovery; digital solutions for sustainable educational programmes related to humanity; agricultural improvement; refugee identity, management, safety and security; disabled and aged population; social media for

[1]Department of Computer Science and Engineering, Military Institute of Science and Technology, Dhaka, Bangladesh

community engagement and building resilience; development of open source web portals for disaster management systems; and the development of humanitarian robotics and drones to provide assistance to people in humanitarian crises and with special need. The possible scopes of ICT-based innovation, research, and development for humanitarian services; technologies used for any ICT-based solutions; and the key qualities that are required to maintain in any ICT-based humanitarian systems are presented in Figure 1.1. Among these, a few ICT-based solutions are briefly introduced next to provide a brief idea about the recent technological innovations for providing humanitarian services.

During any major disaster, ICT-based systems like disaster management system or humanitarian information system facilitate to provide comprehensive, relevant, and reliable information to the responding humanitarian community [3]. On the other hand, the crisis-affected population may also use the ICT (mobile or interactive radio) system to communicate and share their needs to the humanitarian practitioners [4]. Again, robotics and unmanned aerial vehicles (UAVs) are the promising technologies and have a positive impact in providing humanitarian services. For example, the primary uses of the UAVs/drones during disaster situation include taking images from the disaster zones and analysing those to take optimum decisions, mapping the affected area and preparing the route plan to reach the disaster zone, measuring the level of disaster using (chemical) sensor, transmitting the quality information about disaster area, assisting to conduct medical surgery, and coordinating with UAV's network [5]. Tele-operated search and rescue robots that can navigate deep into the rubble to search for victims and to transfer critical field data back to the control console have gained much interest among emergency response institutions. Similarly, robots are used for providing humanitarian services in emergency search and rescue operations like collapsed buildings due to typhoons, earthquakes, weaponry destructions, tornados, and catastrophic explosions where heavy equipment cannot get close enough and thus tele-operated

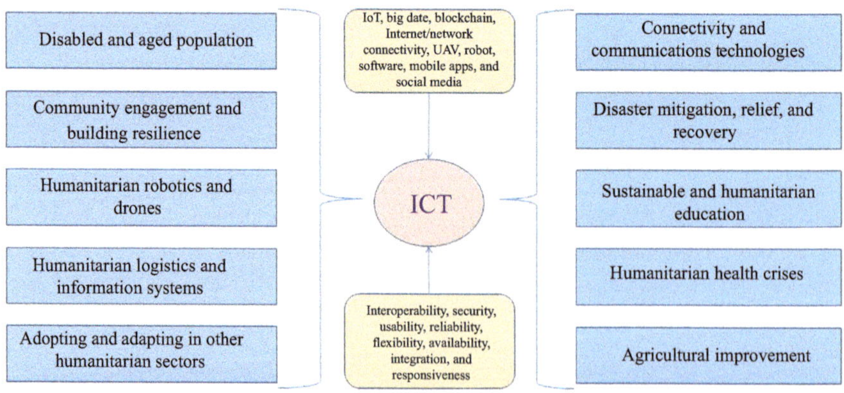

Figure 1.1 Scopes of ICT-based innovation, research, and development for humanitarian services

search and rescue operations are being conducted by the robots [6]. A number of ICT-based assistive tools are developed for the disabled and aged population to survive and fulfil their daily basic needs with less effort and high efficiency – for example, the development of a wearable system to assist elderly people for path finding in indoor environment [7]; Internet of things (IoT)-based wheelchair to assist the people with congenital disabilities and mobility impairments [8]; and a wearable system for the blind people for moving around easier and safer without taking help of others [9]. In another work [10], a design framework is proposed for the humanitarian agencies for adopting the humanitarian principled, namely humanity, neutrality, impartiality, and independence approach to develop humanitarian ICTs. Few other innovations focus to provide health services – for example, developing a web-based system to distribute unused medicine to the poverty-stricken people who need medicine but are unable to buy the prescribed medicine [11]; the development of an intelligent system to provide support for Alzheimer's patients for their survival without others assistant [12]; and the development of humanitarian computing systems using artificial intelligence and social media to aid rapid decision-making during humanitarian health crises [13].

Despite these number of research and innovations, as the world continuously faces humanitarian disasters, there is an ever-increasing need to find ICT-based solutions to overcome complex problems. In other words, though ICTs have changed the lives of people at the operative level, many ICTs' contributions are yet to be uncovered in humanitarian service. Therefore, this book provides new developments, innovations, and research outcomes; case studies and lessons learnt; and other considerations for the creation and deployment of effective ICTs to provide humanitarian services for the resource-constrained and vulnerable populations in the world in order to improve their lives.

1.2 Overview of chapters

The book includes a set of contributing chapters that follow with significant background on the role of ICT in the context of humanitarian-service activities. The primary key to appreciating these chapters is an awareness of current ICT trends in the humanitarian sector. The chapters provide insight into and analysis of various aspects of the humanitarian services with a particular focus on the role of ICTs. The overviews of the remaining chapters included in this book are provided next.

Chapter 2 highlights how the digital engagement (through mobile or other platforms) can significantly impact women and girls in humanitarian settings through freedom of expression and communication, better financial inclusion, more informed decision-making, and improved self-esteem. The chapter provides keen consideration to the digital disparity between men and women in terms of digital engagement in humanitarian settings and offers recommendations to narrow the gender gap through meaningful woman- and girl-centred design of technological solutions.

Chapter 3 presents the analyses of the profound technologies in our contemporary society and shows that profound technologies can contribute to human development based on humanization and inclusion of diversity and supported by reflection on their use and implementation in society, culture, and education.

Chapter 4 presents a dossier of concepts and applications of information, communication, and consciousness science going beyond conventional human and materialistic endeavours of the present-day society. This chapter entails and invokes a systemic holistic perspective towards resolving many issues and challenges that the global society is currently up fronted with.

Chapter 5 intends to study the humanitarian logistics dimension of disaster management so that the vulnerability of the affected area can be reduced. The chapter briefly explains the nature and intensity of the disaster, the role of different stakeholders, and the use of ICT in the disaster cases that happened in India. Based on the review of the literature and the case analysis, a sustainable knowledge framework is proposed for disaster management in India.

Chapter 6 addresses the optimization of the early assessment of the disaster amplitude and features based on data collected through social networks for improving relief planning and for increasing resilience to disasters.

Chapter 7 discusses first the primary building blocks of blockchain technology and then illustrates various use-case scenarios of blockchain technology for humanitarian engineering considering the fields of agriculture, energy, health, finance, education, environment and climate change, and e-governance. A cost–benefit analysis of adopting blockchain technology for humanitarian engineering is also provided in this chapter.

Chapter 8 highlights the potential of big data for disaster management through increasing resilience against socio-ecological vulnerability. The authors emphasize the major principles of big data for their effective use in disaster management like open source tools, strong infrastructure, developing local skills, context-specific data sources, data sharing with ethics, awareness about the right of data, and learning from experience. The chapter also argues that big data is a potential tool for policymakers, administrators, and related stakeholders to take necessary actions during and after disasters like an early warning system, weather forecasting, emergency evacuation, immediate responses, and relief distribution.

Chapter 9 discusses the design and development of a rapid damage assessment system for public elementary school buildings of Albay, Philippines. The chapter aims first to identify the current procedures in conducting rapid damage assessment in public elementary school buildings of Albay; second, to determine the design features of the proposed rapid damage assessment system; and finally, to generate appropriate report formats to represent the rapid damage assessment information. The chapter also presents the findings of the evaluation study that showed the proposed damage assessment system is effective, reliable, useful, and efficient to the end users.

Chapter 10 discusses the technologies for emergency roll-out of broadband public protection and disaster relief (BB-PPDR) communications in humanitarian crisis zones, and proposes a cost-effective and energy-efficient solution for BB-PPDR communication. This chapter also discusses the benefits of the proposed solution for BB-PPDR through an experiment in terms of error vector magnitude, adjacent channel power ratio, and power efficiency.

Chapter 11 proposes a network model by combining delay-tolerant network and content delivery network using TCP/IP (Transmission Control Protocol/ Internet Protocol) and OpenFlow, and with the help of wireless connectivity-enabled devices. The chapter also highlights that the proposed model will facilitate interconnecting the rural villages, which are detached from modern technologies, and give them autonomous solutions so that the village people do not require any costly Internet connection or infrastructure to connect within themselves and with the global Internet. Finally, the author suggests implementing the proposed network model in third-world countries where sometimes technological amenities are scarce.

Chapter 12 proposes a framework for developing an intelligent technology-based adaptive solution that can foster the biological, psychological, behavioural, physical, mental, and emotional well-being and contribute towards active ageing of elderly. The chapter also highlights that the proposed framework is envisioned to pave the way for the future technological intervention in healthcare and social welfare to enhance user experiences and foster elderly care while addressing the varying diversities of the elderly population.

Chapter 13 discusses the design, development, and evaluation of an arcade mobile game (BrickBlaster) for the blind people in Bangladesh for making the mobile fun accessible to everyone. The chapter also discusses the procedure and findings of evaluating the BrickBlaster. The findings of an evaluation study which is replicated with 24 blind people showed that the game is usable and playable by the blind people. The findings of the evaluation study suggest that completion of each level had reasonable difficulty with respect to completion time and the number of attempts. As expected, the difficulty increased as the gamers moved to the next level and no significant differences were observed when analysing the impact of demographics (i.e. gender and high school vs. college) of the players on the playability of the game in terms of *success rate*, *the number of attempts*, *level completion time*, and *player attitude* measures.

Chapter 14 explores the challenges and opportunities of adopting necessary technological innovations in the humanitarian sector in Nigeria. The chapter also proposes a framework based on the features of the innovative technology and the theory of diffusion of innovations for guiding the humanitarian ICT adoption in Nigeria.

Chapter 15 highlights the nature of humanitarian issues affecting the Southern African region and how best adoption and adaption of ICTs can close the gaps of these issues for maximizing the efficiency of the humanitarian sector in mitigation, management, and recovery efforts. The chapter further explores the challenges faced by the Southern African regional countries that are affecting the adoption and adaption of ICTs in the humanitarian sector. Recommendations for adapting and adapting ICTs for the region by the policymakers are also discussed.

Chapter 16 provides concluding remarks with a brief summary of some of the key issues covered in this book, and the ways forward to further research and technological innovations for humanitarian services.

References

[1] Haselkorn M, and Walton R. The role of information and communication in the context of humanitarian service. IEEE Transactions on Professional Communication. 2009;52(4):325–328.

[2] Belliveau J. Humanitarian access and technology: opportunities and applications. Procedia Engineering. 2016;159:300–306.

[3] Van de Walle B, Van Den Eede G, and Muhren W. Humanitarian information management and systems. In: International Workshop on Mobile Information Technology for Emergency Response. Berlin, Heidelberg: Springer; 2008. pp. 12–21.

[4] Chapelier C, and Shah A. Improving Communication between Humanitarian Aid Agencies and Crisis-affected People: Lessons from the Infoasaid Project. Overseas Development Institute, UK; 2013.

[5] Bravo R, and Leiras A. Literature review of the application of UAVs in humanitarian relief. In Proceedings of the XXXV Encontro Nacional de Engenharia de Producao 2015 Oct 13–16. Fortaleza, Brazil.

[6] Ko AW, and Lau HY. Intelligent robot-assisted humanitarian search and rescue system. International Journal of Advanced Robotic Systems. 2009;6(2):12.

[7] Jahan MR, Aziz FI, Ema MB, Islam AB, and Islam MN. A wearable system for path finding to assist elderly people in an indoor environment. In Proceedings of the XX International Conference on Human Computer Interaction 2019 Jun 25 (pp. 1–7).

[8] Hossain T, Sabbir MS, Mariam A, *et al.* Towards developing an intelligent wheelchair for people with congenital disabilities and mobility impairment. In 2019 1st International Conference on Advances in Science, Engineering and Robotics Technology (ICASERT) 2019 May 3 (pp. 1–7). IEEE.

[9] Khan NS, Kundu S, Al Ahsan S, Sarker M, and Islam MN. An assistive system of walking for visually impaired. In 2018 International Conference on Computer, Communication, Chemical, Material and Electronic Engineering (IC4ME2) 2018 Feb 8 (pp. 1–4). IEEE.

[10] Cardia IV, Holzer A, Xu Y, Maitland C, and Gillet D. Towards a principled approach to humanitarian information and communication technology. In Proceedings of the Ninth International Conference on Information and Communication Technologies and Development 2017 Nov 16 (pp. 1–5).

[11] Islam MN, Zavin A, Srabanti S, *et al.* GiveMed: a webportal for medicine distribution among poverty-stricken people. In 2017 IEEE Region 10 Humanitarian Technology Conference (R10-HTC) 2017 Dec 21 (pp. 294–299). IEEE.

[12] Omar KS, Anjum A, Oannahary T, *et al.* An intelligent assistive tool for Alzheimer's patient. In 2019 1st International Conference on Advances in Science, Engineering and Robotics Technology (ICASERT) 2019 May 3 (pp. 1–6). IEEE.

[13] Fernandez-Luque L, and Imran M. Humanitarian health computing using artificial intelligence and social media: a narrative literature review. International Journal of Medical Informatics. 2018;114:136–142.

Chapter 2

Where are the women? How to design information and communication technology to be inclusive of women and girls in humanitarian settings

Kristy Crabtree[1]

Digital engagement (through mobile or other platforms) can significantly impact women and girls in humanitarian settings through freedom of expression and communication, better financial inclusion, more informed decision-making, and improved self-esteem. Yet, the existing gender gap in low- and middle-income countries is worsened in humanitarian settings where gender inequalities are heightened. Owing to systems of patriarchy that have practical consequences on women and girls' mobile or other technological solutions' access and use, women are falling behind. To correct this disparity, programmers must understand the state of access in their setting, real and perceived safety risks to women and girls' engagement, and choose meaningful woman- and girl-centered design. Therefore, the objective of this chapter is to call attention to the digital disparity between men and women in terms of digital engagement in humanitarian settings and offer recommendations to narrow the gender gap through meaningful woman- and girl-centered design of technological solutions.

2.1 Introduction

Forced displacement has effects on safety, health, power, livelihoods, and education of the women, girls, men, and boys living in a state of humanitarian crisis. Information and communication technologies (ICTs) can aid in addressing these consequences by facilitating the dissemination of information, collection of data, management of health, reporting of incidents, access to education, and poverty reduction.

These electronic tools, including communication channels and mobile or digital technology, provide valuable technological solutions. Convenient, adaptable and inexpensive, ICTs are especially prized in humanitarian settings, where

[1]Violence Prevention and Response Unit, International Rescue Committee, Kristy Crabtree, MS Seattle, USA

resources are limited, connectivity is unreliable, and scalable solutions are needed. There's a reason for their rise in popularity. The growth of ICTs can be attributed to multiple drivers: increasing access to the Internet [1], higher mobile device ownership [2], and growing models for use.

Another reason for their prevalence is the applicability of ICTs across humanitarian sectors. ICTs have been deployed for agriculture and crop information, to connect women with safe ways to receive mobile money in markets, to digitize health records, to bring libraries to rural areas, to provide automated pre-natal guidance, and to deliver micro credits, among other examples. These tools can also be used to impact social norms and behavior change while providing a strategic opportunity to scale up projects, reaching significant population groups.

While digital technology brings opportunities and the use of ICTs trends upward, it does not translate into equitable access for all displaced population groups. More than half the globe remains unconnected. Primarily, these billions—female and rural populations—are cut off from participation and the associated opportunities. They are left without the information, leverage, and power that come from being connected online.

There is a disparity in access to and usability of ICTs for displaced women and girls because they are disproportionately affected by gender inequalities heightened during humanitarian crises [3]. This inequality is the result of societal constraints on "approved" activities for women and girls, the associated threat of violence for deviation from expected norms, and restrictions on movement and participation in public life. Social norms, whether conscious or not, dictate what is acceptable for women to do that will maintain the patriarchal status quo. Thus, in these settings where resources, access to services, and opportunities are already limited, women and girls face additional barriers to participation.

The disparity between men and women is reflected in a number of ways in humanitarian settings, including obstacles to digital engagement. The divergence for females from the same levels of access as males is problematic as digital technology gains in popularity and use, as well as employment requirements—thus, widening gender gaps in service access, livelihood opportunities, and technological literacy. The impact is broad, and the lack of access for women and girls influences all areas of their lives—from communication to financial inclusion to health management and decision-making.

So, the question arises—How can we better understand the barriers to active digital participation for women and girls? What can we do to increase their digital engagement with communication or networking platforms, educational programs, and other technological opportunities for empowerment?

For women and girls, there are benefits to the utilization of mobile technology: increased self-esteem, heightened access to information, and lessening of the digital gender divide. However, there are also accompanying challenges. Women and girls in humanitarian settings face additional gender-unique barriers to access and use, including safety risks, such as gender-based violence, when utilizing digital technology. With this lens, it should be asked again, how can the gender gap be intentionally and meaningfully narrowed with the voices and inputs of women and girls globally?

2.2 Know the starting point for women and girls

Effective inclusion of women and girls requires meaningful design. This means seeking out and listening to women and girls. It means understanding their socio-cultural environment, localized concerns or safety risks, and preferred technology modalities and platforms. It means that human-centered design is not enough. It should instead be specifically woman- and girl-centered and follow principles that promote safety and respect and understand the systemic impact of patriarchal systems so as not to further obstruct women's engagement.

While there have been positive trends in women's phone ownership and Internet use, there remain gendered barriers. Knowledge of women and girl's starting point is essential. According to the Global System for Mobile Communications Association (GSMA) Mobile Gender Gap Report 2019, "women in low- and middle-income countries are 10 per cent less likely than men to own a mobile device, which translates into 197 million fewer women than men owning a mobile phone" [4]. The gap increases when considering mobile Internet access with "313 million fewer women than men [using] mobile internet, representing a gender gap of 23 per cent" [4] Women are also more likely to use a smaller set of mobile services and programs as compared to men [4].

From the outset, this inequality in access must be a primary consideration in order to make a meaningful, sustainable impact with ICTs. The reality of gender disparities and experiences must inform and be a central consideration in the design process. ICT project planning and management may fail without understanding existing access patterns and gaps for women and girls.

The trends above reflect the global gender discrepancies in phone ownership and Internet access for women in low- and middle-income countries; however, in humanitarian situations there are additional hurdles. In general, refugees are half as likely to have mobile Internet as compared to the general population, and nearly 30 percent of households have no mobile phone at all [5].

> This is driven by refugee-specific difficulties in affording a device or data plan, a problem caused by factors such as the state of their displacement or by government restrictions on their right to work or move freely from place to place. Owing to a lack of income, refugee households often share phones within the family unit, as well as between families (especially in rural camps). [5]

For both men and women in these displaced settings, device affordability is ranked as the most significant barrier to Internet use, followed by poor literacy, plan affordability, network strength, and a lack of understanding phone plan options [5].

2.3 Systemic barriers to inclusive access

These barriers do not bode well for displaced women and girls' inclusivity in digital technology; seemingly accessible or free to all populations should not be disguised as equal/equitable to all populations. Disparaged in access by their gender and also by displacement, women and girls' struggle for equity puts them at a further disadvantage. They must overcome obstacles in being uprooted as well as insular obstacles, the result of being female.

The digital gender gap can be attributed to two primary issues: (1) patriarchy and the resulting risk of gendered violence and (2) practical barriers to access. Both of these adversities are connected to gender inequality levels. The choice and barriers to using digital technology are influenced by the lived experiences of women. A foundational understanding of the underlying normative ideals and values that create gender disparities is required for inclusive design. Primarily, these "patriarchal systems shape social expectations in both functional and ideological terms to maintain male superiority over women" [6]. These translate into practical barriers as well as restrictive norms affecting practice. These norms have historically affirmed "unequal power relations ... which force women into a subordinate position compared to men" [7].

Put in another way, Women's Protection and Empowerment Technical Advisor at International Rescue Committee Patty Gray says that one manifestation of patriarchy is violence against women. This is rooted in gender inequalities and connects patriarchal systems of oppression with real risks to women's safety [6]. This is crucial to understanding why there are barriers to accessing digital technology. Women must go through safety planning when choosing to either engage or not to engage with technology. Their concern should inform programming and humanitarian interventions.

Access and use of ICTs by women and girls means shifting the power and control of a tangible resource (such as a mobile phone) plus intangible resources (such as knowledge and social networks) away from the status quo—men—to women. It can be a powerful organizing tool and let women know they are together in their thinking around anti-oppression and feminism. Yet, it is because of this shift in power through technology that safety and access must be addressed. Not because women and girls are vulnerable, but because they threaten male power when they have access to such resources. The potential to build social networks and create apps (for example) shakes patriarchy and its roots in racism, inequality, and sexism.

In humanitarian settings, the terminology used most often is gender-based violence (GBV):

> an umbrella term for any harmful act that is perpetrated against a person's will and that is based on socially ascribed (i.e. gender) differences between males and females. It includes acts that inflict physical, sexual or mental harm or suffering, threats of such acts, coercion, and other deprivations of liberty. [8]

Humanitarian crises often intensify risks of GBV, as the crisis creates further imbalance to already-unequal gender norms. This exacerbation can be linked to "increased militarization, lack of community and State protections, displacement, scarcity of essential resources, disruption of community services, changing cultural and gender norms, disrupted relationships and weakened infrastructure" [8].

Incidents of gender-based violence are historically known to be underreported, but it is estimated that one in five refugee women have experienced sexual violence [9]. While GBV can manifest as physical or sexual violence in its most well-known forms, women and girls also experience violence in the form of disadvantages and barriers related to "social power and influence, control of resources, control of their bodies and participation in public life—all as a result of socially determined gender roles and relations" [8]. This includes mental, emotional or psychological harm.

Research shows that there is evidence of changing gender norms in response to humanitarian crises. Disruption caused by crisis has social and economic impacts and alters living conditions where social norms are typically managed, enforced, or created. Some research notes "a tendency for women to take on more and different roles as providers and protectors of families, to draw confidence and determination from these experiences, and to develop their political consciousness and agency. On the other hand, men often find themselves as a loose end, unable to reestablish their position as respected decision makers" [10]. Where men more traditionally held a role as head of household prior to being uprooted, women at times found freedom from social pressures combined with new drivers for economic or social activity. This can place additional risk or burden on female populations, but it can also be freeing. However, this mobility can create tension as gender norms, roles, and relationships change in humanitarian crisis, increasing the incidence of intimate partner violence.

The question about women's digital engagement operates in this context of patriarchal systems, changing gender norms, and the social and economic constraints and drivers that occur during displacement. For digital engagement, barriers for women and girls could include lack of financial means, sociocultural expectations, perceptions of self-value, and risks of violence. For example, in looking at barriers related to mobile device affordability, women can be impacted in their ability to earn by a lack of acceptance of participation in public life (this can also include restricted movement), access to fewer resources, and lack of control over resources. These are gender-unique barriers that limit women's ability to achieve social, political, and economic equality, which then affects women's digital engagement—this being a microcosm of broader gender inequality issues. With restrictions on movement, livelihoods opportunities are limited or nonexistent, making the initial barrier or affordability for mobile devices or mobile Internet insurmountable without intervention.

Another driver of restricted digital access is social approval. If a girl were to obtain a mobile phone and have literacy levels enough to use it, she may not see the relevance or may be restricted from use because of negative perceptions about women and girls' use of digital technology. If it is assumed that women and girls are using mobile technology for romantic relationships, the male figures in their community may restrict use for perceived moral purposes.

2.4 Sample studies on inclusion: Lebanon and Uganda

Acknowledging global patriarchal systems impacting women and girls' digital engagement, two contexts are worth further examination considering the digital gender disparity in humanitarian settings: Lebanon and Uganda*.

In the eight years since the start of the Syrian civil war, Lebanon has hosted 950,000 Syrian refugees, the largest number of refugees per capita [11]—90 percent of which are residing primarily in urban areas [12]. Lebanon has high rates of individual Internet use, at 76 percent [1] as of 2017, and high rates of 3G access as well [5].

While Uganda hosts a similar-size displaced population with 1.2 million refugees and asylum seekers, the population is largely residing in rural refugee settlements [12]. Uganda has significantly lower rates of individual Internet use, at 22 percent in 2017 [1], with most populations limited to 2G coverage at best [5].

In 2017 and 2019, surveys conducted in areas of displacement in Lebanon and Uganda, respectively, aimed at determining levels of access and barriers to mobile device and Internet use for forcibly displaced women and girls. In Lebanon, 257 women and girls were surveyed, and in Uganda 152 females and males responded to the survey. In both Lebanon and Uganda, the survey results revealed trends in gender disparity looking at use, ownership, possession, monitoring, and gendered barriers.

2.4.1 Ownership and possession

For the purpose of comparison, the surveys in both countries asked about the general population's engagement with digital technology. In both settings, perception of women and girls' use of phones was less than that of men's. From a high-level perspective, women and girls are already perceived to use mobile devices less, which may be related to accepted gender norms around use.

Rates of mobile phone usage and smartphone usage were similar in Lebanon at 77 and 74 percent. In Uganda, the rates were 51 percent for mobile phones and just 5 percent for smartphones. In Lebanon, breaking down possession, mobile phones for respondents age 15–17 were owned only 17 percent of the time; for age 18–24, it was 53 percent; and for those 25 and older, phone ownership increased to 57 percent.

Trends in mobile device ownership reflect even more barriers. Only 31 percent of women reported owning a mobile phone in Lebanon. Most often shared or borrowed phones (26 percent) are owned by parents or intimate partners, which could have an impact on the types of services provided/offered and safe communication methods. Phone ownership by women and girls was reported at 44 percent in Uganda. However, in Uganda men rated their ownership 9 percent higher than women's.

*The survey in Lebanon was conducted by International Rescue Committee, ABAAD, Caritas, Danish Refugee Council, and INTERSOS. The survey in Uganda was conducted by International Rescue Committee.

There is a significant difference in how often phones are used. For women in Uganda, the rate of usage every day or almost every day is 39 percent and for men the rate is 58 percent. In Lebanon, 70 percent of respondents reported using their phones every day. However, in Lebanon enumerators expressed doubt about these self-reported statistics sharing, "What doesn't make sense is that most of the women and girls say that they use it every day. However, lots of people are ashamed to share that they do not use the phone everyday as their husbands don't allow them to use it while a seven-year-old boy can use the phone freely."

Other supplemental activities are required for digital engagement, including the purchase of credit and phone charging. In Lebanon, 44 percent of women and girls purchased credit themselves and 24 percent had to ask for credit and share it with someone else. For phone charging 30 percent were unable to perform this task easily. Similarly, in Uganda, 48 percent were able to purchase credit on their own and 19 percent found charging their mobile devices difficult. These two activities could be related to affordability and locations of charging stations.

Even if phones are owned or able to be possessed temporarily, phone use for women and girls is often monitored and subject to approval. Phone monitoring takes the form of looking at the screen during use or reviewing chat and phone logs. For female respondents in Lebanon age 15–17, 33 percent had their phone monitored and 30 percent had to get approval to perform tasks on their phone. For respondents age 18–24, the rate drops slightly to 22 percent monitored, and 31 percent needing approval. For respondents age 25 and older, 19 percent had their phone monitored, and 11 percent needed approval. This has an impact on the ability of women and girls to seek safe service provision and on communication opportunities for potential survivors of family or in-home violence. In Uganda, 15 percent of women had their phone monitored and 18 percent had someone else decide what they could or could not do on their phone.

2.4.2 Barriers to use

When ranked, women in low- and middle-income countries (LMICs) recognized the common barriers to mobile ownership in order from most common as affordability, literacy and skills, safety and security, and relevance [4]. They ranked the barriers to mobile Internet use as literacy and skills, affordability, relevance, and safety and security.

Many of these same barriers are present in humanitarian crises, but there are additional significant barriers for displaced women and girls: disapproval. Restricting beliefs about gender relations and norms exacerbate the barriers women and girls face in accessing ICTs. The hurdle for use is simply higher for displaced women and girls.

In Lebanon, the top barriers to mobile Internet for the general population (male and female inclusive) were perceived as cost of credit and cost of phone. For women and girls overall, they ranked the barriers as prohibition by family, followed by cost of phones and cost of credit. For adolescent and young adult populations 24 years old or younger, the barriers were ranked as disapproval by family, perceptions of appropriateness, and cost (Figure 2.1).

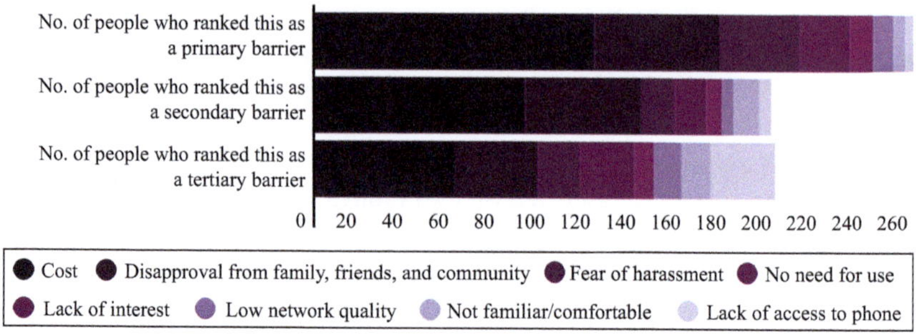

Figure 2.1 Top-ranked barriers to use by women and girls in Lebanon

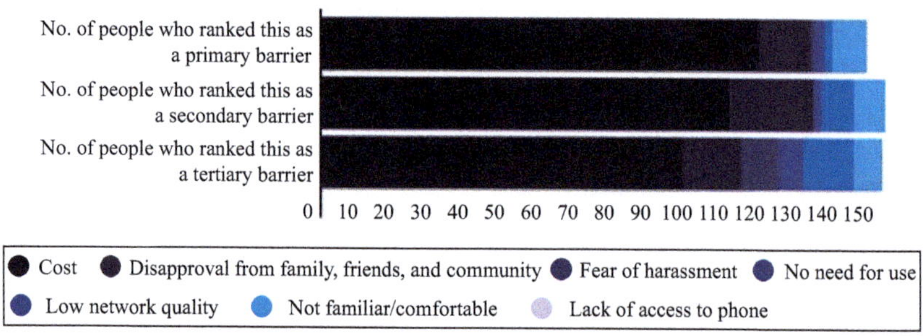

Figure 2.2 Top-ranked barriers to use for women and girls in Uganda

In Uganda, the barriers to usage for both sexes were reported most significantly as the cost. For women, the secondary barrier was lack of interest and disapproval from friends and family. Men rated women and girls' top barriers as cost of phone, cost of credit, and disapproval from the community (Figure 2.2). About 1 in 5 respondents disapproved of women and girls using phones or the Internet.

A solid understanding of barriers to ownership and use and sociocultural attitudes about women's digital engagement is critical to designing interventions or programs that involve women and girls' use of ICTs. Without knowing the barriers and norms affecting use, as well as the compounding issues around barriers (i.e. addressing one barrier may not resolve another barrier), adoption and meaningful uptake of ICTs will not meet the potential. Cost is a primary barrier that prevents use, which then impacts levels of comfort and familiarity or technical literacy. This, in turn, has an impact on perceptions of relevance. So, the barriers women and girls face in digital engagement are circular and compounding. The underlying common barrier are social norms that impact all other barriers. If women and girls cannot readily and fully access mobile phones and the Internet because of negative perceptions around their use and a weakened access to financial resources, their opportunities for digital engagement are at a gendered disadvantage.

2.4.3 *Affordability/prohibitive cost*

Across the humanitarian contexts surveyed in Lebanon and Uganda, as well as LMICs, cost of the device or cost of credit ranked the highest concern and most likely obstacle to mobile device ownership and mobile Internet use. For households in displaced settings with income limitations, "the purchase of a mobile device, credit, and accompanying fees for ownership may limit the household to one device for the entire household or to a device only accessible by borrowing from a community member" [13,14].

Lack of full ownership as a result of cost limits the range of services women and girls can use and their access to mobile Internet. This translates into an impact on digital competence levels. Without the ability to try and test, with full use of the device, digital literacy is suppressed. This disparity can be a "driver of inequity" as restricted or infrequent use can lead to lower levels of confidence [15].

In humanitarian contexts with lower device ownership, there should be some assumption around mobile device procurement necessities or that mobile devices will inevitably be borrowed from others, which has an impact on women and girls' safe use of the devices.

If a program were to design a mobile sexual harassment reporting app, and knowing devices are likely to be shared, it may follow that women and girls would not download the app for fear of stigma or retaliation. Thus, safety has to be a consideration from the perspective of access. If devices are shared or borrowed, programming needs to be designed around this obstacle to do no additional harm with the introduction of ICTs. This could include "quick exit" button or pins to enable access or disguising content under the umbrella of general health information. Creativity is required to ensure women and girls are safe during use.

Beyond the actual cost of device and Internet access, there is a social cost as well. Mobile device ownership increases opportunities (and at times income) and elevates the social status of men. Conversely, this same access or possession has a social cost for women (because of male-dominated society). These restrictive norms cause very real security and safety risks for women and girls.

2.4.4 *Disapproval: restrictive social norms/security and safety*

Perhaps the most difficult barrier to overcome are gendered social norms that dictate what is or is not appropriate for women and girls when it comes to digital engagement. In the settings surveyed, appropriateness was tied to perceived risks that women and girls would face if regularly engaging with mobile Internet, including threats to safety or reputational integrity in terms of morality.

In Lebanon, restrictive social norms, such as the community's disapproval of women and girl's use of phones and the Internet, are a top barrier to digital access. A sense of fear and underlying patriarchal attitudes contribute to this sentiment of concern about what will happen to women and girls if they had increased access, "which contributes to the lack of positive social norms around mobile technology use by females" [13].

In Uganda, there were perceptions (held more often by male respondents) that phones are not useful for girls, that they will lead to school dropouts, and that phones can bring increased opportunities for exploitation and abuse from men. Further, some male respondents noted, "for a wife to be [respected] she should not use a phone," and "girls should not be allowed to use a phone because they [will] not respect her husband or brothers," and "phones are not good because they help women find love and women can get [spoiled]." Just over half the female respondents and 56 percent of men also noted that phones could make women and girls less safe. For women, the reasons they believed they would be less safe were family breakup, exploitation, confusion from men, or love relationships that lead to abuse. For men, the reasons were fear of exploitation or abuse, love relationships, early marriage, familial problems, and school dropouts.

Even if temporary possession or ownership is facilitated for women and girls, there needs to be an awareness of the likelihood that phone use will be monitored by intimate partners or family members. For many, controlling women's use of phones through monitoring is viewed as a necessary activity to ensure women and girls are not pursuing romantic relationships outside of marriage [14].

These attitudes are normative barriers to digital engagement for women and girls. A Harvard evidence review of mobile use by women and girls in India outlined these gender social norms according to pre-marriage and post-marriage expectations of acceptable behavior. Girls are expected to remain "pure" before marriage and subservient to the wishes of their parents. This results in several behavioral consequences including restricted mobility, higher safety concerns, limited interactions, and a limited household role in decision-making [16]. Following marriage, gendered social norms follow a similar trend with expectations of obeying husbands and in-laws as well as following a traditional role as caregiver, managing the house, and prioritizing others' care in the household [16]. The consequences of these social norms dilute women's agency and empowerment and include "decreased ties with natural family, constrained physical mobility, reduced investment of household resources towards daughters, son preference... low labor force participation, [and] inability to leave [the] spouse despite abuse or violence" [16].

While these social norms are specific to India, the surveys from Lebanon and Uganda reflect equivalent trends. Survey results showed expectations for women and girls, fear or concerns about "purity" or romantic liaisons, and disobedience were risks exacerbated by female digital engagement. It reflects a view that men see women in relation to themselves, and they could be threatened by women's digital engagement because of other men. This is shown as a need to control women by limiting or manipulating women's digital engagement. The consequence of this is far-reaching—leaving women behind in the workplace, widening the pay gap, and inhibiting social networks.

Social norms and perceptions about acceptable behaviors influence not only the use of technology but also other barriers to them. These norms can restrict use desire, which consequently limits digital literacy and competence [17], and feelings of digital engagement relevance. They can also complicate/compound the barrier of affordability, as without the financial capability to purchase a device, women and girls have to advocate with income earners for access and ownership. If that decision-maker in the household does not agree with their use, this precludes even the prohibitive cost barrier. Further, if the perception is that women and girls will use phones for a purpose that is deemed not to be culturally appropriate (often related to unequal gender norms), it can complicate use of the device, as these norms may be held by individuals who also believe women's movements should be restricted, making required activities like phone charging difficult or prohibitive.

This restricts women and girls' use of ICTs as well as broader technology engagement such as online governance initiatives and opportunities for participation and voice as well as social networking, communication, and online or mobile educational initiatives. The gap in how women and girls engage with mobile devices or mobile Internet overall is reflective of an overall trend of exclusion from technology, which can have a multiplying effect. Without access to phones, women and girls are hindered in communication, connection, learning, and livelihoods, and their voice is absent from electronic participation. This reproducing effect is worsened through big data, a burgeoning opportunity for research and idea generation that is largely feeding decision-makers, but absent the voice of women and girls who lack equitable access to feed into it.

2.4.5 Relevance and literacy

Women and girls will use mobile phones and the Internet when the risks to use are removed and they are perceived as valuable. However, gender disparities "in education, employment, and income [mean women] lack the exposure to the Internet necessary for digital and information literacy, as well as opportunities to learn and practice computer skills" [18]. With restrictions on movement and interactions in the community, women are further disadvantaged in access and building digital literacy, as exposure to mobile devices and the Internet often happens in public spaces [18]. This is a continuing cycle of low literacy leading to low-value propositions and lack of engagement.

Without steady or regular access to mobile devices or Internet, it can alter perceptions about the usefulness of mobile phones and Internet. The narrative becomes that mobile phones create more problems and don't have added value. This misperception, driven by a lack of access and lack of consultation around problem-solving with women and girls, "can prevent non-owners from seeing the value for money in buying a mobile, even if they can afford one" [4]. Lack of awareness keeps women and girls from accessing the possible benefits of being connected online. Research from Intel's Women and the Web report found "one quarter of non-users expressed a general lack of interest in the Internet and nearly a quarter said they do not believe they [needed] it" [18]. The utility of being connected is not well known, and this can breed what might be seen as technophobia. However, it is really the result of lack of exposure, low technical literacy, and missing nudges to push for more active digital engagement.

Overall, the lower level of ownership, possession, and access and heightened barriers "reflect existing gender inequalities, but also threaten to compound them" [4]. If unaddressed, the gender gap persists and may worsen over time as male populations continue to improve access, literacy, and models for use.

2.4.6 Access matters

Digital technologies remain a beneficial opportunity for women and girls, which is why meaningful design must include consideration of obstacles to access and safety concerns around tech. On both a macro and micro level, there will be positive impacts. From a big-picture perspective, narrowing the gender digital divide can drive economies and become an equalizer as more women transform the online content to their needs and values. The GSMA 2019 Mobile Gender Gap Report estimated that "closing the gender gap in mobile internet use across LMICs could add $700 billion in additional GDP in these countries over the next five years … an additional 0.7 per cent of GDP growth in these countries by 2023" [4].

Equitable digital engagement will bring transformative change not only in economic contributions but also towards sustainable development goals. Access to the Internet drives social and educational change as "women are critical collaborators in the effort to achieve development goals such as reduced child malnutrition and mortality, or increased economic growth, this [digital engagement] gap disadvantages not just women, but their families, communities and countries" [18].

On an individual level, there are positive outcomes as well. Digital engagement raises the possibility of greater expression. With access to information, women and girls can have a greater voice in the exchange of ideas related to household, community, and broader arenas [18]. This contributes to empowerment through feelings of confidence and self-esteem. The micro level also shows more opportunities to research and access needed services, employment opportunities, and beneficial programs. Communication and connection opportunities also develop with digital access, which can lead to expanded social networks and increased social capital.

Digital technology in various formats (communication and networking, e-government, educational or vocational programs, financial inclusion or empowerment, and employment) can heighten access to information for decision-making, increase income-earning opportunities and financial equality, and aid in empowerment that lessens the gender divide both broadly and on an individual level. In Uganda, women noted the value of mobile phones for women and girls as providing communication channels, easier access to humanitarian services, increased understanding of phones, and access to information, news, and empowerment opportunities.

The inverse approach raises the likelihood of women being further disadvantaged: "Women without access to the Internet risk getting left further behind" [18]. To combat this future, exposure is essential. As women and girls, armed with information, observe people outside their community, they can "see themselves and their lives in new ways" [18].

2.5 Recommendations: design for inclusivity

To encourage meaningful participation, there are several key recommendations to design for women and girls' inclusivity in ICTs. From the conceptual to the practical, there are a number of actions the implementers can prioritize to bring women and girls to the table.

2.5.1 Build an inter-disciplinary team

Any ICT implementation will benefit from bringing together sector specialists or service providers who work in areas specializing in services that target women and girls.

One example of this is actors working in the GBV sector. This engagement should not be a one-time requirements collection consultation, but rather a higher-level effort that involves close coordination from problem definition through design, implementation, and maintenance. This will ensure the product delivered is relevant, that it is responding to and addressing barriers, and is woman- and girl-centered.

2.5.2 Woman- and girl-centered design

Human-centered design is simple in principle: taking into account the needs, opinions, and ideas of the end user when creating a product for high usability. As Melinda Gates noted, human-centered design works because "when you let people participate in the design process, you find that they often have ingenious ideas about what would really help them" [19]. While this principle is well-recognized in the design process and an accepted part of ICT implementations, we need to go a step further to really engage women and girls.

Women and girls need to be included, not just to fulfill a gender lens requirement from donors, but to ensure their active and full participation. Women and girls should be at the center of design. *If you want to reach women and girls, they need to be engaged from the start when identifying the problem.* Too often, we start with a technology solution without first having clearly defined the problem from the right perspective. There needs to a shift in the way we deploy projects that are meant to have broad participation. We need to bring in women and girls to define and refine the problem statement through the design process. It's not enough to just have sex-disaggregated data; we should have thoughtful questions to better understand the perspective of women and girls, safety concerns, access barriers, and how the solution fits with the problem they identified.

This is particularly needed in humanitarian settings where harmful or unfair gender roles are heightened as a result of additional restrictions—both physical and time restrictions, expectations, and barriers to agency. As researchers found in the Philippines,

> A person may feel that trying to change the situation that constrains them is futile, that they are not up to the task or that those with power to change the situation will not listen to them .... Engaging such marginalised communities in digital citizenship initiatives may require activities specifically designed to address people's depressed sense of political agency. [20]

Design must also apply a gender lens to raising awareness of risk levels for women and girls' engagement.

> From the perspective of the end user, consider safety concerns, device ownership and access, and the level of risk involved. ICTs should follow the principle of do no harm. This should also include data protection considerations. Adhere to international and national standards for data protection. If data is collected from the user, consider the risks if this information was exposed. If personal information is collected from the end user or about displaced populations, informed consent must be provided in advance of data collection, citing the purpose for data gathering. [13]

Planning for safe use when implementing woman- and girl-centered design includes considering trends in possession, ownership, and monitoring of activities. This requires service providers to consult with women and girls through the design process about real and perceived safety risks. This has an impact on the mobile services offered and overall technology solution in terms of content and function.

> Assuming mobile devices are shared or borrowed should force implementers to consider wrapping, associating, or including sensitive content (e.g., gender-based violence response services) with other general information, for example having a website or Facebook page that highlights women's health services and, as can be also discussed in a masked manner, intimate partner violence. Implementers could also consider ICT tools that can accommodate a "quick escape" button so the user can quickly leave the site if someone starts to monitor their use. Planning could also include safety tips on technical literacy and how to be safe on the Internet. [13]

2.5.3 Consider guiding principles

In design, ICT practitioners would do well to consider the guiding principles of gender-based violence response, notably the principles of safety, confidentiality, respect, and nondiscrimination. Even if ICT deployments do not involve reporting of incidents, violence, or protection concerns, these principles will push implementers to be aware of the safety context of women and respect their needs in ICT use.

The guiding principle of safety [21] places the safety of women and girls as the primary consideration. Because women and girls face additional safety risks, ICT implementers should include this guidance in their design and implementation. How will this ICT product or tool affect women and girls' safety? Will it introduce additional potential harm or opportunities for stigma or retaliation?

Given that mobile devices in humanitarian settings are less likely to be owned and more likely to be borrowed, shared, or monitored, there must be a strong consideration for confidentiality. This is a requirement if seeking support services for GBV or choosing to report acts of violence, but confidentiality could also be prized in situations outside of incident reporting. In some settings, involvement in activities and programs in public spaces can bring safety concerns and therefore would be aided by privacy or confidentiality.

As with any project or tool introduction, respect should be a central guiding point. Minimum standards in the field of gender-based violence response call on service providers to respect "the choices, wishes, rights and dignity of the survivor" [21]. This follows the trend of woman- and girl-centered design.

Finally, we can think about the principle of nondiscrimination in the context of inclusion of marginalized populations. This may include overlooked and often unconnected groups like rural or pre-literate women, who are not included in assessments, design, and participation. This has an effect on them being further excluded from digital engagement.

2.5.4 Mind the gap (assess locally)

There are numerous reports on the gender gap in digital engagement for women in LMICs, but research is infrequent with this same population in humanitarian settings. Surveying one setting and extrapolating or generalizing the results to other humanitarian settings will not serve women and girls well in this nascent field of research. What works in Iraq may not be transferrable to Cameroon or Bangladesh. For this reason, local ICT assessments in the setting where you work are crucial to ensuring your technology integration is feasible, acceptable, and informed by local realities. Knowledge and understanding of the gender gap will "help operators know where to target efforts and identify the most important barriers preventing women from owning and using mobile phones in their market" [22].

Another complication of applying statistics from setting to setting is that while overall female access to mobile phones and the Internet may be increasing, it could remain stagnant or worsen for overlooked populations: rural, pre-literate, or traditional nonusers. Those least connected should be a focused part of assessments.

Any research or assessment should focus on use/access, ownership and possession, patterns of use, common barriers, and desired uses. This can be done through an ICT assessment that looks broadly at availability, affordability, awareness, ability, and agency [20]. More specifically, this can include demographics, general usage trends and gendered trends, levels and practices of personal use, personal access and barriers, and access to information and mobile phones. An initial simple survey can be conducted in a setting to understand these influencing factors.

As a final more advanced recommendation, implementers can also use machine learning models to understand larger-scale trends of gendered use. GSMA has a Gender Analysis and Identification Toolkit (GAIT) that "uses machine learning to estimate the gender of unknown customers based on the usage patterns of a sub-sample of customers with a known gender" [23].

2.5.5 Address the barriers (do not treat male culture over women's as the accepted norm)

For women and girls to participate, the barriers to access and use need to be addressed: safety, accessibility, affordability, usability, relevance, and preferences. Primarily, service providers and practitioners need to ensure that women and girls feel safe when using mobile devices or the Internet. For this, involving women and girls in planning, implementation, and feedback are critical.

The service providers also need to address the imbalance in women and girls' access and ownership of mobile devices and access to the Internet. Accessibility is a primary reported barrier to use and must be addressed based on unique settings and the context of women and girls there. Just providing a mobile device may be too brusque an approach, as gender dynamics in the home could impact access, even if provided a mobile device.

Affordability also impacts access. To utilize mobile Internet and full digital engagement, women and girls need to be able to afford or possess devices, data plans/credit, and related fees for transaction. With additional barriers to income-earning, livelihood opportunities, or restricted movement, this presents a barrier that can be addressed through provision of devices (with careful assessment), or library models.

Once the barriers of access and affordability are addressed, usability is another possible barrier to consider. This includes building technical literacy and confidence where it may not currently exist.

> Multiple studies over the years have shown that formal skills training—be it in agricultural production or ICT applications—helps women facing severe gender or class discrimination improve their self-esteem. Training helps women overcome the intimidation that they sometimes feel in the face of new technology. [18]

Relatedly, women and girls need to feel that technology solutions are relevant for them and meet their self-identified needs. One way this can be accomplished is through the production of tailored content and utilization of preferred communication, or other platforms. ICT practitioners should evaluate preferred modes of delivery and the likelihood of utilization [24]. "The mode of ICT delivery will have an impact on who participates. Assessing for device types, levels of comfort and likeliness to utilize, in light of safety concerns, is crucial" [24]. Women and girls need to feel this is relevant in the marketing for the product/tool. In promoting women and girls' use, "you need to put women front and centre in your marketing" [24]. Representation matters, and women are drawn to messages that reflect them.

2.5.6 Focus on women and girls' participation

Encouraging true participation of women and girls means asking a new type of question. Currently, the question we're asking is—"How can we get more women and girls to use ICTs or engage with technology?" The question instead should be—"What are the choices women and girls see in how they participate technologically?" [25].

Agency is the answer. Going forward, we should be trusting women's agency, believing they know what is desirable, safe, and sensible in terms of technological participation. To encourage that they need to see models for safe and effective use by other women and girls, and they need to see tech that is built with them and for them. In talking about getting girls into coding, Dr. Jane Abbate, Associate Professor of Science and Technology in Society at Virginia Tech, says the way to encourage women and girl's participation is not to tell them we want to train them on coding necessarily, but instead to offer another "in" through trying to solve an identified problem and using technology as a tool to do that [25]. The same could be said for digital engagement, focus with women and girls on the problem statement first and plug in technology as a tool to achieve change.

2.6 Conclusion

ICT has great potential for women and girls in humanitarian settings. Women and girls' digital engagement is a requisite part of the gender equality process. Women and girls will fall behind without access to information, the ability to communicate, inclusion in the digital economy, or encouraged agency. However, the benefits of engagement need to be poised against the gendered challenges and barriers for women and girls to have equitable access and use.

More research is still needed, but looking at LMIC and survey results from Lebanon and Uganda are a clue into access points, known barriers, and opportunities for creative problem-solving. What is clear is that gendered disparity in digital engagement is worsened in humanitarian settings. Planning and precautionary measures are required for implementation to be inclusive of women and girls and their lived experiences. Without this, they will be excluded from digital inclusion.

The Inter-Agency Standing Committee Gender-Based Violence Guidelines call on service providers to promote gender equality and address the contributing factors that lead to gender-based violence by reducing risk, promoting resilience, and aiding in recovery. ICT should be an aid in this process. ICTs present an opportunity for inclusion, raising the voices of women and girls and lessening the digital divide between men and women.

Thoughtful and well-designed technology solutions and integrations in humanitarian aid can scale valuable interventions, access hard-to-reach populations, and strengthen efficiencies. The onus is on service providers and ICT practitioners to enable women and girls' participation, input, and feedback from the problem definition stage through design and implementation. Ask women and girls. Consult with women and girls. Listen to women and girls.

References

[1] World Bank Group (USA). *Individuals Using the Internet (% of Population)* [online]. 2017. Available from https://data.worldbank.org/indicator/IT.NET.USER.ZS [Accessed 12 March 2019].

[2] GSM Association. *The Mobile Economy 2018* [online]. 2018. Available from https://www.gsma.com/mobileeconomy/wp-content/uploads/2018/05/The-Mobile-Economy-2018.pdf [Accessed 28 February 2019].

[3] United Nations Office for the Coordination of Humanitarian Affairs. *Global Humanitarian Overview 2019* [online]. United Nations. 2019. Available from https://www.unocha.org/sites/unocha/files/GHO2019.pdf [Accessed 15 March 2019].

[4] GSM Association. *The Mobile Gender Gap Report 2019* [online]. 2019. Available from https://www.gsma.com/mobilefordevelopment/wp-content/uploads/2019/03/GSMA-Connected-Women-The-Mobile-Gender-Gap-Report-2019.pdf [Accessed 20 February 2019].

[5] United Nations High Commissioner for Refugees. *Connecting Refugees: How Internet and Mobile Connectivity Can Improve Refugee Well-being and Transform Humanitarian Action* [online]. Geneva. September 2016. Available from http://www.unhcr.org/publications/operations/5770d43c4/connecting-refugees.html [Accessed 2 April 2019].

[6] Namy S, Carlson C, O'Hara K, *et al.* "Towards a feminist understanding of intersecting violence against women and children in the family." *Soc. Sci. Med.* 2017;184:40–48. (Originally cited in Dobash R (eds.)). *Violence against Wives: A Case against the Patriarchy*. New York: Free Press; 1979.

[7] United Nations General Assembly. *Declaration on the Elimination of Violence against Women* [online]. United Nations. 1993. Available from http://www.un.org/documents/ga/res/48/a48r104.htm [Accessed 15 March 2019].

[8] Inter-Agency Standing Committee. *Guidelines for Integrating Gender-based Violence Interventions in Humanitarian Action: Reducing Risk, Promoting Resilience and Aiding Recovery* [online]. Inter-Agency Standing Committee. 2015. 5. Available from https://gbvguidelines.org/wp/wp-content/uploads/2016/10/2015_IASC_Gender-based_Violence_Guidelines_full-res.pdf [Accessed 2 April 2019].

[9] Vu A, Adam A, Wirtz A, *et al.* "The prevalence of sexual violence among female refugees in complex humanitarian emergencies: a systematic review and meta-analysis." *PLoS currents.* 2014:6.

[10] El-Bushra J. "Gender and forced migration: editorial." *Forced Migration Review*. 2000. Available from https://www.fmreview.org/sites/fmr/files/FMRdownloads/en/gender-and-displacement.pdf [Accessed 2 March 2019].

[11] United Nations High Commissioner for Refugees. *Lebanon* [online]. February 2019. Available from http://reporting.unhcr.org/sites/default/files/UNHCR%20Lebanon%20Fact%20Sheet%20-%20February%202019.pdf [Accessed 2 March 2019].

[12] United Nations High Commissioner for Refugees. *Urban Refugees* [online]. Available from http://www.unhcr.org/uk/urban-refugees.html [Accessed 3 March 2019].

[13] Crabtree K., and Geara, P. "Safety planning for technology: displaced women and girls' interactions with information and communication technology in Lebanon and harm reduction considerations for humanitarian settings." *Int J Humanitarian Action*. 2018;3:3. Available from https://doi.org/10.1186/s41018-018-0031-x [Accessed 27 Feb 2019].

[14] Santosham L. *Bridging the Gender Gap: Mobile Access and Usage in Low and Middle-income Countries* [online]. *In: Connected Women*. GSMA. 2015. Available from https://www.gsma.com/mobilefordevelopment/wp-content/uploads/2016/02/Connected-Women-Gender-Gap.pdf [Accessed 16 Nov 2016].

[15] *The State of the World's Children 2017: Children in Digital World*. New York: UNICEF. 2017.

[16] Barboni G, Field E, Pande R, Rigol N, Schaner S, and Moore CT. *A Tough Call: Understanding Barriers to and Impacts of Women's Mobile Phone Adoption in India* [online]. Harvard Kennedy School. October 2018. Available from https://epod.cid.harvard.edu/sites/default/files/2018-10/A_Tough_Call.pdf [Accessed 15 April 2019].

[17] Anid N, Cantileno L, Monique JM, and Rahilla Z (eds.). *The Internet of Women: Accelerating Culture Change*. Denmark: River Publishing. 2016. p. 181.

[18] Intel. *Women and the Web: Bridging the Internet Gap and Creating New Global Opportunities in Low and Middle-income Countries* [online]. Intel. 2012. Available from https://www.intel.com/content/dam/www/public/us/en/documents/pdf/women-and-the-web.pdf [Accessed 2 April 2019].

[19] Rope C. "The human element: Melinda Gates and Paul Farmer on designing global health." *Wired Magazine*. 11 December 2013. Available from https://www.wired.com/2013/11/2112gatefarmers/ [Accessed 11 April 2019].

[20] Roberts T, and Hernandez K. *Digital Access Is Not Binary: The 5'A's of Technology Access in the Philippines* [online]. Institute of Development Studies. 1 January 2019. Available from https://www.ictworks.org/wp-content/uploads/2019/02/5a-technology-access.pdf [Accessed 20 March 2019].

[21] United Nations Population Fund. *Minimum Standards for Prevention and Response to Gender-based Violence in Emergencies* [online]. UNFPA. 2015. Available from https://www.unfpa.org/sites/default/files/pub-pdf/GBVIE.

Minimum.Standards.Publication.FINAL_.ENG_.pdf [Accessed 2 February 2019].

[22] Lindsey D. *"Understand the Gap": How Mobile Operators Can Better Understand the Gender Gap in Their Customer Base*. GSMA. 20 November 2018. Available from https://www.gsma.com/mobilefordevelopment/programme/connected-women/blog-1-understand-the-gap-how-mobile-operators-can-better-understand-the-gender-gap-in-their-customer-base/ [Accessed 21 February 2019].

[23] GSM Association. *The Gender Analysis & Identification Toolkit* [online]. GSMA. 2018. Available from https://www.gsma.com/mobilefordevelopment/wp-content/uploads/2018/09/GSMA-Gender-Analysis-and-Identification-Report-GAIT-August-2018.pdf [Accessed 20 April 2019].

[24] ICT Works, *Four Surprising Ways Mode and Gender Can Transform Your Survey Data*. [online]. ICT Works. 13 March 2019. Available from https://www.ictworks.org/mode-gender-survey-data/#.XOgdPxRKjIV [Accessed 14 March 2019].

[25] Crabtree K. "Girl Code." *The Riveter*. 2016;5(3):19–20.

Chapter 3

Profound technologies: towards exploring the technological disruption and the challenges for a more humanized and inclusive education*

Marcelo Careaga[1], Laura Jiménez[1], Juan Molina Farfán[2], María Graciela Badilla[1], José Luis Carrasco-Sáez[3] and Luis Díaz[4]

Technologies exist because man exists. It is not possible to understand them unaffected by the meaning of their existence and how, through the history of cultures, human intelligence has been creating new forms of adaptation that have offered unique solutions to man's problems.

The technologies are not neutral. They reflect the man himself expressed through his talents and from his deepest contradictions and defects. Technology can be built to improve living conditions or to destroy and kill. Therefore, the sense of technological creation and the use of the manifestations of the *techné* is debated between the exercise of virtue, emanating from the permanent values that allow us to know how to be, the classic *areté* of virtuous life, and through the know-how related to *techné* (from Greek *τέχνη*, art, technique or trade) also understood as a skill to do something.

Technology, as an awesome achievement that aims to serve mankind, is a tool that can be employed either virtuously or unethically. These technologies need to be incorporated into the daily life of men embedded within processes of humanization and inclusion, which consist of an ongoing reflection on how the technologies should be used. They should be understood not only as consumer goods, but as both responses to and symptoms of times of change, where the creators of these technologies are responsible for what they create. They are humanized in the sense

*This chapter is written in the context of Regular Fondecyt Project 2019 N°1191891.
[1]Faculty of Education and CIEDE researcher, Universidad Católica de la Santísima Concepción, Concepción, Chile
[2]Faculty of Education, Department of Language Science and Literature, Universidad Católica de la Santísima Concepción, Concepción, Chile
[3]Doctoral Program in Education, Faculty of Education, Universidad Católica de la Santísima Concepción, Concepción, Chile
[4]Doctoral Program in Education, Faculty of Education, Universidad Católica del Maule, Talca, Chile

that these technologies are designed for a specific user as well as to fulfil a specific purpose and function.

The objective of this chapter is to analyse the profound technologies in our contemporary society, the process of humanization and inclusion as well as life forms based on coexistence, solidarity and the distributed participation of those who teach and learn through technologies.

Profound technologies arise within a certain anthropological and cultural context; they are designed to support the dynamic and critical interaction of users, and are thus humanized such that ethical and social tendencies are built into their use. Without an educational system that takes care of these new demands of culture on a human scale, of its interactions with cultural singularities and with the individuals assumed as managers of identity, technological opportunism will continue to deepen the great macro consequences inherited from the modern world, delaying ethical, moral and ecological growth. Profound technologies, based on humanization and inclusion of diversity, can contribute to human development, supported by reflection on their use and the implementation of individual, social and cultural content and practices.

3.1 Analysis of technology as cultural anthropology

The technique is by definition a set of effective knowledge accompanied by knowledge of the reasons or causes for which the procedure is effective. Aristotle [1,2] systematically approaches the definition of *techné* and its comparison with other types of knowledge. In *Book I of Metaphysics* and in *Book II of Physics* (fourth century B.C.), they are only a sample of the writings where this concept is considered as a type of specifically human knowledge linked to its cognitive capacity. The concept of *techné* is associated with the forms of knowledge of the human being and is related to classical science.

Aristotle considered that whoever builds a rudder knows the form it should have and how it should work, but whoever possesses the navigation technology knows how and why it should have that form and not another, or knows what type of material it should be made of. *Techné* reveals itself as a type of specifically human knowledge related to its rational capacity. This consideration specifically affects man's relationship with nature. As Adorno and Horkheimer [3] mention, what men want to learn from nature is how to use it to completely dominate it and other men.

According to the above, we cannot consider techné as a synonym of technique, nor vice versa, since one must be cautious when relating technique, in its anthropological and instrumental definition, with knowledge (episteme, knowledge) or with intelligence since, being the fruit of a human activity, it transforms the I can (dynamism, power, force) into a specific function of man (ergon, work). The technique, in short, is man's creation of an artificial environment for life. The artificial will be understood as a world built through the intentional intervention of man. The technique seeks to produce a useful object, while the techné prioritizes identifying a perfect act that makes it possible to obtain an ethically good and politically just life [4].

Foucault relates a juridical, negative and restrictive vision of power, towards a positive conception of power, a vision that he calls a technology of power. In order to analyse it, questions are posed: How was it possible that our society, western society in general, conceived power in such a restrictive way, so poor, so negative? Why do we always conceive power as rule and prohibition, why this privilege? [5, p. 2]. On the contrary, Foucault, using some ideas coming from Marxism, argues that there is not only one power, but that there are several forms of power, which he understands as meshes of power that operate as forms of domination. For example, there are forms of subjection that operate locally: in an office, in the army, in a slave-type property or in a property where there are servile relations. These are always local, regional forms of power that have their own modality of operation, procedure and technique. All these forms of power are heterogeneous. We cannot then talk about power, if we want to make an analysis of power, but we must talk about powers or try to locate them in their historical and geographical specificities [5, p. 4]. What he does, in his analysis, is to apply a historical gaze to classify different families of power technologies, including feudal power, monarchic power, adding modern productive power, which is exercised so that individuals are constituted as part of a population destined to be machines of wealth production, to produce goods and to produce other individuals, emerging a new technology of power that he calls biopolitics.

Ferré, when considering a postmodern epistemological theory, which focuses from an ecological vision of the world, establishes relations with the relevant presence of values in the conformation of this vision. In his book *Philosophy of Technology* [6], he makes a study on the meaning, purpose and results of the technological changes experienced by modern society. He analyses the topics and methods that philosophy uses to address problems related to the impact of technologies on human society, assuming that the relationships between philosophy and technology are dynamic and permanently modified. It attends epistemological problems, which are based on a philosophy applied to technology, specifying its relationship with intelligence in its theoretical and practical dimensions. In order to analyse these reciprocal relations, it establishes links between the development of advanced technology and modern science. Ferré also analyses the problems that have accompanied technological development, confronting more optimistic visions of modern existence with more pessimistic or critical visions regarding the role of these technologies. To support and complete his analysis, he incorporates ethical notions about the new phenomena associated with the automation of work, the presence of computers in everyday life, the use of nuclear energy, the development of genetic engineering and the characteristics of the Third World. Ferré includes debates on the mutual influences between technology and religion, technology and metaphysics, which end up forming a philosophical position on technology.

Stiegler, in his renowned works, *Technique and Time, Vol. 1 The Sin of Epimetheus* [7] and *Vol. 2 Disorientation* [8], confronts the development of technology as an intrinsically anthropological element, intimately linked to the development of man as a species and a sustainable future. Rejecting the technique as object of thought, the technique is the unthinkable [7, p. 9]. He raises that, initially, philosophy isolated the episteme, understood as pure knowledge, of the techné, assumed as ability, as effective procedure. The essence of the technical entities, in general, is enunciated on the inheritance of that conflict in which the philosophical episteme fights against the sophistic technology, devaluing with it any technical knowledge [7, p. 13]. Philosophical thought considered technique as the opposite of knowledge and study, because it was associated with a trivial technical instrument. Technology was considered as a means to achieve objectives previously attained by reason. In *Vol. 1 The Sin of Epimetheus*, it's used the image of Prometheus (Προμηθες, foresight, the one who sees in advance, the foresighted), from Greek mythology, to associate him as the father of technology and technological progress. Prometheus forgets the figure of his twin brother Epimetheus (Επιμηθες Epimetheus, the one who reflects later, the neglected or forgetful, the one who sees too late, once things are consummated), who possesses the sin of omission, a sin that leaves man naked and unarmed in the face of nature. Man's vulnerability compels him to create the technique in order to conquer space and time, and thus his inferiority in the face of nature.

Stiegler develops a critical thesis concerning the scope of technique. He proposes a human form of technology, which allows humanity to develop and evolve, rather than a form of strange and inhuman technology, which does not care or care about people. He categorically states that the technique is original and should not be understood as an increase in the very fact of human existence. In this view, the thesis does not directly relate technology with technical objects, but as the environment built from the human, man being the one who programs the environment and makes it habitable for humans. His vision stems from a latent malaise with civilization, where people have become consumerists, because they need to fill an emotional void left by the sadism of the technological system. A system that alienates man enslaves him, exiles him and transforms him into an object, objectifying him. Dealing with technique and time constitutes a profound reflection on the nature of the technical object and on the characteristics of contemporary technology. Stiegler defines technique as that vital tool for the transformation of the human being, but which is complemented with the essence of the human being, the individuality, ingenuity and creativity that each person has.

Stiegler highlighted the risks that citizens can fall into when they are immersed in a technological environment. There are valid tools available for processing, sharing and storing information, but, he critically asserts, he is wary of those tools that replace processes associated with human reasoning. Technologies modify established mental structures and have changed the way knowledge is shared. Stiegler's critical vision compels him take back his life and recover the sense of his own existence. New technologies allow us to have access to time and light, and facilitate the realization of complex processes instantaneously. The technology that emerged in the nineteenth century generated a progressive euphoria, where lifestyles could only improve. The techniques developed by man were taken to industrial levels; however, the idea of human beings without technology was not possible. In fact, it is the technologies that elevate the position of human beings within their environment and educate them. For this reason, according to Stiegler, the technique is an element of hominization; it is a component of man's very existence.

The techné is required for the constitution of the identity of the Anthropos. The human being acquires additional qualities that are born from the artificiality created by him. Its naturality would originally be secondary. The essence of man, his history, his purpose as a species, would be rooted in the artificiality created. In his analysis of technique and time, he postulates that the role of technology has been repressed. There would exist a kinematic constitution of consciousness, which would consist in the fact that the technological instruments form the material substratum of memory and, through these, the objective fixation of the experience lived by man is established.

Deleuze, in his book *Qu'est-ce que la philosophie?* [9] and the Post-scriptum on control societies [10], formulates ideas developed from the concept of enclosure centres raised by Foucault. Deleuze affirms that these centres of power – prisons, hospitals, factories, schools, families – find themselves in a crisis, located as they are within a growing mood of decadence. In antiquity within the society of sovereignty (Monarchy), men operated simple machines of mechanical type; in the disciplinary society (modern State), men were equipped with machines that worked with energy; the control society (Postmodernity) began to operate using computers, which it understands as third-type machines. The crisis occurs because the purposes and functions that were proper to this type of societies were, in turn, totally different.

Today's society is called the control society and it is exercised fluidly in open spaces, in a deterritorialized way, through psycho-pharmaceuticals, television consumption, marketing, private indebtedness and consumption, among other modalities. What is essential in them are the fluctuating and interchangeable figures such as those that show the value of one currency in the others, the incessant movement of surf that replaces slow and strategic sports such as boxing. Factories are replaced by companies, which are ductile and changing formations, simple machines by computerized systems of production and control. Individuality is replaced by external, computerised and computerizable divuales that move in a virtual space [10, p. 1]. It is not necessary to appeal to scientific fiction to conceive a control mechanism capable of providing at every moment the position of an element in an open environment, whether an animal within a reserve or a man in a company (electronic necklaces). Félix Guattari imagined a city in which everyone could leave their apartment, their house or their neighbourhood thanks to their electronic (dividual) card through which they erected barriers; but there could be days or hours in which the card was rejected; what matters is not the barrier, but the computer that signals the position, licit or illicit, and produces a universal modulation [10, p. 4].

Latour [11], carrying out studies about society, science and technology, proposes that scientific or technical work be immersed in the invisibility of its own effectiveness or success. When a scientific discovery is embedded in the laws of society or a certain technology works efficiently, the subjects who use such knowledge or machines tend to concentrate on the benefits they bring more than on the internal complexity that allows them to work (for example, nobody wonders about how a computer or a car works – they are only used).

These theoretical debates highlight the need to reflect on the use of technologies and thus overcome the technological opportunism, reducing their appropriation to the sole use of them. This endeavour necessitates a new concept. Profound technologies refer to the conceptual and ethical approach that emanates from the confrontation between modernity and postmodernity to deconstruct [12] modernity from a conception of technology as cultural anthropology, understanding it as extensive manifestations of human intelligence more than as available resources. It's a call to influence the humanization of society and to contribute to the conformation of a more inclusive, supportive, collaborative and expansive culture.

3.2 Modernity and postmodernity: an incomplete cultural transition

In order to define deep technologies, it is essential to incorporate into the analysis the confrontation between modernity and postmodernity, the latter understood as an incomplete cultural transition.

Toulmin states that modernity had two different origins. The first was the literary or humanist phase, which, going back to Petrarch, would have been dominant in the fifteenth and, above all, sixteenth centuries, with Erasmus and Montaigne. The second, the scientific and philosophical phase, which would have prevailed in the seventeenth century and which could be considered a counter-revolution, as society turned its back on many truly humanist themes and attitudes. The second beginning of modernity, led by the philosophers of the seventeenth century, wiped out the old preoccupations of Renaissance humanism.

Toulmin [13] tried to systematize this process by distinguishing four types of essential changes in the modes of humanistic knowledge, which are summarized in the transition from the oral to the written, the transition from the particular to the universal, the transition from the local to the general and the transition from the temporal to the timeless. The dismantling of the modern worldview was postponed until after the post-war period of the Second World War, reaching its maturity in the 1960s. It is a resumption or aggiornamento of the process of revision and critique of anti-humanistic modernity, initiated in the Enlightenment, radicalised at the end of the nineteenth century by authors such as Schopenhauer, Mauthner and Nietzsche, and frozen from 1914. Thus, during the 1960s and 1970s, the abstract and rationalist formalism of the sciences and arts was once again abandoned; a new style of writing about science, less exclusively logical and more open to historical questions began to be written about [14, pp. 15–34].

Modernity represents a historical period of profound changes, which took place since the Enlightenment and which represents part of a cultural transition, which some identify as postmodernity, others as a society of knowledge, cybernetic culture, liquid society or global culture. Above these conceptual disquisitions, it is necessary to confront what characterized modernity and its macro consequences, with the symptoms of change associated with these drastic modifications of the way of being intelligent of men. Therefore, the new ways of solving their problems, to elucidate whether these solutions in the individual, cultural and human scale dimensions are constituting symptoms of humanization and inclusion or are deepening modern macro consequences.

3.2.1 Main characteristics of modernity

Among the most relevant characteristics of modernity, the following can be identified:

1. The pre-eminence of thought systems that will see their maximum expression in ideological–cultural models linked to liberalism and nationalism and, later, to neoliberalism and neopositivism.
2. The development and deepening of scientific knowledge and technologies, and processes that influenced the fragmentation of knowledge and the hyper-specialization in human endeavour.
3. The development of economic liberalism, expressed in the recognition of the legitimacy of private property, the consequent accumulation of wealth and the accumulation of capital, free competition in markets, the subsidiary character of the State which is restricted to a regulatory role without involvement in the planning and development of economic activities.
4. The development of industrialization, through two Industrial Revolutions [15], the first developed between 1750 and 1840 and the second between 1880 and 1914. In these processes, the bourgeoisie and the business class are strengthened and the wage-earning working class emerges, which ends up being the genuine expression of a new type of social organization, typical of Capitalism, which will be expressed through new contradictions linked to the ownership of the means of production and social demands. These revolutions involved social, economic and technological transformations that meant a rapid transition from a rural economy based on agriculture and linked to trade, to an urban economy, which experienced increasing industrialization and mechanization.
5. The establishment of a new political system which tended to put an end to absolute monarchies. It was organized on the basis of the division of state powers, the predominance of constitutional systems governed by a Magna Carta (Constitution), the prevalence of constitutional monarchies in which kings hold sovereignty and the people hold supreme power, exercised through representatives participating in Congresses, Chambers or Assemblies, in which political decisions are made that affect rulers and citizens. Also, systems such as enlightened despotism and democratic-representative systems are implemented.
6. The sustained development of the sciences, based on positivist principles and linked to the economy, leading towards a fragmentation of reality and to a more diverse and specialized understanding of natural and human phenomena,

with the conception of theories such as Theory of Evolution [16], Microbial Theory [17] and Atomic Theory [18], among others.
7. The development of thought which will be expressed in new trends related to idealism, dialectical materialism, nihilism and nationalism, among others.
8. Along the social dimension, the conditions for the great social movements will be created. The working class and the political class will undertake their demands and struggles, which will lead to liberal, bourgeois and proletarian revolutionary movements.
9. In modernity, the phenomenon of imperialism develops, as the maximum expression of colonialism, due to the fact that several European countries overflow their geopolitical borders, conquering countries of Africa, Asia and other latitudes, with the purpose of providing themselves with raw materials for industrial development and of improving their commercial capacities.

3.2.2 Positive macro consequences of modernity

Among the main positive macro consequences of modernity, we can identify the advances that were made at this stage of history for humanity:

1. Cultural progress, with the trend towards mass education as a form of leverage for the progress of society.
2. Advances in medicine and the significant increase in the life expectancy of human beings.
3. Scientific and technological development, with the undeniable improvement in living conditions.
4. Diversification of the production of goods and services, improving the quality of life of men.
5. The production of energy generated by man, with the implementation of energy grids to support industry, transport and provide domestic energy.
6. The rapprochement between men and between singular cultures, by the increase in efficiency of transport systems as well as information and communication technologies (ICTs).
7. The tendency towards the installation of democratic systems of government.

3.2.3 Negative macro consequences of modernity

However, modernity is beginning to be exhausted as a historical stage and as a cultural expression, and is also associated with negative macro consequences:

1. The fragmentation of knowledge, due to the development of sciences and hyper specialization, which resulted in the partial implementation of solutions to human problems.
2. The irrational exploitation of natural resources, without sufficiently considering the sustainability of using resources that are not renewable.
3. Ecological disaster, through the generation of unclean energies, the inadequate treatment of industrial wastes, the production of domestic waste and a social tendency towards apathy and a lack of awareness concerning environmental issues.

4. The exhaustion of meta-stories, ideologies, doctrines and explanatory models, which sought to solve the great problems of modern man, because neither liberalism, socialism, social democracy, communism, anarchism nor revisionism, to name but a few proposed systems of social organization.

   Only the neoliberal system remained dominant, which crossed ideological frontiers installing itself with its forms of life based on individualism, competitiveness and consumption, bringing into its system social effects that prevail until today and, in some cases, tend to deepen, such as the following:

   (i) Social inequality, increasing the segregation between a minority population that controls political and economic power and the majority of people relegated to work systems that operate as forms of subsistence, without the material and subjective (emotional) recognition they deserve.
   (ii) The regressive distribution of income, which means that the rich population earns ever more and the poor tend to be poorer, a gap that can also be noticed between countries, as the great powers of the world determine the present and future of all peoples.
   (iii) The deepening of the irrational exploitation of resources that has led to our current ecological crisis with economic factors consistently being prioritized over respect for nature and the environment.
   (iv) The exclusion of minorities and intolerance towards diversity of gender, sex, ethnicity and beliefs.
   (v) The prevalence of individualism and consumerism, creating a market of superfluous needs, which induce an implacable neurosis of consumption which, rather than causing happiness, causes permanent dissatisfaction, both to those who consume, because its spiral of consumption is unfinished, and those who cannot consume because they do not have sufficient resources and who, therefore, carry a strong dose of frustration and social resentment, which manifests itself in social maladjustment, neglect, delinquency and drug addiction, among other individual and social consequences.

3.2.4 The idea of postmodernity

In this context, postmodern thought arises, which we can situate in a phase of cultural transition. With the recognition of this historical phase as such, it is possible to avoid opening a conceptual debate to try to define this stage, which is one of profound cultural, social, political and economic change, and which manifests itself in the individual dimensions, of singular cultures and of global culture on a human scale. In this way, it will not yet be necessary to assume a position to say whether it corresponds to the Information Society, Global Culture, Knowledge Society, Cyber Culture, Liquid Society, Relative Society or any other meaning that may appear from the clear definition that characterizes this new phase of search. We will only say that we are in a cultural transition from Modern Culture to Postmodern Culture which, in an applied sense of the dialectic of Hegelian history, would be the antithesis of the previous one.

Therefore, taking ideas from reference authors plus some of our own, we can try to identify general characteristics of postmodernity and of the new thought forms that begin to incubate [19–21], among those that can be identified.

The indisputable withdrawal of ideologies, understood as the meta-stories, guided human activity. Modern ideologies have been revised (revisionism), reformed, relativized and deformed, renouncing their pure and initial conceptions. In some cases, it has definitively collapsed. This is explained, besides their own historical wear, by the tendency to the loss of the interest in the theoretical, on the part of the postmodern man and by the little practical applicability that these ideologies had. They lost their reliability, as a means to take strategic control of the State to obtain the political power and guarantee the transformation of society.

The reality for the postmodern man has ceased to be an object of study exposed to the inquisitive gaze of men who build science from it. The relationship between man and his reality has experienced a strong reductionism. It has been limited to a utilitarian link, in which everything that surrounds man possesses an exchange value, which depends on the extent to which it satisfies his needs and contributes to his life and progress.

In postmodern man, the concepts of use value and exchange value acquire maximum relevance. In this idea, Lyotard [19] proposes that the old principle that the acquisition of knowledge is inseparable from the formation (bildung) of the spirit, and even of the person, falls and will fall even more into disuse such that these concepts themselves lose their use value.

From the epistemological and axiological point of view, in postmodernity there is a tendency to the loss of rational rigour. The creation of knowledge as well as ethical and moral behaviour no longer depends directly on rational capacity. Taking positions on one thing or another becomes an expression of an extreme liberality in human behaviour. Ethical conduct, derived from this kind of knowledge, is not justified or sustained in great guiding principles of human conduct, but rests on a relativism of customs and on the anxiety of generating useful knowledge to increase efficiency in the production of goods and services and to persistently improve macroeconomic variables. This provokes erratic collective behaviours, importing only what accommodates the particular vision of individuals. Post-modern men are essentially concerned with casuistry, which provides background for solving concrete problems. They abandon, with this, the analysis of the great principles or idealistic theories.

The tendency towards hedonism, characterized primarily by an open addiction to consumerism, with consumer products idolized as a direct source of satisfaction and pleasure. An evident symptom of change is the tendency to live on credit, as opposed to the modern emphasis on the accruement of savings as an economic discipline.

There is a particular option and a specific focus on problems, without there being a great concern to specify the theoretical or practical aspects that characterize human action or to visualize reality as a complex whole. There is a clear renunciation of holistic theoretical conceptions, such as those proposed by modern theorists such as Leibniz, Kant, Hegel and Freud. This characteristic constitutes one of the paradoxes of postmodernity, where there is a clear tendency towards the globalization of culture and, nevertheless, the particularity of cosmogonies prevails.

There is also the immediate vision of human life, replacing the transcendent existential concern with the sustained will to improve its quality in pursuit of joy. According to Roa [22], this is a transition from the ethics of duties (Aristotelian and Kantian) to the ethics of rights. Ethics was always a discipline concerned with discriminating between those things that one wants to do and should do. The ethics of duty are, for example, the classical ethics of Kant, the categorical imperative, and the fact that man worships duty over the worship of will and power led him to say that the beauty of the moral order could only be compared to that of the starry sky in a serene night. However, such Kantian ethics, which would be one of the dynamic axes of modernity, and the same as another type of ethics of duties, would be what today appears as simply an anachronism.

Following the ideas raised by Lipovetsky [20], we would be in the times of the post-duty epoch. All human rights are proclaimed and defended, but the duties associated with these rights are not necessarily emphasized. It proclaims the right to life, to individuality, and to sexuality. Postmodernity maximizes the empire of these rights. However, there is a complacent and benevolent tendency towards the fulfillment of duties, weakening the ethics of duties.

One of the most significant changes lies in the essential modification of the epistemological linear subject–object paradigm, which was typical of modernity. One moves towards a new epistemological relation, characterized by the decodification of the subject in his relation with the object, a question that will be very present in the formulation of the anthropological–philosophical paradigm when the emergent *Homo ciberneticus* is characterized and its development as a three-dimensional (3D) man.

In the light of the modern concept, the subject was a researcher, an inquisitor, an investigator, an impartial observer of the reality (this reality understood in two dimensions: the objective reality of things and the immanent reality of ideas), which was the object of his study. However, in postmodernity, with the development of technosciences, such as computer science and cybernetics, the object becomes active, dynamic and interactive, turning over to the subject through the modification of the subject, its identity, its modes of action and its customs.

With the loss of importance of modern theories, a new perspective on social theory appears, [21] differentiating it from sociological theory, which has led to a reflexive theory, centred on epistemological and ontological questions over sociological ones [22, p. 164; 23]. When man's subjectivity acquires value, new spaces are built. Differences are produced in traditional sociological categories. A new type of knowledge acquires importance because in the post-modern contingency new tensions appear between the roles of individuals, their singular identity proper to the culture to which they belong in their interaction with elements of culture on a human scale. Thus a relationship is established between scientific knowledge and its effect on people's daily lives. New categories are emerging for the understanding of social and cultural space, which demands a new theory that goes beyond modern theories.

3.3 The context of technological disruption and the challenges of humanization

At present, the cultural transition is undergoing the vertiginous change of technological disruption, which is projected with notable modifications to human life over the next two decades. The concept of disruption implies a sudden interruption of something, a sudden rupture, in this case in the dominant culture that is modernity. This disruption is associated with innovation and its impacts, but, at the same time, it implies a prospective vision of future trends. Any situation, element, activity, behaviour or decision that produces a sudden rupture with reality or the stable environment leads us inexorably to a process of change, whether physical or not [24]. In terms of technology, (τεχνολογία, téchnē and logia, λογία; meaning techné, τέχνη), is art, technique, craft or skill, concepts derived from Plato and Aristotle.

Technology is man's creation of an artificial environment for life. Therefore, technological disruption is the dynamic process of advancement in the artificial solutions that man manages and whose technological effects provoke sudden changes. As these changes are projected towards the near future, a drastic modification in man's way of life is projected, which will bring new forms of social behaviour and work. There will be many advances associated with innovation, the important thing being to achieve effective performance, anticipation and foresight. The Fourth Industrial Revolution is taking place, which is a new productive model that will change the world in a way that we can only glimpse today. It will change not only what we do, but also what we are (Klaus Schwab, President of the World Economic Forum).

3.3.1 Main characteristics of disruptive technologies

Among the characteristics of disruptive technologies, one can visualize the following:

1. The irruption of artificial intelligence (AI) in daily life and in the production of goods and services. AI is a process by means of which a machine can be made intelligent [25] having applications, for example, in the natural processing of language, artificial vision, augmented reality, robotization applied at different levels of human performance, etc.
2. Massive macrodata, also known as big data, consisting of the application of data intelligence for large-scale data processing, with characteristics such as volume, processing speed and variety of data (structured, semi-structured, unstructured).
3. Robotization, applied to production and domestic life. Robotization, in principle, is the artificial reproduction of movement, but, with the development of AI, has experienced an expansion of its application at different levels of operation, tending to replace many functionalities exercised by humans, such as production processes, automated services and use of technological products, in telehealth, in transport systems (e.g. trains and airports), etc.
4. The Internet of Things (IoT) [26] is based on the interconnection of sensors and transmitters in computer systems and networks, being able to monitor and

manage the state of things and living beings that are connected and learn from that connection, for example, domotics (intelligent houses), Internet for me (systems that learn from and facilitate my life), intelligent clocks (heart rate, disease control) and Smart Cities (intelligent cities with intelligent traffic lights, traffic management, services, security, etc.). Among the most important challenges of the IoT is the security of things vs. ransomware of things, the latter consisting in the manipulation or computer attacks directed at interconnected things.

5. Autonomous vehicles for transport and industry, based on Lidar technology, which allows sensors to emit electromagnetic waves and receive them to make decisions about the objects around them, for example, taxis, trucks in mining and freight transport. Full automation of transport is planned for the near future. Among the negative effects, there is disruption to the labour market.
6. The digital financial market, known as blockchain (Satoshi Nakamoto), which consists of the decentralized registration of encrypted, transparent and secure transactions (sales, real estate, vote counting, donations, immediate payments, etc.), which incorporates online financial management, transaction control and crypto currencies, such as Bitcoin, ICOs (Initial Coin Offerings) and token.
7. 3D printing (consumer 3D) and four-dimensional (4D) printing (nanotechnology with physical and biological materials): 3D printing allows the manufacture of objects by printing layers of material, based on digital models, which impacts on current supply lines, develops the concept of Consumer 3D (production in the home, e.g. consumer goods and household objects), and 4D printing, which consists of applied nanotechnology, will allow printing of objects with physical and biological materials, automated production, with self-assembly, biomedicine, etc. – among other multiple technological applications associated with technological disruption, which is presumed to drastically modify human life in the next two decades and following stages of this postmodern cultural transition.

3.3.2 A deeper look into disruptive technologies

The ideas of (i) Technological Disruption, (ii) Intelligent Cities, (iii) Collaborative Economies and Work Automation and (iv) Artificial Intelligence and Education will be analysed from a deeper perspective.

3.3.2.1 Technological disruption

Since its origins, human beings have invented various technologies that have penetrated their social organizations and the way in which they conceive their reality. For Capra [27], technology is older than science, because man had to invent it for his subsistence. The greater or lesser acceleration in the use of a technology is known as multiplier factor of the technology [28], Moore's law [29] or law of accelerated performances [30]. According to Negroponte [28], the technology multiplication factor relates to the number of times a technology can improve the function for which it was created. In the mid-1960s, Moore [29] discovered that it was possible to double every 24 months the number of transistors that could be assembled in an integrated circuit, increasing their speed, since electrons had to travel shorter distances. The law of accelerated yields attempts to explain the acceleration and exponential growth of the fruits of a complementary biological and technological evolutionary process. It is based on some of the following principles associated with technological disruption: (a) evolution is sustained by positive feedback (the next stage of the evolutionary process will be built with those methods that were successful in the previous stage); (b) an evolutionary system is an open system whose yields increase over time, at least exponentially (beyond Moore's intuition); (c) a specific paradigm will grow until its potential is exhausted and replaced by another, so that growth continues exponentially.

Negroponte [28] and Kurzweil [30] coincide in that the technological revolutions or changes of paradigm have had dissimilar times of propagation. The first sketches of technology created by *Homo sapiens* (wheel, fire and stone tools) took thousands of years to perfect and be fully adopted. An important difference of the digital revolution with respect to the previous technological revolutions is that its multiplying factor (of the order of one billion) and its process of evolution have been very superior. Only 70 years ago, the first electronic computer (The Electronic Numerical Integrator and Computer (ENIAC)) was introduced, whose development was the preamble to technologies that changed the way human beings relate to each other, build knowledge and expand the limits of what they consider to be reality. On the other hand, agriculture is assigned a multiplication factor of the order of 100; 10 because of the invention of the plough in the Neolithic revolution; and 10 because of the subsequent use of chemical fertilizer [28]. In the case of the industrial revolution, the multiplication factor would be 1,000.

Kurzweil [30] suggests that from the twentieth century to the present, the development of computing can be understood through five exponential growth paradigms: (a) electromechanics, (b) the relay, (c) the vacuum tube, (d) the transistor and (e) the integrated circuit. According to Moore's law [29], the maximum performance curve of the integrated circuit should give way to a sixth paradigm: 3D molecular computation, which using as reference to the human cortex and cerebellum, will combine two-dimensional (2D) systems with 3D structures, based on nanotubes and computation with DNA, among other technologies. Advances in digital development allow the rise of high-speed mobile Internet networks that promote the rise of virtual worlds, social network platforms and video streaming, IoT, augmented reality, big data and AI, among other technologies. With the IoT, the information available on the Internet will come from humans and from any device connected to the network. The rise of social networks and video platforms, such as YouTube or Vimeo, will continue to change the way we communicate, with videos capable of delivering greater impact than the written word. Augmented reality will provide us with a mix of hologrammatic, physical and virtual worlds. Our entire 2D footprint will tend to be tracked by various big data algorithms capable of making predictions about our tastes, using interaction with our digital devices (connected to other sensors and devices), which are currently capable of correctly interpreting almost 80% of our queries [31].

3.3.2.2 Intelligent cities

One of the social expressions of the transition to postmodernity is reflected in the concept of Smart City. According to Harrison *et al.* [32], it is defined as follows:

1. A city instrumented, through the use of sensors, and personal devices to capture data from the real world.
2. Interconnected, through the integration of data captured in computer platforms.
3. Intelligent, which includes the analysis, modelling and visualization of information to make decisions. Although this definition bases the concept of city intelligence rather on computer decisions, it does not necessarily mean that its inhabitants carry out all their management online [33].

There are differences between a digital city and an intelligent city, as the former seeks to provide a broadband computing infrastructure to meet the requirements of businesses, citizens and governments [34], while the latter appears as a result of the intersection between the Knowledge Society and the digital city [35]. It was a group of IBM researchers who first coined a definition for Smart City [32], situating it as an instrumented city (use of sensors and personal devices to capture data from the real world), interconnected (integration of data captured on computer platforms) and intelligent (analytical, modelling and visualization of information to make decisions). It is interesting to note that this concept has gradually become more humane. For example, the European Commission, in its report Mapping Smart Cities in the EU, states that 'a Smart City is a city that seeks to solve public problems through technology-based solutions in the framework of a partnership between different participants, both public and private' [36].

In this same perspective, the Digital Country Foundation of Chile defines Intelligent Cities as Cities that, through the application of technology in their different fields, become more efficient localities in the use of their resources, saving energy, improving the services delivered and promoting sustainable development, solving the main problems faced by citizens. In this way, the public's quality of life is improved [37, p. 3].

The verification of generating inclusive spaces within cities that avoid differences between people allowed the idea of coining the concept of Smart Human City, which incorporates a more humanized vision than the Smart City, putting at the centre the rights of citizens rather than the use of ICT [38].

The trend towards the sensorization of cities is based on the analysis of large amounts of data in real time, using machines and different algorithms, which can lead to an increase in automation and polarization of work in various areas of knowledge, as explained next.

3.3.2.3 Collaborative economies and work automation

Each technological revolution has resulted in societies debating its effects on the economy and employment. In 1767, the economist Sir James Steuart argued that the introduction of machinery could have negative effects on employment rates at the time [39, pp. 33–34].

The first formal movement of resistance to technological advances and their impact on people's lives, known as ludism, was led by English artisans, who protested between 1811 and 1816, destroying the machines that could replace them in their jobs [39, pp. 34–35].

Since the mid-twentieth century, rapid technological changes and advances in robotics and AI have once again raised concerns about the consequences of automation. A kind of neoludism, fostered by intellectual distrust towards the advancement of science, founded under a dominant social structure, protected by the new digital economy and the knowledge society [31,40,41].

Automation is understood as the computerization of works that can be carried out by computers and algorithms [42], related to the use of robots, AI and machines. Automation can have an impact on the reduction and polarization of employment. Studies show that the impact of technological development and technological adoption in the workplace fosters job polarization, causing low- and high-skilled jobs to gain market share at the expense of medium-skilled occupations [43,44], among others.

Routine tasks can be fully coded and automated. Examples of such tasks are company accounting, office administrative work or the execution of repetitive physical operations [45].

In contrast, non-routine tasks and abstract tasks cannot be easily coded. This phenomenon, analysed from the irruption of virtuality as a new postmodern human dimension, is understood according to Lévy [46], as a Moebius effect that confronts the classic works of physical spaces, with the new dynamics of telework, questioning the notions of time and the boundaries between the private and the public.

In short, the trend towards the automation of work causes tensions in the world of work. Jobs at medium level could be in danger of disappearing, dividing employment between those of low qualification and those of high qualification (which require critical thinking, emotion or solving the unpredictable), which still cannot be replaced by machines.

The automation and boom of digital platforms encourage the emergence of gig economies or economies on demand, and circular or collaborative economies. Both tendencies are embedded within the economic modifications (within the capitalist system) of postmodernity, as a consequence of technological disruption. At the end of the seventeenth century and until the middle of the twentieth century, industrial society operated under the rhythm of clocks that led man's time [47].

However, technological disruption has caused this model of industry to open the way to a new type of globalized and post-industrial capitalism [48]. Gil García [49] argues that from the 1970s onwards, capitalism began its neoliberal transformation, not only invading the economic sphere, but also becoming a political form that expands the limits of the exchange of goods, invading and pressuring people, society and states to live in permanent competition, capable of converting any behaviour into economic behaviour [50–52].

The trend towards the automation of work raises the need to increase the growth rates of countries through the development and implementation of technologies that expand the supply of products and services.

An OECD study [44] describes some technologies that could open up new opportunities, especially for developing countries:

1. Sensors combined with the IoT that predict possible problems on multiple devices.
2. More accurate predictions of new demands for goods and services.
3. 3D printing.
4. Autonomous machines, big data and AI.
5. Nanotechnologies, among others. In this scenario dominated by the neoliberal model, the concept of a collaborative economy that makes extensive use of ICT to produce, exchange and consume goods and services emerges.

There are new models of companies that seek to improve their efficiency and the relationships they have with their customers to maintain their competitive advantage. Its advocates suggest that this model is a source of innovation that causes positive social and environmental impacts through efficient economic processes, which also allows citizens to access extra income by introducing underutilized goods to the market [53,54].

Their detractors argue that they are companies that emerge in deregulated markets, generating precarious work with mercantile practices hidden behind social or environmental principles [55–57].

Collaborative economies could become a hidden expansion of capitalism using technological platforms [58], promoting the expansion of digital monopolies in different parts of the world. A situation that some authors suggest can become a neo-feudalism of cognitive capitalism [59], in which the subject is compelled to constantly improve their competitive position by resorting to ventures in any area of his life, becoming producer and consumer of his own image. Under this perspective, the main objective is to work on itself in order to transform itself permanently, to improve, to become increasingly effective [50].

3.3.2.4 AI and education

ICTs are key elements of our society, generating changes in the economy [60], the way we communicate [61] and the acquisition of knowledge [62,63]. Education has not been oblivious to these changes; currently, the teacher has a variety of ICT tools for the processes of teaching, management, administration and research [64–67].

For the students, the use of ICT should be motivating such that the student is fully engaged with the learning process, interacting with reality and observing the results of this interaction, developing critical and creative thinking skills, integrating and facilitating the understanding of what has been learned in a dynamic way, in order to achieve more effective learning [68]. It should be noted that the use of ICT does not ensure an increase in learning levels. The design and the way in which ICT is incorporated can make it accessible or inaccessible [69–71]. For Gil *et al.* [72], inappropriate use of ICTs can lead to isolation or exacerbate phenomena such as loneliness and depression, as well as addiction, violence and loss of privacy.

The use of technologies requires, as a priority, to be attended by education, since through academic accompaniment the leisure time of young people is rationalized and reoriented towards the global requirements that the knowledge society is demanding. ICTs represent the communication channel most used by young people, occupying a fundamental role in student development. Teachers play a fundamental role in the use of ICTs; they must know and assimilate new technologies with pedagogical intent, guiding students towards the knowledge required by the information society [73].

Faced with this scenario, teachers must focus on new cultural practices, where the important thing is the person, and technology is an instrument that facilitates the teacher's work in the process of generating knowledge [74]. Educational institutions not alien to this change make efforts to modernize learning processes in the classroom; in this situation, ICTs play a fundamental role in this area, rethinking educational techniques in which the student is the centre of the learning process [75].

At this point, techniques and paradigms of tools such as AI appear as alternatives to the problems that arise when trying to introduce computers to support different learning strategies. AI includes not only the development of intelligent machines, but also the way in which these systems affect our learning [76]. AI in learning environments offers effective support in teaching processes, reflecting aspects of behaviour and capacity to assimilate knowledge in each student; these advanced pedagogical tools provide individualized learning experiences, allowing teaching processes to be more adaptable [77].

These new technologies facilitate the generation of advanced pedagogical tools, since they not only allow the creation of procedures based on modern knowledge management processes, but also facilitate the generation of new environments and modalities in training [78]. Intelligent tutors based on AI systematically construct and update an instructional plan based on the needs of the student, and they are designed to identify and determine the most suitable methods for acquiring knowledge [79–81]. AI learns from patterns of interaction and associates them with groups of pedagogical relevance so that the teacher can better understand the ways in which students think and where they can be effectively guided. AI can also provide such diagnostic data to students to reflect on their metacognitive approaches and possible areas for development. Therefore, AI will have important potential in learning diagnostics, analysis and extraction of educational data [82]. AI-based tutors adapt pedagogy, materials and instruction to provide an additional learning scaffold in the context of the student's assimilated perception.

Real-time assessment for understanding information provides a powerful tool for adaptation [83]. Educational institutions face a set of possibilities and challenges regarding the use of AI, promoting the permanent learning of a model of self-instruction; the use of this type of tool must motivate and stimulate students to become involved in the educational process, interacting with reality, developing critical and creative thinking skills in order to increase the skills that will become enduring competencies in twenty-first century students [68,83].

3.4 Two reference models to explain the dynamics of the globalized world

In order to understand the disruptive changes and thus attain satisfactory levels of humanization and inclusion, it is necessary to propose new paradigms that outline and explain these future trends.

The basic question is, how does the world move today? What is different about the cultural transition? An initial answer is the coexistence of realities.

Today the world moves interconnected between the space of places and the space of flows [84]. It is the development of the global village anticipated by McLuhan and Powers [85]. This is the physical or objective space, referred to as the temporal and space categories. And the one that coexists with cyberspace is referred to as the virtual category. These circles or dimensions intersect epistemologically in a border that is characterized by two phenomena: (i) information management and (ii) knowledge management.

3.4.1 The pedagogical circuit model of knowledge management

The pedagogical circuit model of knowledge management [86] seeks to establish how this epistemological frontier is resolved, specifying that it is the mediation of language that establishes the difference between the consumption of information, characterized by the technological opportunism that consists in accessing and representing available information, with the management of knowledge, which consists in the creation of intellectual, theoretical or practical constructs which are managed by means of some language with the intention of making them transferable, linking individual intellectual capital with the conformation of collective social capital (see Figure 3.1).

3.4.2 The 5R model

This model establishes the dynamic way in which men under cultural transition link and move. This model (see Figure 3.2) consists of an incremental circuit that begins in the social networks (R1), which are the virtual environments in which, at present, social interactions between people are dynamized, establishing communication between the physical space and cyberspace; these networks are conformed from networks of contacts (R2), in which the first communication links are established; these contacts evolve into collaborative networks (R3), when the members establish common interests and co-elaborate among themselves. That is to say, they are capable of building something that is common to them; when co-elaboration exists and endures in time these networks are transformed into networks of trust (R4), which corresponds to a higher stage of connection in which permanence and joint solutions work because deep and reciprocal links were established that are capable of overcoming contingencies and eventual differences, in order to support virtual communities; and, finally, when those who trust manage to coalesce an epistemological, social, communicational, artistic, organizational, religious, political or

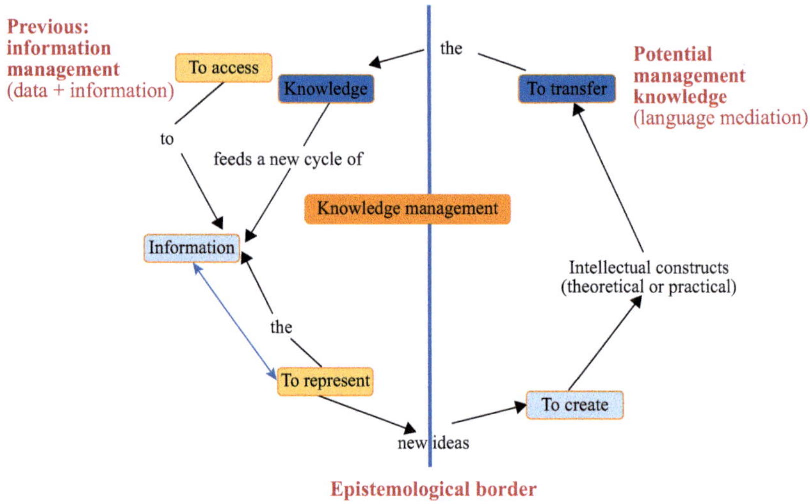

Figure 3.1 Pedagogical circuit model of knowledge management [86]

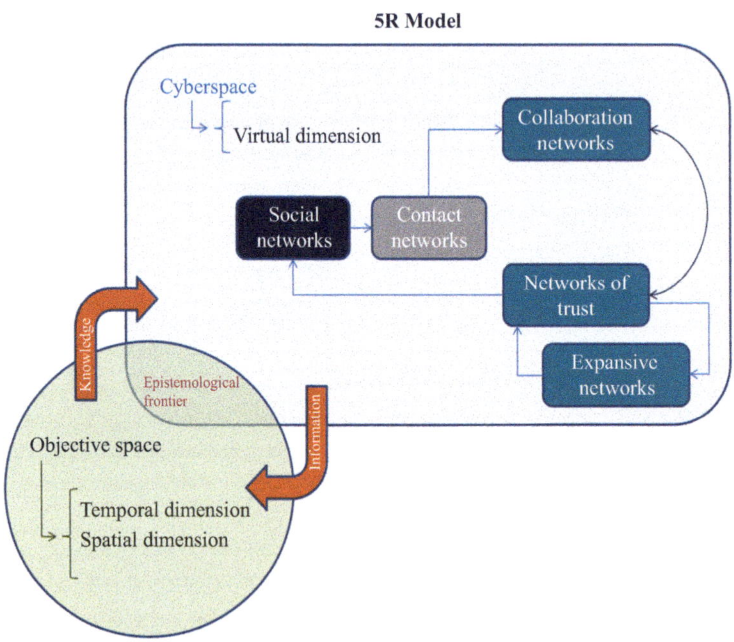

Figure 3.2 The 5R model [87]

cultural referent, a dynamic of globalized escalation is produced characterized by the configuration of expansive networks (R5).

3.5 The 3D man and the new pyramid of needs

Postmodern man is 3D. The man of the cultural transition broke with the limits of time and space to invade the virtual world as a 3D man who moves in physical space (time and space) and, simultaneously, into the virtual space (cyberspace). In this context, projected towards disruptive technological trends, new needs emerge, which are not fully represented in Maslow's traditional Pyramid of Modern Needs.

So, we put forward an inverted pyramid that complements it, considering from the base the new needs of 3D subjects, the needs of the cultural singularities to which the agents of culture belong and the needs associated with culture on a human scale. Among these needs, we can highlight the following:

1. Individual needs (subjects as cultural agents): digital literacy, 2D identity (real + virtual), holistic-eclectic citizen, autonomous learning, self-regulated learning, access to information, information management, information representation and development of talents.
2. Specific needs (local culture).
3. Global needs (culture on a human scale): needs emanating from culture on a human scale, transculturation, knowledge transfer, the need to rethink the known, to adapt to new contexts, to contextualized digital inclusion.

This new pyramid of needs gives an account of the coexistence of two worlds, the face-to-face and virtual worlds, that work differently and that, both, require processes of humanization and inclusion that account for the full satisfaction of those needs added since the introduction of deep technologies (Figure 3.3).

3.6 Reflections on a humanized and inclusive appropriation of technologies in society, culture and education

Technologies emerge from man himself and are essentially the artificial dimension of human life. They emerge as adaptive manifestations to the environment and have developed throughout history to improve the living conditions of humanity. However, technologies are not neutral, they constitute dialectical entities; human beings carry in their essence man's own contradictions, and they are made to provoke well-being or to produce harm.

The history of man has been debated between contradictions that have shaped his culture, with technologies being one more expression of these contradictions. The most evident tension lies in constructing technologies almost exclusively for progress, which is material, since as they are oriented towards the purpose of producing well-being, under this excuse they can cause harm.

Figure 3.3 Pyramid of needs for the digital citizen [88]

The welfare state is circumscribed to a spiral of consumption, a process in which the demand for new technologies is endless, since the satisfaction of emerging needs is unfinished. This provokes an overflowing and thoughtless dynamism in the production of technologies, exposing them to a late discovery of their possible harmful effects. For example, Japan is currently banning the use of microwave ovens because their harmful effects on health were confirmed after millions of people around the world used them. The alternative is to design and produce technologies for development, which implies putting them at the service of human, social and cultural development; this should lead them to be factors that contribute to the common good.

Technologies are humanized in their essence when they acquire anthropological depth, when they are understood as a manifestation of human intelligence and when they have been made to be used under an ethical conduct that orients their action so as not to affect or cause harm to people or the environment. A simple device such as a knife can be used to process food or to kill. It is in its use that it is humanized and acquires its essential meaning. When technologies are made and used to cause harm, they lose their meaning and become dehumanized. War machines will never be an expression of technological humanization, since in their essence they have been built to destroy. It is purpose and use that humanizes technologies.

Technologies are a fundamental part of a cultural transition, between modernity and a new culture that is still in definition. This process, of profound transformations on a human scale, is defined from different theoretical conceptualizations: postmodern culture, information society, global culture (village), knowledge society, liquid society, relative society, end of history and cyber culture, among others. Above these conceptual definitions is the need to discover the cultural keys of change, to establish the demands of humanization and inclusion. Among these keys, it is possible to identify the following:

1. The emergence of a virtual epistemology, where it is necessary to establish the border between the administration of information (access and representation) with the management of knowledge (creation and transfer).
2. Cybernetization and cybernetic consciousness. It means in practice that more and more people unleash complex networks of automated decisions based on very simple decisions, which activates decisions and digital directions that operate to achieve a certain purpose, without people needing to know how the systems operate (cyber black box). In new generations, this operates as an intuitive cyber consciousness, as children are born manipulating computer and cyber artefacts. The problem lies in the fact that the tendency is situated in the consumption of information, assumed as access to pre-made knowledge, rather than as an opportunity to produce new knowledge mediated by technology and the use of language as a means to make it transferable.
3. Role modification, which implies that individuals have modified their individuality, to become global 3D citizens, who live their lives coexisting in the objective world of space and time with the cyberspace that makes up the digital world.

4. New human needs. The cultural transition demands the satisfaction of new needs, emanating from individuals as cultural managers, from the uniqueness of the human groups to which they belong, and from culture on a human scale.

The dynamic that moves today's society is the movement of networks which operate as open systems. These new dynamics act through the power of social networks which mobilize individuals and human masses; networks of contacts which make up the preliminary instances of virtual coordination; networks of collaboration which are based on the sharing of common interests; networks of trust which deepen in the sustained management over time of common purposes; and expansive networks which are the maximum expression of the successful management of purposes, by generating the capacity to mobilize networks of networks. In this dynamic, human inclusion is above ideological, political, religious, gender, social and even cultural differences, since they go beyond the limits of time and space, invading unlimited virtuality. It encloses the new dimension of inclusion in the cultural transition, which goes beyond the limits of modernity.

With technological disruption, new needs emerge:

1. New needs of the Subject as cultural manager: digital literacy; information management; development of talents; self-regulated and autonomous learning; holistic and eclectic skills; 2D identity development.
2. New needs of local culture from the singularity.
3. New needs of global culture: digital inclusion; adaptation to new contexts; rethinking the known; knowledge transfer; culture on a human scale.

In education, the demand is paradigmatic, and it is necessary to rethink education. The student of the cultural transition develops from a 3D digital identity: (i) identity of the subject as manager of his talents; (ii) identity from the cultural singularity of the human group he belongs to; and (iii) identity on a human scale. The 3D student learns in physical spaces (time and space) and simultaneously learns in cyberspace (virtuality). It is necessary to complement a curriculum located with a distributed curriculum, whose main characteristics are the following: (i) virtual epistemology; (ii) students with 3D identity; (iii) flexible school, where autonomous, self-regulated and lifelong learning is promoted; (iv) knowledge management, to link individual intellectual capital with the conformation of collaborative social capital; (v) knowledge from the cultural singularity to manage knowledge in culture on a human scale; (vi) integration of contents in learning networks, to overcome the technological opportunism of information consumption; and (vii) ecological and lifelong education, as an existential option to form happy students in learning.

3.7 Conclusions

Technological disruption is projected into the next two decades, as a very rapid and abrupt historical, social and cultural change. It means that the ways of solving human problems will become cybernated; that is to say, more and more people, from very simple automated decisions, will unleash complex networks of digital decisions.

Progressively more people will have many automated devices applied to everyday life and work. There will be various transhuman creatures (Kurzweil, 2005), known as cyborgs, that will assist men in solving many of their needs; this added to the IoT, virtual commerce, nanotechnology, biotechnology, intelligent air and land vehicles, and 3D and 4D printing, among other postmodern technologies. All of them, as a whole and in their progressive massification, will drastically modify human life.

This forward-looking panorama contains ethical–ontological demands unprecedented in the history of humanity. A new anthropological–philosophical definition will be needed to differentiate and, at the same time, make human life and intelligence coexist with life and AI. If we assume that technology is what man is, it means that technology is born in anthropological notions that define and determine man himself.

Technology contains in itself man's own contradictions. Man, as a natural intelligence, creates an AI as an extension of himself, so technology is man, expressed through his talents, his ethical convictions and in his virtues and defects, his evils and hatred.

Deep technologies constitute a necessary and urgent effort, of an ethical and ontological type, that we human beings must make, from today and in the future, in order to assign good contents to the development and use of technologies. The basic problem is what technologies are built for, what contents we human beings provide them with and what we use them for. In their usability appear the contents of technological re-humanization.

Different is the reality of the Deep Internet or Deep web [89] which is an invisible virtual space, which is hidden from most users, whose contents are not linked by conventional Internet search engines. It is estimated to have a volume of information 45 times larger than the open Internet. It contains a large proportion of cyberspace, which is configured for the operation of private forums, intranets of institutions and companies and hidden publications and is also used to develop links to local and international crime, to coordinate terrorist actions, to traffic drugs and to virtually manage other manifestations of the most abject interests and passions of the human being. This disruptive context questions the conscience of men, demanding definitions that tend towards the re-humanization of technologies. As long as good men exist, knowledge will be generated and technologies built for a better life, and as long as knowledge and technologies are created for noble purposes, there is still hope.

References

[1] Aristóteles. Libro I de la Metafísica. S. IV a.C.
[2] Aristóteles. Libro II de la Física. S. IV a.C.

[3] Adorno, T. and Horkheimer, M. Dialéctica de la Ilustración. Fragmentos filosóficos Madrid. Trotta; 1944/1998.

[4] Sparza, J. Curso General de Disidencia. Apuntes para una visión del mundo alternativa, Madrid: Ediciones El Emboscado/col. Metapolítica; 1997.

[5] Foucault, M. Las redes del poder. Conferencia dictada a Facultad de Filosofía de la Universidad del Brasil. Revista Barbarie. 1976; 4.

[6] Ferré, F. Philosophy of Technology. USA. Athens: University of Georgia Press; 1995.

[7] Stiegler, B. La técnica y el tiempo. Vol. 1 El pecado de Epimeteo. Paris: Editions Galilée; 1994.

[8] Stiegler, B. La técnica y el tiempo. Vol. 2 La desorientación. Paris: Editions Galilée; 1998.

[9] Deleuze, G. Qu'est-ce que la philosophie? París: Ed. Minuit; 1991.

[10] Deleuze, G. Post-scriptum sobre las sociedades de control. Polis, Revista de la Universidad Bolivariana, vol. 5, núm. 13, 2006, Universidad de Los Lagos Santiago, Chile. http://www.redalyc.org/articulo.oa?id=30551320

[11] Latour, B. Nunca fuimos modernos. Ensayo de Antropología simétrica. Argentina: Siglo XXI; 2007. 12.

[12] Derrida, J. Una filosofía deconstructiva. Zona erógena, 35; 2007. Available at: www.jacquesderrida.com.ar/textos/filosofia_decontructiva.htm

[13] Toulmin, S. Conceptual Revolutions in Science. In: Cohen, R.S. and Wartofsky, M.W. (eds) A Portrait of Twenty-five Years. Boston Studies in the Philosophy of Science; 1967. pp. 58–74, Springer, Dordrecht. DOI https://doi.org/10.1007/978-94-009-5345-1_6

[14] Castany, B. Humanismo, modernidad y posmodernidad. Una reflexión sobre el doble origen de la 'modernidad' a la luz de 'Cosmópolis' y 'Regreso a la razón' de Stephen Toulmin. Cartaphilus: Revista de Investigación y Crítica Estética, ISSN-e 1887-5238; 2018 (16): 11–35. Available at: https://revistas.um.es/cartaphilus/article/view/356821/256051

[15] Scott, J. and Baltzly, A. Readings in European History since 1814. New York: Appleton-Century-Crofts, Inc; 1930.

[16] Darwin, C. The Different Forms of Flowers on Plants of the Same Species. New York: Appleton and Company; 1877.

[17] Pasteur, L. Note sur la cristallisation du souffre; 1848. *OP1*, 31–32.

[18] Dalton, J. A New System of Chemical Philosophy, Part 1. 1808, London.

[19] Lyotard, J. F. La condición postmoderna. Madrid: Cátedra; 1989.

[20] Lipovetsky, G. La era del vacío. Ensayos sobre el individualismo contemporáneo. Anagrama. Colección argumentos; 1983.

[21] Seidman, S. The Postmodern Turn: New Perspectives on Social Theory. UK: Cambridge University Press; 1994.

[22] Roa, A. Modernidad y posmodernidad. Santiago de Chile: Andrés Bello; 1995.

[23] Noya, J. (2004). Teoría e investigación en la sociología española contemporánea. Política y Sociedad. 2004; 2(41): 163–174. Available at: http://revistas.ucm.es/index.php/POSO/article/view/POSO0404230163A

[24] Bayón, F. Disrupción; 2018. Available at: http://www.eoi.es/blogs/fernandobayon/2014/02/24/disrupcion/
[25] McCarthy, T. La Teoría Crítica de Jürgen Habermas. Madrid: Tecnos; 1987.
[26] Ashton, K. That Internet of Things thing; 1999. Available at: http://www.rfidjournal.com/articles/view?4986
[27] Capra, F. La trama de la vida. Una nueva perspectiva de los sistemas vivos. Segunda edición. Barcelona: Editorial Anagrama; 1999.
[28] Negroponte, N. Ser digital. 1st ed. Barcelona: Ediciones B, S.A; 1995.
[29] Moore, G. Cramming more components onto integrated circuits. Electronics. IEEE Solid-State Circuits Society Newsletter, 2006; 11(3), pp. 33–35. Available at: https://ieeexplore.ieee.org/document/4785860
[30] Kurzweil, R. La singularidad está cerca. Cuando los humanos trascendamos la biología. España: Cimpapres; 2005.
[31] Giraldo Ramírez, M.E., Sánchez Arias, V.C., Paquienséguy, F., et al . Futuro de los sistemas y ambientes educativos mediados por las TIC. México; Universidad de Guadalajara, Sistema de Universidad Virtual; 2017. Available at: http://148.202.167.116:8080/jspui/bitstream/123456789/1833/1/Futuro%20de%20los%20sistemas%20y%20ambientes%20educativos%20mediados%20por%20las%20TIC.pdf
[32] Harrison, C., Eckman, B., Hamilton, R. et al. Foundations for smarter cities. J. Res. Dev. 2010, 54, 1–16.
[33] Colado, S., Gutiérrez, A., Vives, C. and Valencia, E. Smart City. Hacia la Gestión Inteligente; Marcombo: Barcelona, Spain, 2014.
[34] Ishida, T. and Isbister, K. Digital Cities: Technologies, Experiences, and Future Perspectives; Springer: Berlin, Germany, 2002.
[35] Albino, V., Berardi, U. and Dangelico, R. Smart Cities: Definitions, Dimensions, Performance, and Initiatives. J. Urban Technol. 2015, 22, 3–21. Available at: https://www.tandfonline.com/doi/full/10.1080/10630732.2014.942092
[36] Arizmendi, S., Navío, J. and Portilla, J. Smart cities: ¿Cómo determinar el estado de desarrollo de una ciudad inteligente? Telos Cuad. Comun. Innov. 2016, 105, 49–65.
[37] Cohen, B. and Obediente, E. Estudio Ranking de Ciudades Inteligentes en Chile. Centro de Estudios Digitales, Fundación País Digital. 2014. Available at: http://dg6223fhel5c2.cloudfront.net/PD/wp-content/uploads/2014/06/Ranking-Ciudades-Inteligentes-en-Chile.pdf.
[38] Olmedo, E. and López, A. De la Smart City a la Smart Human City. Inclusión digital en aplicaciones. Revista Fuentes. 2015 (17), 41–65. Doi: http://dx.doi.org/10.12795/revistafuentes.2015.i17.02
[39] Mokyr, J., Vickers, C. and Ziebarth, L. The history of technological anxiety and the future of economic growth: is this time different? Journal of economic perspectives. 2015; 29(3): 31–50.
[40] Fukuyama, F. The end of History and the lastman. Nueva York: The Free Press (edición española traducida en Barcelona por Editorial Planeta) 1992.

[41] Argüello, S. Los efectos de la automatización sobre el trabajo Desempleo tecnológico, polarización del mercado laboral y políticas públicas. Biblioteca del Congreso Nacional de Chile | Asesoría Técnica Parlamentaria; 2019. Available at: https://www.bcn.cl/obtienearchivo?id=repositorio/10221/26941/1/Los_efectos_de_la_automatizacion_sobre_el_trabajo.pdf
[42] Frey, C. and Osborne, M. The future of Employment: How Susceptible Are Jobs to Computerisation? Working Paper Oxford martin School University of Oxford; 2013.
[43] Fundación Chile. Automatización y Empleo en Chile REPORTE FINAL; 2017. Available at: https://www.comisiondeproductividad.cl/wp-content/uploads/2018/11/Nota-T%C3%A9cnica-8-Automatizaci%C3%B3n-y-Empleo-Fundaci%C3%B3n-Chile.pdf
[44] OCDE. Future of work and skills. Paper presented at the 2nd meeting of the G20 employment working group; 2017. Available at: http://bcn.cl/28mxu
[45] Author, D. Why are there still so many Jobs? The history and future of workplace automation in Journal of Economic Perspectives. 2015; 29(3):3–30.
[46] Lévy, P. ¿Qué es lo virtual? Barcelona: Ediciones Paidós Ibérica; 1999.
[47] Foucault, M. Vigilar y castigar, Buenos Aires: Siglo XXI; 1989.
[48] Sibila, P. El hombre postorgánico. Cuerpo, subjetividad y tecnologías digitales. Buenos Aires: Ediciones B, Barcelona; 2005.
[49] Gil García, J. Desigualdades, límites y posibilidades para la trasformación del capital en las economías colaborativas. Revista de Estudios para el desarrollo social de la comunicación. 2017 0(15), 32–67. Available at: http://revista-redes.hospedagemdesites.ws/index.php/revista-redes/article/view/505/534
[50] Foucault, M. Nacimiento de la biopolítica: curso del Collège de France (1978–1979). Ediciones Akal. 2009; 283.
[51] Laval, Ch., y Dardot, P. La nueva razón del mundo. Editorial GEDISA; 2014.
[52] Brown, W. Undoing the Demos. Cambridge, MA: MIT Press; 2015.
[53] Comisión Europea. Una Agenda Europea para la economía colaborativa; 2016.
[54] Rodríguez, S. Los modelos colaborativos y bajo demanda en plataformas digitales. Sharing España y Asociación Española de la Economía Digital; 2017.
[55] Slee, T. What's Yours Is Mine: Against the Sharing Economy; New York and London: Or Books; 2016.
[56] Scholz, T. Platform Cooperativism. Challenging the Corporate Sharing Economy. New York, NY: Rosa Luxemburg Foundation; 2016.
[57] De Rivera, J. Sociología de las instituciones digitales y consumo colaborativo. Tesis inédita UCM; 2017. Available at: http://sociologiayredessociales.com/textos/la-teoria-del-consumo-colaborativo.pdf
[58] Morozov, E. 'La fiebre de las plataformas'. Tribuna de El País; 2015. Available at: https://elpais.com/elpais/2015/06/25/opinion/1435251058_955249.html
[59] Kostakis, V. and Bauwens, M. Network Society and Future Scenarios for Collaborative Economies. Basingstoke: Palgrave. Macmillan, 2014. DOI: 10.1057/9781137406897.0012.

[60] Lombardero, L. Trabajar en la era digital. Pixel-Bit. Revista de Medios y Educación. 2016; 49, pp. 243–244.
[61] Roig, D. Socionomía. ¿Vas a perderte la revolución social? Bilbao: Deusto; 2012.
[62] Gairín, J. Gestión del conocimiento y desarrollo organizativo: formación y formación corporativa. Wolters Kluwer Barcelona: 2012; 68.
[63] Sánchez, J. and Ruiz, J. Recursos didácticos y tecnológicos en educación. Madrid: E. Síntesis; 2013.
[64] Garcia, I., Peña, I., Johnson, L., Smith, R., Levine, A. and Haywood, K. Informe Horizon. Edición iberoamericana; 2019.
[65] Durall G., Gros S., Maina M., Johnson L. and Adams S. Perspectivas tecnológicas: educación superior en Iberoamérica. Austin, Texas, The New Media Consortium; 2012.
[66] Johnson, L., Adams Becker, S., Cummins, M., Estrada, V., Freeman, A. and Ludgate, H. NMC Horizon Report: Higher Education Edition. Austin, Texas: The New Media Consortium; 2013.
[67] Sharples, M., Adams, A., Ferguson, R., *et al.* Innovating Pedagogy. Exploring new forms of teaching, learning and assessment, to guide educators and policy makers. The Open University; 2014.
[68] Esteve, F. and Gisbert, M. El nuevo paradigma de aprendizaje y las nuevas tecnologías. Revista de Docencia Universitaria. 2011; 9(3): 55–73.
[69] Plomp, T., J. Pelgrum, W. and Law, N. International comparative survey of pedagogical practices and ICT in education. Education and Information Technologies. 2007; 12: 83–92. Available at: 10.1007/s10639-007-9029-5.
[70] Unión Europea, versión consolidada del Tratado de Funcionamiento de la Unión Europea, 13 de diciembre de 2007, 2008/C 115/01, disponible en: https://www.refworld.org/docid/4b17a07e2.html [consultado el 26 de junio de 2019].
[71] Barrera O. and Leigh L. El uso y mal uso de las computadoras en la educación: evidencia de un experimento aleatorio en Colombia. Documento de trabajo sobre investigación de políticas del Banco Mundial. 2009.
[72] Gil D., Vilches A., Edwards M., Praia J., Marqués L. and Oliveira, T. A proposal to enrich teachers perception of the state of the world: first results. Environmental Education Research. 2003; 9(1): 67–90.
[73] Upegui M. El consumo tecnológico de los jóvenes... Un desafío para los docentes. Colección Académica de Ciencias Sociales. 2017; 4(1): 51–60.
[74] Rifkin, J. La sociedad de coste marginal cero: el internet de las cosas, el procomún colaborativo y el eclipse del capitalismo. Barcelona, Paidós; 2014.
[75] Cabero, A. Los retos de la integración de las TICs en los procesos educativos. Límites y posibilidades. Perspectiva Educacional. Pontificia Universidad Católica de Valparaíso, Chile. 2010; 49(1): 32–61.
[76] Dima L., Urdaneta R., Sosa G., Esperón G. and López D. Búsqueda de la calidad del aprendizaje mediante herramientas de inteligencia artificial. Ciencia y Tecnología. 2012; 1(57). Available at: 10.18682/cyt.v1i12.654. DOI: 10.18682/cyt.v1i12.654

[77] Caro, M., Josyula, D. and Jiménez, J. Modelo pedagógico multinivel para la personalización de estrategias pedagógicas en sistemas de tutoría inteligentes. Dyna Medellin. 2015; 82. 185–193.

[78] Gisbert, M., González M. and Esteve, F. Competencia digital y competencia digital docente: una panorámica sobre el estado de la cuestión. Revista Interuniversitaria de Investigación en Tecnología Educativa; 2016.

[79] Ainsworth, S. and Fleming, P. Evaluating authoring tools for teachers as instructional designers. Computers in Human Behavior. 2006; 202.

[80] Jimenez, J. Un modelo de planificación instruccional usando razonamiento basado en casos en sistemas multiagente para entornos integrados de sistemas tutoriales inteligentes y ambientes colaborativos de aprendizaje. Universidad Nacional de Colombia. Tesis Doctoral; 2006.

[81] Rodríguez, G. La inteligencia artificial en la educación superior. Oportunidades y Amenazas. INNOVA Research Journal. 2017; *l* 2(8): 412–422.

[82] Tuomi, I. The Impact of Artificial Intelligence on Learning, Teaching, and Education. Policies for the future, Eds. Cabrera, M., Vuorikari, R & Punie, Y., EUR 29442 EN, Publications Office of the European Union, Luxembourg, 2018, doi: 10.2760/12297, JRC113226.

[83] Popenici, S and Kerr, S. Exploring the impact of artificial intelligence on teaching and learning in higher education. Research and Practice in Technology Enhanced Learning; 2017. Available at: https://doi.org/10.1186/s41039-017-0062-8

[84] Castells, M. La era de la información: economía, Sociedad y cultura. Alianza, Madrid, 1997.

[85] McLuhan, M. and Powers, B.R. La Aldea Global. Transformaciones en la Vida y Los Medios de Comunicación en el Siglo XXI; Editorial Gedisa: Barcelona, Spain, 1989.

[86] Careaga, M. and Barnes, S. Actualización Circuito Pedagógico de Gestión del Conocimiento. 2015. Available at: http://marcelocareaga.blogspot.com/

[87] Careaga, M., Fuentes, C., y Molina, J. El Modelo 5-R. 2018. Available at: http://marcelocareaga.blogspot.com/2018/12/modelo-5r.html

[88] Carrasco-Sáez, J.L., Careaga Butter, M., and Badilla-Quintana, M.G. The new pyramid of needs for the digital citizen: A transition towards smart human cities. Sustainability; 9, 2258. 2017. Available at: https://www.mdpi.com/2071-1050/9/12/2258

[89] Bergman, M. The deep web: surfacing hidden value. Journal of Electronic Publishing. 1994/2001; 7, 1–17. Available at: http://www.press.umich.edu/jep/07-01/bergman.html

Chapter 4

New frontiers of human wisdom: information, communication and consciousness empowered decision-making under the broader realm of reality

Sanjay Bhushan[1]

The history of human creative and innovative pursuits is, at times, highlighted with instances of some most intellectually accomplished minds' biggest regrets about their own intuitive wisdom and scientific discoveries, ultimately proving to be unwarranted during some future course of modern observations and scientific developments. The most notable of all is Albert Einstein's self-admitted, suppressing his own intuitive conviction, the so-called 'greatest blunder ever made in life' in 1917 about his failure to accept the validation of his own equations of the 'cosmological constant' (usually denoted by the Greek capital letter Lambda: Λ) in his original field equations of general relativity, which ultimately turned out to be in fact quite consistent with current observations in view of the fundamental role that this cosmological constant now appears to be playing in the modern cosmology and closely associated to the concepts of dark energy and quintessence. This instance reinforces the very deep-rooted, fundamental and pioneering role played by intuitive consciousness as a guiding instinct or force in majority of pure creative enterprises and pursuits that the humans have been engaged with since the time immemorial. In this context, the present chapter aims at presenting a dossier of concepts and applications of information, communication and consciousness science going beyond conventional human and materialistic endeavours of the present-day society. It entails and invokes a systemic holistic perspective towards resolving many issues and challenges that the global society is currently upfront with. Systems perspective of cognitive consciousness holds that cognition can be explained by means of a continuous dynamic system in which all the elements are interrelated.

The domain or system of cognitive consciousness can be hypothesized to be one such system where numbers of endogenous and exogenous variables have

[1]Department of Management, Faculty of Social Sciences, Dayalbagh Educational Institute (Deemed University), Dayalbagh, Agra, India

dynamic interplay impacting the physiology, minds and meditations of individuals and consciousness practitioners. Hence, understanding the non-linearity of this form of evolutionary spiritual Cognitive–Physical dynamic system should be paramount. In the same manner, Quantum consciousness can be understood as a Neuronal–Spiritual phenomenon which is inseparably associated and in communion with the dynamic Genetic-physical existence of a human being. Probing further deep also reveals a unique phenomenon of *super causality* as a sum total of all creational causalities and retro-causalities. It can be said that complex dynamical systems are neither completely rigid nor fully random; instead, they display a unique balance of 'integration', cohesion and robustness at the global level and at the same time, differentiation and multiple realizability at the component levels. Hence, in the domain of consciousness too, we see downward causation, in which upstream determinants at the higher-order consciousness (Spiritual) levels influence and regulate events at lower levels of human cognitive consciousness.

It was only as a result of cutting edge physics, particularly quantum physics, the very fundamentals of all existence (perceived) so far that has confirmed again and again that Consciousness is in fact the very ground of Being. There is nothing else apart from consciousness, from which arises mind as does matter. Physicists, of late, have also come to realize that the universe is interconnected in much subtler ways than had once thought. In quantum physics, the observer and the observed can no longer be separated and the whole is more fundamental than the part. According to this view, consciousness imposes 'downward causation'. In other words, our free will is real as intention.

When we act in the world we really are acting with the causal power of our intentions. The conception of downward causation conveys the proposition that while the whole is to some degree constrained by the parts (known as upward causation), at the same time, the parts are to greater degree constrained by the whole.

So, the behaviour of the parts (down) is determined by the behaviour of the whole (up), and thereby, determination moves downward instead of upward. In other words, the global level of integration (the result of 'upwards causation') may produce 'downwards' effects, acting eventually upon the local level of the parts/zones. This view, however, does not deny that matter also has causal potency – it does not deny that there is causal power from elementary particles upward, so there is upward causation – but in addition, it insists that there is also an integral play and overwhelming impact of downward causation.

Methodology. Documented under various sections of the chapter, starting from decision-making in professional business setting to move into the least chartered territories of quantum consciousness, neural and cognitive sciences, consciousness, genetic algorithm (GA), super causality and finally the loop integral dynamics of information consciousness will help the readers visualize the intriguing phenomenon of connectivity and causality which is inherent in the most basic framework of the broader realm of nature and reality. It provides an exposition about the systems perspective of information, communication and consciousness sciences which hold that the various forms of realities can be conceptualized as a continuous dynamic system in which all the elements and sub-elements are closely or distantly interrelated and numbers of endogenous and exogenous variables have dynamic interplay impacting the Physics, Physiology and Psychology together.

4.1 Neural network-assisted information system dynamics for business decision-making

Application of information systems science in business has lately emerged as an exciting area of applied research. It aims to study the behaviour of complex business systems and analyse the convergence of the two powerful approaches of system approximation in form of artificial neural network (ANN) and system dynamics which are based on rigorous modelling and simulation processes. It also envisages a dynamic possibility of data and decision interface generated between the two in order to improve the overall efficiency of a representational system design and decision-making (Figure 4.1). Along with system dynamics, in the area of soft computation-based dynamic modelling researches, artificial neural network – popularly coined as ANN or NN (neural nets) – has also emerged as a computational model as neuro-morphic systems based on biological neural network. As a consequence of this interface, neural network can be used for enriching the SD model in a way that it would assist an SD modeller in policy stabilization of input variables by maintaining a value of target variables within a desired level.

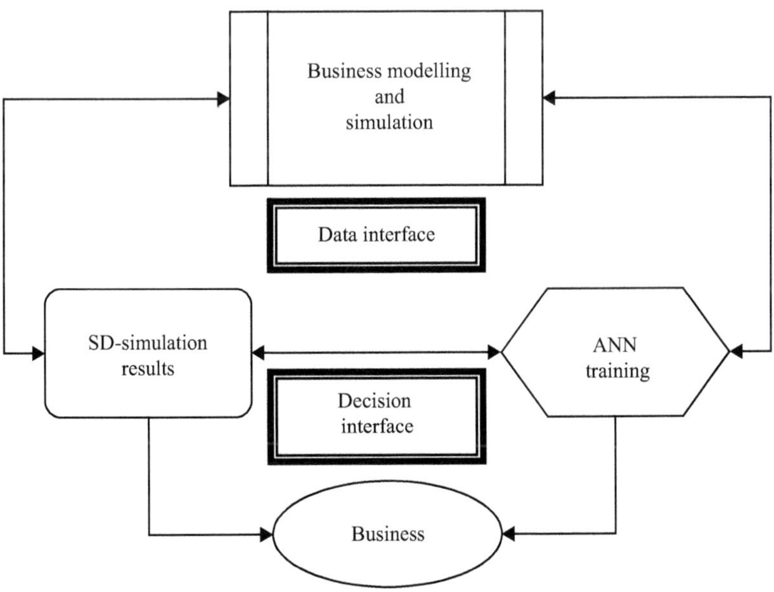

Figure 4.1 System dynamics and neural network interface

Here, the luxury available to the experimenter is of the enormous amount of data generated through SD simulation which would be used for training and testing the working model of a neural network. The goal is to leverage this added understanding into the design and implementation of more efficient business processes and policies.

4.1.1 Data interface of SD–ANN

The dyadic interface and exchange of data inputs between SD and ANN simulation environments promises to facilitate a mutually beneficial enrichment of system design for efficient business decision-making. A successfully trained ANN model is prudent to obtain quick response outputs within acceptable error for values of input covered by training input space. Hence, the ANN model provides excellent opportunity to store the knowledge contained in the variety of SD simulation results which would have otherwise been discarded after analysis. Fortunately, this interface can be facilitated by neural net software products which give access to the power of neural network environment while working within a familiar spreadsheet environment, which in turn is back-interfaced to the SD simulation environment (Figure 4.2). Certain commercially available software products increase the range of applications even further by being easy-to-use, well integrated with Microsoft Excel, and requiring no prior expert knowledge of neural networks to perform complex analysis. In this case, as we are performing SD simulation and generating a mass of data results, the data are simultaneously transferred to a spreadsheet environment which also has a built-in functionality of neural network module. After tagging SD simulation generated data sets under the classified columns of Input and Output (Desired), we get ready to initiate neural modelling and simulations.

4.1.2 Decision interface

Simulation-based experiments performed by Sterman [1] (John D. Sterman, Professor of Management, MIT Sloan School of Management and Director of

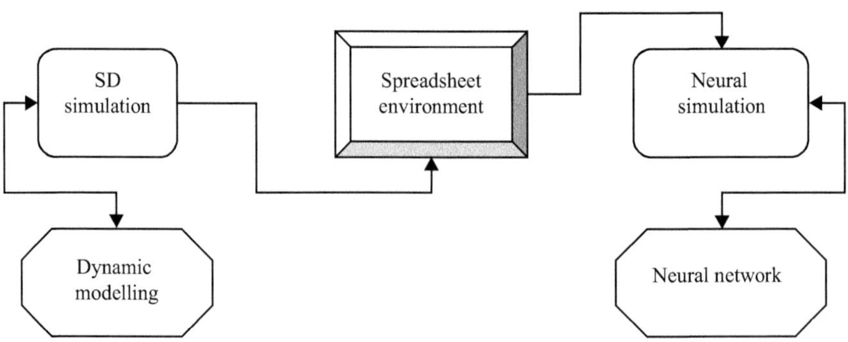

Figure 4.2 SD-spreadsheet–ANN interface diagram

MIT's System Dynamics Groups) showed that under system dynamics, role-players' performance was improved slowly as their experiences are accumulated. The slow learning process highlights the potential power of human intelligence or heuristics. This potentiality can be improved multifold by replacing human decision-maker with a neural network. An ANN model can be successfully trained and used as a quick response model for fast feature extraction of the dynamics of the integrated system such that the outputs are within reasonable acceptable error for values of inputs covered by the input space of training patterns. Feedforward processing of ANN allows the control system to adjust to the change in decision structural response much faster than is possible if adjustments are based solely on spectral averaging in the main feedback loop.

Thus, it greatly solves the response time problem of a feedback loop and in fact could act as a quick response model as an expert decision-making system. In this order, it is appropriate to design a scheme to investigate the results generated by system dynamics simulations in order to globally recognize systemic patterns exhibited by a complex system based on the performance of either categorized or randomly distributed large databases. Such a mechanism would be dynamic to adjust itself according to the user's query (by activating the query module in the neural software model) and the characteristics of the underlying databases. Therefore, the suggested SD–ANN interface could certainly be transformed into an artificially trained decision-support system for handling complex business scenarios even containing incomplete information. This network will accomplish the decision-making task in a massively parallel manner and it is a subsystem which learns and constructs a knowledge base for quick response business decisions.

4.1.3 Direction for future research

The above-deliberated interface of ANN into SD modelling offers a tremendous opportunity of developing an experiential expert system for business policy control, strategy formulation and decision-making. Moreover, the need of such systemic integration is particularly pronounced on account of the easy compatibility offered by these two system research domains. As a consequence of this interface, neural network can be used for enriching the SD model in a way that it would assist an SD modeller in policy stabilization of input variables by maintaining a value of target variables within a desired level. Here, the luxury available to the experimenter is enormous amount of data generated through SD simulation which would be used for training and testing the neural network. Probably, the challenge here is not in the neural network architecture itself, but instead in the choice of variables and the information used for training. Certain exciting possibilities exist in seeing the entire SD–ANN interface described above happening as a real-time (also, online) simulation interface on a novel framework of data exchange between SD and ANN environments which can significantly improve simultaneous convergence and also reduce the overall computational time of simulation and neural training. Moreover, the present study has only considered with multi-layer perceptron (MLP)-neural architect and a simplified SD model of innovation. Other architects of neural network can also be well integrated and experimented with more complex SD models and that too with varying parameters in order to produce better insight on controlling complex business system decision-environment.

4.2 Science of cognitive consciousness and decision-making: a convergent systemic interface to ANN and GA

The domain or system of cognitive consciousness can be hypothesized to be one such system where a number of endogenous and exogenous variables have dynamic interplay impacting the physiology and psychology (minds and meditations) of individuals or decision-makers. Hence, understanding the non-linearity of this form of evolutionary cognitive–physical dynamic system could be of paramount significance. System dynamics, ANN and GA together have the potential of approximating the behaviour of extremely complex systems such as of spiritual–physical system.

This section conceptualizes systemic interfaces between SD, ANN and GA in order to explain the dynamics of cognitive–physical systems through their connectionist approaches that may also be quite instrumental in developing an experiential learning system for attaining consciousness of high order in the personal and professional decision-making activities.

4.2.1 Dynamics systems approach towards cognitive consciousness

Dynamic systems approach of cognitive consciousness holds that cognition can be explained by means of a continuous dynamic system in which all the elements are interrelated. One can safely infer that dealing, controlling and optimizing only 'segmented' variables or components of a cognitive–physical system would lead to suboptimization of the whole system and that too through an erroneous way of consciousness awakening among decision-makers. Hence the requirement is of a holistic conception which would enable construction of a dynamic model for the variability and uncertainty of a non-linear diversified cognitive–physical system representing complex interdependencies between various influencing interest flow or stock variables and help realistically analyse the performance trade-offs associated with different consciousness awakening decision-making assumptions. Such endeavour will also take into account the time evolutionary dynamics endogenously created by such cognitive–physical system structures stretched over a dynamic time track. Otherwise, leaving the constituent interest actors to strive to optimize their individual performances will definitely lead to a weak or suboptimal decisions and consciousness awakening solution.

Hence, system dynamics framework-based consciousness-related decision-making is preferably essential in order to provide the basic building blocks necessary to construct such system models that help in developing our fundamental understanding about the impact dynamics of such multilevel–multivariable complex systems over human minds and their corresponding efforts to awaken their consciousness. For a practitioner or conscious decision-maker, the goal would be to leverage this added understanding to design and implement effective consciousness-awakening solutions.

Now, a much simplified 'single stock-multiple flow' order cognitive–physical system dynamic model can be conceptualized as demonstrated by Bhushan [2], where 'Consciousness Awakening' is treated as a stock variable as it indicates the level or accumulation of consciousness in terms of high consciousness quotient whose value can only be changed by flows in and out (the double-lined arrows). These flows are controlled by a (+) positive feedback loop emanating from the stock of 'gained consciousness' eventually reinforcing the entire consciousness dynamic system by positively triggering the values of inflow variables conceptualized here as 'consciousness genome' and 'consciousness neurons' under the dynamic evolutionary channel variables of ANN and GA.

The above model also envisages dynamic reinforcement loops (+) mutually created between and amongst the subsystem variable components of ANN, GA, consciousness neurons and consciousness genome themselves. Thus, it provides a holistic view of the system by graphically showing the causal relationships between different elements of the system and the system as a whole (Figure 4.3). It has to be reiterated that another key element of the system dynamics approach is the time evolutionary view which allows the representation of the behaviour of the system as it evolves through time, giving a dynamic rather than a static view of the system.

Going beyond the prevailing scientific paradigm considering information processing inside the central nervous system as occurring through hierarchically organized and interconnected neural networks (for instance, the visual information

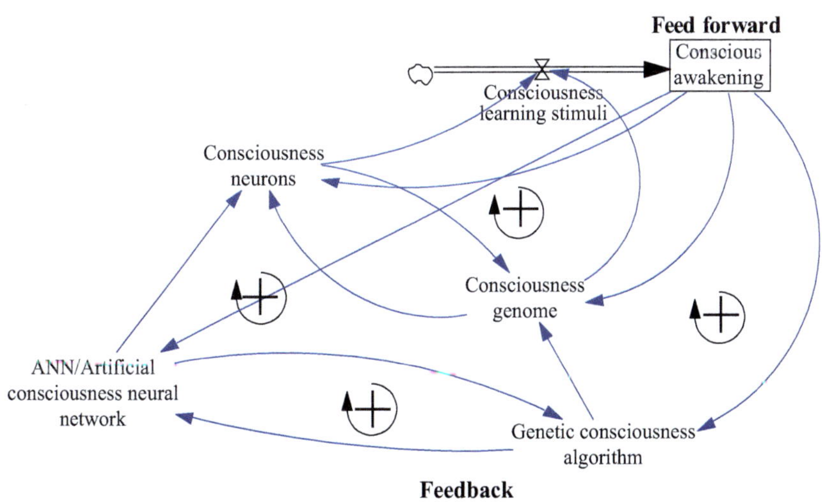

Figure 4.3 Feedback–feedforward-enabled consciousness system dynamics stock-flow diagram

is first hierarchically processed at the level of retina from the photoreceptor rods and cones, to the ganglion cells to be then hierarchically proceeded within the levels of primary, secondary and tertiary sensory and interpretatory cortical regions), we could have the similar conception of structural interconnectedness generalized to understand and describe the spiritual–physical system.

A structural neural pattern might be conceived to exist for a cognitive–physical system in the human brain where the theoretic existence of 'conscious neurons' can be conceptualized in the grey and white matter of human brain, which could have a hierarchical pattern of interconnectedness within and outside the cerebral cortex of a human brain. The hypothesis of conscious neurons is based on the metaphoric premise of existence of electromagnetic force in the physical world. As electromagnetic fields by contrast are regarded as real and independent parts of our physical world and even though one may sometimes be able to specify the values of such a field by appeal to the behaviour of particles in it, the fields themselves are regarded as concrete constituents of reality and not merely as abstractions or sets of relations among particles. Similarly one could regard 'consciousness' as referring to a component or aspect of reality that manifests itself in conscious states through 'conscious neurons', but is also more than merely the abstract nominalization of the adjective 'conscious' applied to them.

It can be safely inferred here that as an external stimulating technology (which can be coined as artificial consciousness neural network – ACNN), ANN offers remarkable opportunity for developing a feedforward enabled experiential learning system of a complex system like one conceived here as conscious–physical system by mimicking the neural patterns of human brain. Specializing in feedforward assisted experiential knowledge; a neural network (ACNN) can store and expand its knowledge base via strikingly human training routes through a synthetic learning process and information storage involving interconnection strengths of synaptic weights.

In this regard, one of the most significant concepts in the field of neural networks is the Hebb's biological learning law, according to which the more frequently activated synapses strengthen the connection while those ones less frequently activated weaken it. One another very significant advantage of externally inducing or activating consciousness synapse can be explained through the feature of 'summation process', where the extensive mutual networking of neural synapses of an activated bunch of consciousness neurons and their corresponding synapses having the possibility of awakening even 'weak neurons' (with dormant or very low synapse and hence, not reaching to the threshold for action potential initiation). This can be stimulated by (both through deep concentration and meditation activity) a simultaneous firing or bursts of action potentials of trained and activated consciousness neurons initiating 'awakening impulse' in their cells where neural synapses are weak, dormant or even dead.

However, understanding of non-linear dynamics in neural circuits by theoretical, computational and experimental means is essentially required for the purpose. This type of study should also concern with the non-linearity at synaptic, cellular and network levels. Moreover, the role of synchronization as carrier of information with respect to global activity patterns in the brain should also be addressed. These good properties of cognitive–physical ACNN neural networks can inspire many investigators to offer solutions for most problems with sufficiently big network and adequate training by finding corresponding network topology and training rules for every particular task. Thus, ACNN can significantly serve as an adaptive but artificially stimulated consciousness networking system which can change the neural structure based on external or internal information inputs that flow through the network during its learning phase stretched over a dynamics time track. It can thus be inferred that the use of ACNN in order to understand, construct and train a conscious–physical neural system can be greatly advantageous to emulate brain functioning going metaphysical (i.e. beyond physical neural constraints) towards offering an explanation for the dynamics of awakening of dormant conscious forces contained within.

4.2.2 GA: consciousness gene

As stated and speculated by D. Jones [3], a prominent genetic scientist, in *Nature* that

> if consciousness is a definite, inheritable characteristic, it must have had survival value for it to evolve. It then follows that consciousness must be encoded somewhere in our genes. Only a single gene may be enough, for consciousness seems to be an uncomplicated phenomenon ... we can determine for certain if any of the *lower* animals are also conscious.

Genes are virtually responsible for every step of the neurotransmitter cycle, including the formation, transport, pre-synaptic expression and post-synaptic reception of the transmitter. Genes thus operate at every level of the neural process. They can indeed be treated like fundamental building blocks for both the structure and the functioning of the brain. They set the stage for how neurons and functional groups of neurons act in response to different inputs. An SD-segmented diagrammatic representation of such interconnection can be shown as a causal model. Genes are therefore fundamental for the way we experience, think and behave. Therefore, based on the above postulate, we can safely move from ACNN into the domain of GA-another affiliate of systems science. DNA has been reported to emit electromagnetic radiation (photons) in the microwave spectrum (1–4 billion cycles per second derived from the electricity that is flowing through our nervous system and spinal cord) and hence it may also carry a message to the consciousness neurons having photon receptors or neurotransmitter receptors through 'electrical synapses' which exist much less in number than chemical synapses in the neural arrangement but are more reliable (depicted as genome-neuron loop in the consciousness system dynamic Stock-Flow Model 3; Figure 4.4). This would conceivably build a tunnel of DNA–photon exchange through cognitive–cerebral gateway. On the other hand, crossover is the basis of GAs but the possibility of mutation is also very exciting as it can greatly reduce the transition time taken to evolve a super genome through a natural selection and crossover period. In fact, the

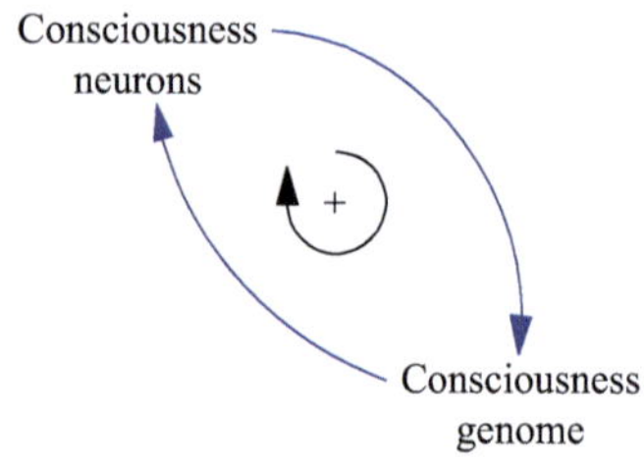

Figure 4.4 Genome–neuron exchange

desired solution may happen not to be present inside a given genetic pool, even a large one. Mutations allow the emergence of new genetic configurations, which, by widening the pool, improve the chances to find the optimal solution.

Thus, all in all, system dynamics, NN and GA together have the potential of approximating the behaviour of extremely complex systems such as being encountered in cognitive–physical system. The above deliberated system approach based interfaces between SD, ANN and GA offer a tremendous opportunity of explaining the dynamics of such system through their connectionist approaches that may also be quite instrumental in developing an experiential learning system for attaining cognitive consciousness of high order.

4.3 Quantum consciousness, neural genetic correlate and quantum computation interface: new frontier for decision-making

Human cognitive consciousness, in some difficult to define sense, is also quantum holistic in character. This gives hope that it can finally be explained in 'quantum' terms on the lines of quantum mechanics. Classical mechanics, on the other hand, does not naturally accommodate and adequately explain the dynamics of cognitive consciousness whereas it is lately believed that cognitive consciousness appears to exhibit 'quantum' like qualities. This realization can come only when we look at the subatomic scale of cognitive consciousness where all regular solid structures disappear or dissolve into waves or waves of probabilities. The way we perceive the reality around cognitive consciousness as of now is basically an illusion or 'perceived reality'. A number of theories propounded in this regard aim to subscribe to the idea of quantum neuro-physics explaining how the classical world of consciousness can originate from quantum processes inside the human mind/brain. Moreover, by introducing consciousness, one obtains a philosophically elegant resolution of the paradox of quantum measurement which says that probabilistic occurrences in the quantum world can be replaced by definite occurrences when they enter into the realm of consciousness.

Quantum consciousness can be understood as a neuro-physical-consciousness phenomenon which is inseparably associated and in communion with the dynamic genetic–physical existence of a human being. The present section postulates a dynamic hypothesis and scientific exploration of quantum consciousness; it's neural-genetic correlate and the prospect of quantum computation apart from proposing a thrilling prospect of genetic engineering in order to evolve a perfectly compatibility human physical system in communion with the 'quantum conscious spiritual system'. Quantum consciousness is one of those multidisciplinary areas of integrative systemic research where we borrow the principles of quantum physics and quantum mechanics to find answers for some of the most intriguing questions pertaining to cognitive consciousness. Most of the nineteenth-century science was founded upon a 'Newtonian Absolute Physics' which provided a description of the world as interplay of forces obeying immutable laws and following a pre-determined trajectories pattern. This was considered as the 'billiard ball' view of the world. This comfortably solid and deterministic world view of the materialists has however been undermined by the new physics, and in particular through quantum science. Discovering the properties of subatomic matter, an element of probability had to be introduced into the physicists' calculations, and each subatomic event was characterized in itself inherently unpredictable and uncertain – one could only ascribe a probability to the outcome. In the quantum picture of the world, each individual event cannot be determined exactly, but has to be described by a 'wave' of probability. Thus, the simple billiard ball model collapses at the subatomic level. Quantum thinkers are generally aware of the famous equation of quantum theory embodying Heisenberg's Uncertainty Principle expressed as Planck's constant = (uncertainty in energy) × (uncertainty in position).

4.3.1 Science of quantum consciousness and neural design

Seen from a distant macro perspective, we may get the impression about the pre-structured cognitive consciousness expressed as exact and precise predictable way, but, in fact, cognitive consciousness does also exhibit, albeit at the quantum scale level, 'quantum' like qualities. According to the quantum explanation, even the formation of the so-called deterministic consciousness structures cannot be achieved through a linear deterministic route. A growing number of investigators believe that the first step towards a science of consciousness towards decision-making is to discover the neural correlates of consciousness. In this way, cognitive neuroscience can be found to be a branch of the scientific study of biological mechanism underlying cognition, with a specific focus on the neural substrates of mental processes and their behavioural manifestations. It addresses the questions of how psychological and cognitive functions are produced by the neural circuitry. One of the strongest proponents of a theory of consciousness founded on quantum theory, Sir Roger Penrose [4,5], a leading British physicist (in association with his American associate Prof. Stuart Hameroff) [6,7] opined that consciousness must be a quantum phenomenon because neurons are too big to account for consciousness. According to him, inside neurons there is a 'cytoskeleton' (also known as Corticons as primitive substance than neurons), the structure that holds cells together, whose

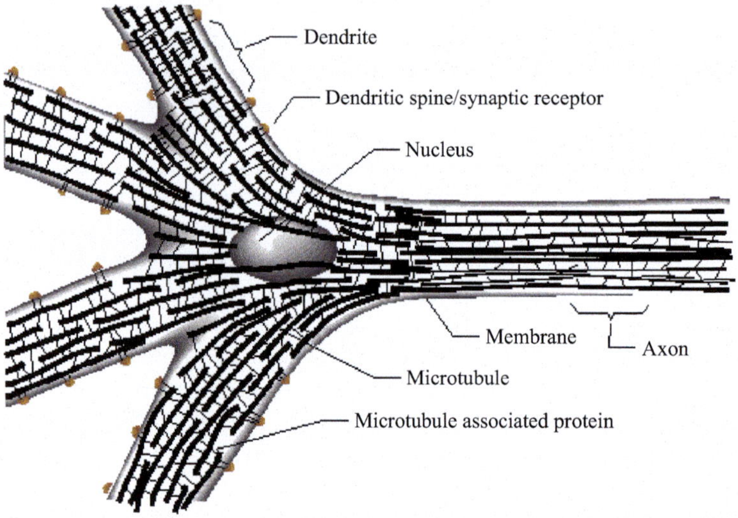

Figure 4.5 Microtubular neural structure

'microtubules' (hollow protein cylinders 25-nanometres in diameter) control the function of synapses (Figure 4.5).

The hypothesis propounded by Penrose is also affirmed by the Japanese physicist Kunio Yasue and the American physicist Gordon Globus who claim that brain substrates uphold second-order quantum fields. Their idea is derived from the quantum field theory developed in the 1960s by another Japanese physicist Hiroomi Umezawa [8] and on his concept of 'Corticons' as more primitive substance than 'neurons'. We can call it 'quantum conscious neuron' and hypothesize it to be like a light or photon like entity – a superluminal (photogenic/photonic) particle having dynamic quantum motions inside the grey-white matter of cerebral cortex.

Hence, the above hypotheses primarily aim to subscribe to the idea of 'quantum neurophysics' explaining how the classical world of consciousness can originate from quantum processes inside the mind/brain. Quantum neuro-physicists also believe that several layers of the brain can host quantum processes, whose quantum properties can effectively explain the dynamics of cognitive consciousness. They conceptualize the human brain as a macroscopic quantum system fundamentally made up of subatomic structures such as microtubules which lie inside the neuron and which contain quasi-crystalline water molecules that again lend themselves to quantum effects.

The possibility of quantum neural excitation and transformation is firmly founded on the postulate of the quantum neural states (superpositioned wave-like entities) having their own space-time geometry coined as 'objective reduction' (Figure 4.6). Under special circumstances such as in the case of electrical or

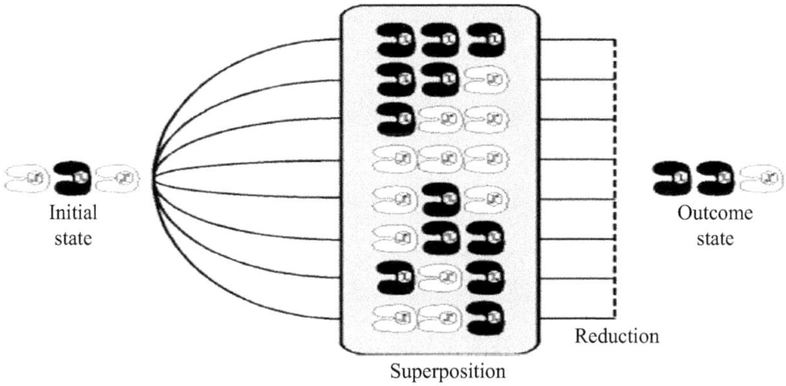

Figure 4.6 Objective reduction collapse propounded by Penrose–Hameroff

concentration–meditation-induced excitation, which microtubules of quantum neurons are suitable for, the separation of space-time geometry reaches a point known as the quantum gravity threshold where the system must choose one state or collapse into other state. Hence, when a quantum neuron system collapses from, say, a quantum superposition state of low consciousness to a state of high consciousness during the concentration-meditation phase without any external environmental intervention or assistance, by objective reduction criterion, we sense out an internal intervention of a non-computable element in it causing the self-collapse.

4.3.2 Objective reduction postulate of quantum consciousness

According to Robert Penrose [4,5], consciousness is an actual physical process, a sequence of quantum state reductions connected by $E = h/t$ to an objective threshold inherent in space-time geometry. He further stresses that underlying structure of the quantum world embeds platonic information as precursors of life and consciousness. As we all also know that Einstein had also stated that matter is equivalent to curvature in space-time, which also stands true not only at large macro scales, but at infinitesimally tiny scales too – the basement level of the universe at the Planck scale, i.e. 10^{-33} cm. Penrose thus observed that quantum superposition – an object in two places/states simultaneously – equated to two space-time curvatures in different directions: a bubble or separation in the fabric of reality. But the separations are unstable, and after a time t (given by $E = h/t$) will spontaneously self-collapse to one curvature or the other as depicted in Figure 4.7.

Enhancing and reaching to the higher quantum neural potential above the quantum gravity, threshold level can be closely associated to the state of 'Trance' or very high-degree consciousness referred in various esoteric but scientifically interpreted religious literature, particularly in the eastern tradition of Saints (Sant Mat) in India.

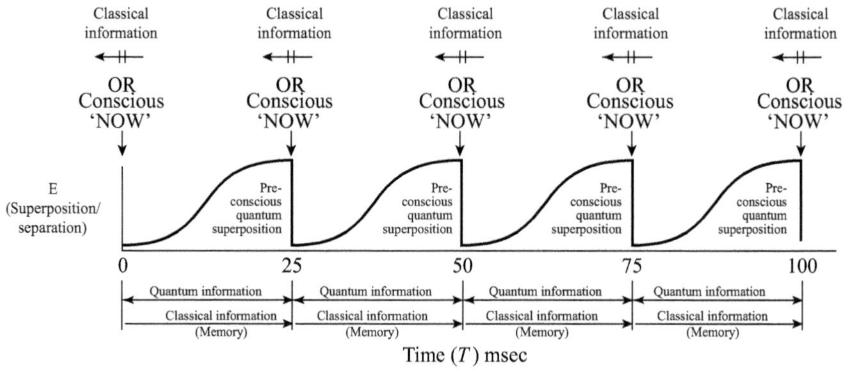

Figure 4.7 Sequence of preconscious–conscious events

The state of trance, as explained more exquisitely in the eastern religion of Saints – 'Sant Mat' in India, is described as the elimination or the cessation of the functions of the physical media resulting in human spirits being endowed with higher powers of senses, foresight and access to a plane which is different from the planes of the three dimensions. This may also be corroborated to the quantum spin carrying quantum information and different configurations of quantum spin geometry ultimately leading to varieties of quantum coherent 'high-order consciousness experiences'. This is probably why quantum consciousness has yet not been measured applying any physical measurement or it is not even conceivably measurable by the quantum computers. Note that in the case of external environmental intervention, as it turns out; the systems in quantum state can perform massively parallel and efficient computing and may collapse to the solution. This is the basis for so-called quantum computing which is currently a very live topic.

On the above lines of the consciousness-excitation postulate, in 1986 John Eccles, the British neurophysiologist who discovered neurotransmitters, had also speculated that synapses in the cortex respond in a probabilistic manner to neural excitation, a probability that could well be governed by quantum uncertainty given the extremely small size of the synapses 'microsite' that emits the neurotransmitter. Also, as a quantum equivalent of the neural summation process, the 'self-collapse' of quantum neural state results in particular 'conformational states' that may further regulate neural processes.

These conformational states can interact with neighbouring states to represent, propagate and process information and each self-collapse corresponds to a discrete quantum conscious event taking place under cortical brain. Sequences of such events then give rise to a 'stream' of enhanced quantum consciousness. In nut shell, the quantum neural phenomenon controls the cognitive consciousness operation of the brain through its underlying quantum coherent state which is speculated by neurophysicists and scientists to be arranged systemically in space-time geometry at the fundamental Planck scale and that a self-organizing Planck-scale process results in quantum consciousness in humans that may eventually lead to high-order conscious decision-making in them.

4.3.3 Quantum consciousness and genetic framework: genetic correlate to quantum conscious neurons

Every quantum behaviour of cognitive consciousness shall have a dual characterization as self-governing superpositioned coherent entity grounded on the genetic platform. The conscious decision-making is therefore made up of two parts, one that deals with the quantum expression of cerebral neurons and the other that deals with how and why those expressions originate. This dualism proves that consciousness is an expressed and potentiated feature of human brain which is somehow related to its genetic properties. It also states a tight relationship between the quantum structure of consciousness and its genetic or functional organization in the form of Genome. From these principles, it follows that manifestation of quantum consciousness is due to the genetic and functional organization of the brain. What it also suggested is that quantum consciousness doesn't come simply from the act of putting cerebral neurons together by magic, although it is a fundamental property of human neurons, and its quantum conductivity depends on the proper circumstances to take place or say, genetic constitution to build in the human genome in order to derive the optimum results.

The neuro-genetic thesis of quantum consciousness proposed here can further be validated on the Darwinist theory of human evolution. If we assume that a similar law of evolution is responsible for all living phenomena, from the creation of species to the physiological system, and we admit that mind is one of them, then, a possible scenario of neural-genome interface emerges, which is compatible with our hypothesis. In nutshell, quantum consciousness engrained in human mind is a mental equivalent of a genetic thread. Similarly, if quantum consciousness comes from a fundamental property of matter or say, human genome, then, we can study why and how, under special excitation circumstances (as described in the earlier section), that very property enables a particular configuration of matter (the cerebral neuron) to exhibit quantum consciousness.

The basic postulate here is, as we know, genes predispose to various ailments and diseases much before their actual and physical manifestation in the life of a human being and this has been proved substantially and authoritatively through the research findings of the human genome projects carried out in number of countries where scientists have succeeded in sequencing the human genome. Following this line, we can also say that cognitive consciousness might also be coded somewhere in the human genome knowing the genetic–neural gateway and transmission mechanism.

This theory, however, admits the existence of quantum consciousness as separate from the physical properties of neural matter as we know them, but, at the same time, we also need to accept quantum consciousness, at the least, influenced or moderated by the physical genetic property of humans who behave in a fundamentally different way from the physical genetic properties of other low conscious species. So, in a sense, quantum consciousness is not a 'physical' property, but it is still a property derived or at least moderated from its constituting genetic physical matter. It can be reiterated here that although neural wiring in human brain is carried out by processes that are derived from the genome, but the information for this wiring is not entirely contained within the genome itself. Rather, a human brain develops by a progressive sequence of steps each involving interplay between genetic programmes and environmental influences as propounded in the Darwin theory of evolution.

4.3.4 Quantum uncertainty and genetic certainty

The famous equation of quantum theory embodying Heisenberg's Uncertainty Principle emphasizes that in the quantum picture of the world, each individual event cannot be determined exactly, but has to be described by a 'wave' of probability. However, if we trace the evolutionary placement of neurons in the human brain migrating and developing axon connections from early childhood to maturity in a human being as also mapped now under the human genome project, we are compelled to infer that the Nature also appears to be highly deterministic in various instances of specificities including the arrangement of cerebral neurons firing quantum pulses of human consciousness. Nonetheless to say that certain genes in the human genome have a unique function in space and time in developing quantum neural consciousness in the brain but that would also be moderated under the dynamic evolutionary influence of the environmental factors. Genes are therefore considered as fundamental building block for they constitute the basic neural framework and leading the way we experience quantum consciousness.

It is no wonder that the DNA has been reported to emit electromagnetic radiation of photons in the microwave spectrum (1–4 billion cycles per second derived from the electricity that is flowing through our nervous system and spinal cord) and hence it may act as a carrier or the 'connecting medium' of message to the quantum-conscious neurons having photon receptors or neurotransmitter receptors through 'electrical synapses' which exist much less in number than chemical synapses in the neural arrangement but are more reliable.

This conceivably builds a tunnel of DNA–photon exchange through cognitive–cerebral gateway. Going by the same premise of Microcosm-Macrocosm theory elaborated in the faith of saints from eastern tradition, it is suggested that the human physical body possess those genotype–phenotype patterns of DNA, which can emit a radiation of higher order, building an inseparable communion with the consciousness photons (neurons), thus evolving into a quantum conscious neuron-genome gateway.

4.3.5 Computation of quantum consciousness: the brain as measuring device

At the quantum level, the computation of quantum consciousness is possible through the protein qubits orchestrated by synapses inside the microtubule arrangements of neural circuitry. Hence, the brain is in effect treated as a Heisenberg-type quantum measuring device, according to which, the mental life of each human being is representable as a subsequence of the full sequence of Heisenberg events. The neural wave function enfolds superposed possibilities, and then consciousness intervenes to choose one classical branch and annihilate the others.

However, some even propose that *consciousness is cybernetic* where the evolution of the quantum wave function of neural circuitry is not random, but optimized under a principle of 'least neural action'. In this way, the random effects of quantum consciousness can be replaced by a 'cybernetic' quantum consciousness operating as a free-willing agent.

This also gels well with the earlier idea, giving rise to a remarkable and dynamic possibility of existence of genetically engrained quantum neural phenomenon in the human brain. Under the cybernetic approach, consciousness participates in an interaction. According to this, consciousness is a quantum eruption offering possibilities to match with sensory input and thus with reality.

Therefore, mental states are not randomly chosen in mental acts but conserve real symmetry and evolve under optimal control (i.e. minimization of the neural action). Whichever ways, by introducing consciousness, we can conclusively recognize that brain substrates uphold second-order quantum fields, and so should not be treated as ordinary physical measuring devices. One definitely obtains a philosophically elegant resolution of the paradox of quantum measurement which says that 'probabilistic occurrences in the quantum world are replaced by definite occurrences when they enter consciousness'.

This implies that Schrodinger's cat (a quite infamous paradox of 'half dead–half alive cat' propounded by Schrodinger to explain quantum uncertainty) becomes definitely dead or definitely alive when a conscious being sees it. In other words, coherent super positions, the multifaceted quantum waves, exist in the transcendent order until consciousness brings them to the world of appearance with the act of observation, and in the process, consciousness chooses one facet out of two, or many, that are permitted by the mathematics of quantum mechanics. Hence, consciousness is not about doing something to objects via observing, but consists of choosing among the alternative possibilities that the wave function presents, recognizing the result of choice.

This conscious 'selection' is a continuous process happening inside human brain, which also has a measurable physical channel of manifestation. Collaborating to this quantum manifestation of human brain, it has long been known that the global electrical activity of the brain is characterized by distinct oscillatory components at different frequencies and that they correlate well with such large functional states as wakefulness and sleep, and during the period corresponding to rapid eye movement (REM) sleep (in which a subject, if awakened, reports having been dreaming), 40-Hz oscillation similar in distribution phase and amplitude to that observed during wakefulness is observed.

Thus, it proves that quantum consciousness has some direct physical effect and one may then hypothesize that the entropy reduction of arrays of possible states by human consciousness is correlated with those changes in the states of conscious systems which correspond to conscious acts of decision-making. This further connects to the assertion of the present idea that consciousness is a process which is part of the dynamics of certain physical systems under 'an intelligent design' expressing itself physically through neural-genome gateway.

4.4 Integrative and causal epistemology of consciousness

If we probe further deep into the dynamics of quantum consciousness, it would reveal another unique phenomenon of '*super causality*' as a sum total of all creational causalities and retro-causalities. It can be said that complex dynamical systems are neither completely rigid nor fully random; instead, they display a unique balance of 'integration', cohesion and robustness at the global level and at the same time, differentiation and multiple realizability at the component level. Aristotle had also identified four types of causal relations: *material, formal, efficient* and *final*. In the area of consciousness too, we see downward causation, in which upstream determinants at the higher-order consciousness levels influence and regulate events at lower levels of human cognitive consciousness. It was only as a result of cutting edge physics, particularly quantum physics, the very fundamentals of all existence perceived so far that has confirmed again and again that consciousness is in fact the very ground of Being.

Consciousness also has this dualistic aspect, i.e. combination of both the physical act of state reduction and the correlated psychological intentional act or tendencies. The focus of this paper has been to provide exposition on the phenomenon of top-down super causality by examining and synthesizing a galaxy of theories and postulates propounded by eminent scientists and thinkers namely Stapp [9,10], Godel, Penrose [4,5], Hameroff [7], Laszlo [11], Cayce and Campbell [12,13], and also based on some fundamental spiritual proclamations in the eastern tradition of religion.

Quantum physicist and consciousness scholar Stapp's model of consciousness [9] is tripartite. In his views, reality is a sequence of discrete events in the brain. Each event is driven by three processes that operate together, perhaps in synchrony:

- The '*Schrodinger process*' is a mechanical, deterministic process that predicts the state of the system where the Schrodinger's equations describe the state of a system as a set of possibilities in contrast with Newtonian one certainty principle.
- The '*Heisenberg process*' is a conscious choice where we know something only when we ask Nature a question. This implies, in turn, that we have a degree of control over Nature. It has 'quantum Zeno effect', in which we can alter the course of the universe by asking questions repeatedly and rapidly.
- The '*Dirac process*' gives the answer to our question. Nature replies, and once she has replied, we have learned something by increasing our knowledge. This necessitates a change in the state of the universe, which directly corresponds to a change in the state of our brain which can be corroboratively referred to as Penrose–Hameroff [6,7] process where occurs an objective reduction of the wave function compatible with the fact that has been learned.

Hence, summarily we can infer that the state of the universe is represented by a wave function which is a compendium of all the wave functions that each of us can cause (Although Stapp highlighted the Individual case of an observer, it can have the attribute of collectivism too) collapse with our observations. That is why it is a collection of subjective acts, although an objective one.

If we probe further deep into this conception about quantum consciousness, it would reveal a unique phenomenon of *super causality* (as a sum of all creational causalities and retro-causalities), and the above three stages appear to be in the grand scheme of this most fundamental causal framework inhabiting this creation. And, the most visible physical evidence at the material plane of it could be the present state of the world and the conditions under which men live that can be regarded as the reciprocal causal image of their own individual and collective states of consciousness.

In other words, it won't be naive to say that the material world around us is nothing but the physical impression of the collective cognitive minds of the humanity (global consciousness) which may in turn be the causal impression of the higher-level universal mind. This is further reflected upon in the book *Mindful Universe* by Henry Stapp [10] (which is a must read for people interested in consciousness from a scientific, philosophical and personal point of view) where in his words '... according to the new conception, the physically described world is built not out of bits of matter, as matter was understood in the nineteenth century, but out of objective tendencies – potentialities – for certain discrete, whole actual events to occur'.

Each such event has both a psychologically described aspect, which is essentially an increment in knowledge, and also a physically described aspect, which is an action that abruptly changes the mathematically described set of potentialities to one that is concordant with the increase in knowledge. According to him, this coordination of the aspects of the theory that are described in physical/mathematical terms with aspects that are described in psychological terms is what makes the theory practically useful.

4.4.1 Downward causation and the elements of super causality

The conception of downward causation conveys the proposition that while the whole is to some degree constrained by the parts (known as upward causation), at the same time, the parts are to greater degree constrained by the whole. So, the behaviour of the parts (down) is determined by the behaviour of the whole (up), and determination thereby moves downward instead of upward. In other words, the global level of integration (the result of 'upwards causation') may produce 'downwards' effects, acting eventually upon the local level of the parts/zones. As first articulated and demonstrated by Donald Campbell in 1974 [12,13] and Calyton and Davies [14], the notion of downward causation describes how higher organizational levels influence lower levels in hierarchically organized biological systems, such that all processes at the lower levels of a hierarchy are restrained by and act in conformity at the laws of the higher levels.

4.4.1.1 Brain

The history of brain research over the last few centuries contains two conflicting theories of how the human brain works. One theory stresses that the brain consists of a vast collection of distinct regions each localizable in the cerebral cortex and each capable of performing a unique function. The other school of thought looks upon the brain not as a collection of specialized centres, but as a highly integrated and mutually interacting organ. In this view, no single function can be the sole domain of any unique part of the cortex. Obeying the old dictum, the holistic brain is greater than and different from the sum of its parts.

4.4.1.2 Mind

According to the Austrian philosopher Karl Popper and the Australian neurophysiologist John Eccles, the mind inhabits its own world, that of subjective states. This world follows its own rules and regulations that are not the laws of physics. The conscious mind – made out of some sort of metaphysical ectoplasm – imposes its will onto the brain by affecting the way neurons communicate with each other in the part of the cerebral cortex concerned with the planning of movement and action by adjusting the synaptic release probabilities. According to one school of thought, by promoting synaptic traffic between these neurons in one location and preventing it in another, the conscious mind imposes its will onto the material world. For those raised with a faith in 'Free Will', this seems to reconcile a religious point of view with a scientific stance. Having said that, it is also true to state that understanding human freedom is the most important and most difficult long-term challenge facing the neurobehavioural sciences.

4.4.1.3 Consciousness

In the area of consciousness too, we see downward causation, in which upstream determinants at the higher-order consciousness levels influence and regulate events at lower levels of human cognitive consciousness. It was only as a result of cutting edge physics, particularly quantum physics, the very fundamentals of all existence (perceived) so far that has confirmed again and again that consciousness is in fact the very ground of Being.

There is nothing else apart from consciousness, from which arises mind as does matter. Physicists have come to realize that the universe is interconnected in much subtler ways than had once thought. In quantum physics, the observer and the observed can no longer be separated and the whole is more fundamental than the part. In this view, consciousness imposes 'downward causation'. In other words, our free will is real as intention. When we act in the world we really are acting with causal power. This view does not deny that matter also has causal potency – it does not deny that there is causal power from elementary particles upward, so there is upward causation – but in addition it insists that there is also downward causation.

4.4.2 Role of tendency in creating actual and subsequent conscious events

Probing further into the causal dynamics of consciousness, it is noteworthy that Heisenberg's notion of the quantum potential is synonymous to the notion of 'tendency', which relates to the situation before measurement happens. So, it expresses the idea of a reality independent of measurement. Quite interestingly,

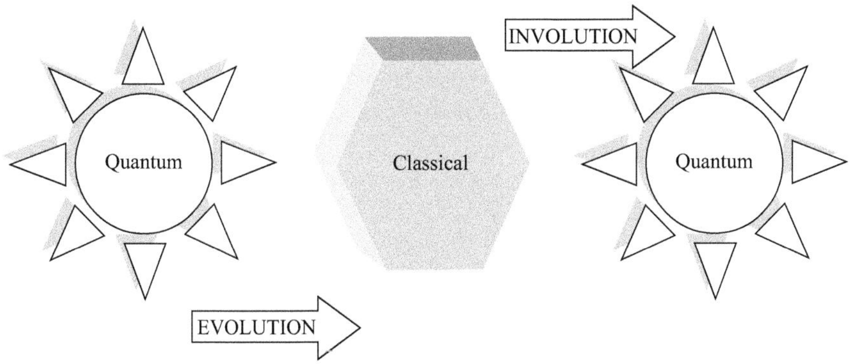

Figure 4.8 Transient dynamic stream of quantum-classical consciousness events

immediately after its actualization, each actualized event further holds the tendency for the impending actualization of another, subsequent actual (conscious) event in a stream of evolution and involution. This clearly reinforces the time-driven transient dynamism of our universe (or, many other universes under the creation) (Figure 4.8).

4.4.3 Modelling super causality

As most exquisitely demonstrated by Prof. P.S. Satsangi, the world renowned system scientist from India [15], a causal model is an abstract model that describes the causal mechanisms of a system. The model must express more than correlation because correlation does not imply causation.

Following the structural dynamics between brain, mind and consciousness, if we have to model consciousness at the bottom physical level of a human brain, then D will represent the set of brain causal variables whose values are determined under the influence of factors from the level of Mind (C), Spirit (B) and Supreme Source (A), i.e. outside (exogenous) the structure model of physical Brain (D). Similarly, values of endogenous variables represented shall be determined by factors from the model structure of physical Brain (D) itself. Underpinning the significance of downstream interventions in consciousness, we can further invoke feedback-oriented dynamic modelling framework for capturing the inherent causal dynamics of downward hierarchization of consciousness ranging top-down from A to D. An advantage of clearly modelled consciousness hierarchies with demonstrated links between top-down levels is that they allow us to visualize and also predict the impact of interventions aimed at one higher level on factors downstream from that level.

Therefore, designing and targeting interventions based on established relationships has the potential to heighten the effectiveness of interventions in enhancing low-order cognitive consciousness to high-order spiritual consciousness in humans. Hence, it can be interpreted that Newton's classical physics and Plank's quantum physics provide necessary conditions (but not the sufficient conditions) for what happens at the material (physical–psychological) plane; they provide the possibility space for what happens, but does not determine the outcome. Top-down super causation allows higher-level causes to manifest their effective causality over the levels below.

Therefore, 'super causality' context is the key to all physical–psychological outcomes, and multiple causation across the hierarchy is always at work. Moreover, all causal phenomena have a teleological existence and are based on purposeful activity and free will, enabling the mechanism of grand evolution and involution that lies beyond the explanatory scope of physical science.

The focus of this section has been on top-down super causality explanation because it is the mode of causation that is least considered at the present-day modelling for consciousness towards decision-making. Extending the postulate propounded by Nancy Murphy and others in the context of neurobiology of free will and downward causation, it can also be emphasized *that bottom-up, same-level, and top-down causation all occur at the same time, in an orchestrated concert or synchrony, enabling the emergence of complex dynamics of consciousness based in modular hierarchical system.*

Broadly speaking, same-level causation is where the action is (psychical consciousness); bottom-up causation enables it to happen (physical consciousness); top-down causation decides what should happen (spiritual consciousness) and the prime cause of all these three to manifest their individual and collective properties is the source of super causality, i.e. God consciousness (Figure 4.9).

4.5 Information consciousness within the realm of creation and its mathematical interpretation

At the cosmic level, the physical analogy of information consciousness idea is best corroborated by an excellent depiction of 'Penrose diagram', demonstrated by Sir Roger Penrose [16], which is conformably equivalent to the actual metric in space-time. This conformal factor is chosen such that the entire infinite space-time is transformed into a diagram of finite size where every point corresponds to a two-dimensional sphere.

Consciousness does not speak the language of mathematics, but it is true that one can endeavour to model it comprehensively in the language of mathematics. Descartes was also, as was so often the case, well ahead of his time by describing continuity and dimensionality, the factors that define his view of space as an actual vector space accessible to mathematical and physical analysis. We can also use mathematics to abstract, idealize and describe these phenomena in order to make some predictions that would be very different otherwise. Drawing parallel to the law of conservation, the information force field of consciousness that is believed to exist and pervade in every region of creation including our universe is believed to be inhabited by information as its fundamental constituent. It evolves and continuously interacts with other fields and matter particles and engage in the exchange of mass–energy with them.

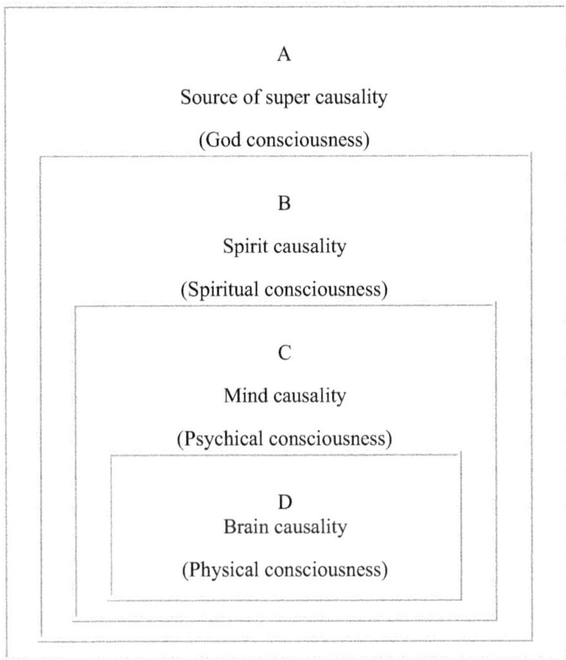

Figure 4.9 Hierarchy of causality

This process is also consistent with the existing postulates of physical mass–energy phenomenon explained by laws such as Higgs boson field and the emergent phenomenon of gravity. Mathematically, information consciousness can be described as a function of transformation of potential information consciousness into kinetic consciousness, i.e. $f(C_0 + C_1)$ which can further be assumed to be a divergence operator explaining how much the consciousness field gradient tends to spread radially outward or converge inward; a kind of behaviour observed under the Gauss' law of electric field lines charging or spreading out from charges in every direction.

4.5.1 *Mathematical interpretation of information consciousness*

In the following section, a mathematical interpretation to express the dynamics about the origin and characteristics of information consciousness is proposed. It does not intend to delve into specifics of micro detailing as there are existing approaches propounded by various people in pursuit to understand the same, notably, Orch-Objective Reduction Theory propounded by Hameroff and Penrose [6,7], Integrated Information theory (Guilio Tononi) and others.

In other way, consciousness field can be fully characterized by its divergence and curl where potential consciousness energy is termed to be self-absorbed (C_0)

whereas kinetic energy is a manifest form of information self-awareness as C_1. We describe this single bit of quantum information (qubit) using two real numbers involving infinitely many decimal points in between them. So, $\Delta C^{0\leftrightarrow 1}$ movement from zero to one also includes the fundamental property of infinitely big or small quantities as infinity $\frac{\infty}{\infty} = 1$ and one can always add a subsequent '1' to any number or quantity to make it further bigger or larger.

And, once we put $\Delta C^{0\leftrightarrow 1}$over $(S, T)^{\infty}$ i.e Space-time, things start demonstrating divergence as $\frac{\Delta C^{0\leftrightarrow 1}}{(S,T)^{\infty}}$.

In graph theoretical terms, (*S*,*T*) is an ordered pair comprising a space set *S* of vertices or nodes or points or regions representing Cosmos made up of infinite galaxies or multiverses bounded together with a time-set *T* of edges or arcs or lines or motion/movement between the vertices stretching to infinity. It can also be visualized as a Cosmic Topology.

Explanation for ∞

Simply stating, it is an infinity of infinite number of reference frames of time and space. $(S, T)^{\infty}$: The bracket represents a boundary condition of an individual reference frame or set of infinities within infinity, a space-time within a space-time. It also indicates singularity as a singularity (in the context of gravitation) is simply a 'point' or a 'region' in space-time where the values of certain quantities predicted by our theories blow up to infinity. For the singularity to be real, this must be true irrespective of the choice of the coordinate system of description itself. There are apparent singularities that vanish with an appropriate choice of coordinate system.

4.5.2 Information consciousness as vector field and explanation for $\Delta C^{0\leftrightarrow 1}$

Vector fields are often used to model, for example, the speed and direction of a moving fluid throughout space, or the strength and direction of some force, such as the magnetic or gravitational force, as it changes from one point to another point similar to the idea presented in the earlier section. This representation of a vector field depends on the coordinate system, and there is a well-defined transformation law in passing from one coordinate system to the other.

More generally, vector fields are defined on differentiable manifolds, which are spaces that look like Euclidean space on small scales, but may have more complicated structure on larger scales like the scale of universe. In the realm of mathematics, the elements of differential and integral calculus extend naturally to vector fields. When a vector field represents force, the line integral of a vector field represents the work done by a force moving along a path, and under this interpretation conservation of energy is exhibited as a special case of the fundamental theorem of calculus. Since, vector fields can usefully be thought of as representing the velocity of a moving flow in space; this physical intuition leads to notions such as the divergence (which represents the rate of change of volume of a flow both in positive and negative directions) and curl (which represents the rotation of a flow).

Moreover, treating information force field as an 'object' of study as emphasized in Michael Faraday's field theory, information consciousness field can be visualized as a vector field to find a mathematical expression for such force fields; the elements of integral calculus in the form of line integral of a vector field of consciousness would represent the work done by this force moving along a path. Accordingly, '*C*' (information consciousness) is as an integral expressing change of consciousness from 0 (global source – potential energy) to 1 (local sink – localization of kinetic energy) of a spiritual entity.

Therefore, information consciousness can be proved being coded in the binary digits of information bits of 0 and 1 and can explain very authentically the entire mechanism of creation or evolution resulting due to a continuous transition, change or shift of the information consciousness between 0 and 1 as denoted by $\Delta C^{0\leftrightarrow 1}$or $\oint_0^1 C$. The reason that forces meeting these conditions can be called conservative is that if all of the forces on an object are conservative, we can define a quantity called potential energy, denoted $U(r)$, a function only of position, with the property that the total transformation energy is constant, i.e. conserved. To define the potential information energy, we must first choose a reference point r_o, at which U is defined to be zero (for gravity, we typically choose the reference point to be ground level). Then $U(r)$, the potential energy, at any arbitrary point r, is defined to be

$$U(r) = -W(r_o \to r) = -\int_{r_o}^{r} F(r') \cdot dr'$$

In words, $U(r)$ is minus the work done by F when the information moves from the reference point r_o to the point r and as the information flows along the path, it starts transforming inherent potential energy (mind-matter tendency) C_0 into kinetic information consciousness energy C_0 while attaining the level of 1, and then may return back to reservoir. These information particles go along the trajectory and they exhibit the transition dynamics of potential and kinetic energy states.

Now, supposing we have some two-dimensional consciousness vector field, $F(x,y)$, and a closed information consciousness curve C indicated by $\oint_0^1 C$ wandering through this field, then the journey of information along a trajectory C in the consciousness field F can be visualized intuitively as the tiny segments which make the whole curve C, each with an amount of mass-energy exiting or entering through it per unit time. Therefore, these movements or transitory changes as the consciousness can be added up and represented best as a line integral or integral curves or trajectories or flow lines. For an information object moving along a closed loop path C in a consciousness vector field $F(x,y)$, the total work done by the field on the object is obtained by summing up the differential work done in moving from one point to another through a function which return the outward unit normal vector at some point on the curve C. This gives the line integral concept of information consciousness and can be expressed as

$$\oint_C F.dr$$

The classic problem for it is to calculate the work done on a particle travelling along some curve in a force field represented by a vector field. If one considers a line integral along a *closed* path, the symbol $\oint$ is sometimes used to emphasize that the path is closed.

Inspired by this physical intuition about information consciousness, we are led to understand the notions of divergence furthermore. The reservoir of consciousness as the primary and the most primitive source of consciousness acts as an end from where the force field emanates and spreads throughout the creation, i.e. Positive Divergence; whereas, at the point of singularity when all these information meet back at their source, it can be termed as Negative Divergence or Convergence, as the divergence at a point represents the degree to which a volume around the point is a source or sink for the vector flow, a result which is made precise by the divergence theorem. The divergence can also be defined on a Riemannian manifold, that is, a manifold with a Riemannian metric that measures the length of vectors.

4.5.3 Line integral loop dynamics

It is important to depict the whole trajectory of movement of information consciousness as 'caustics' which are non-local characteristics of states that can be determined and understood only if the whole trajectory is known. This topological depiction plays a decisive role in quantum description of reality at macro cosmic level.

As we have concerned ourselves with the idea that information consciousness always takes values between 0 and 1, it may also be interpreted as the value transiting between 0 to 1 back and forth. So, in this case as we know that the line integral is path independent between any two points, the line integral along the first information consciousness curve $C - 1$ as the forward path moving from 0 to 1, i.e. C_1 of $f \cdot dr$ is going to be equal to the line integral of returning information consciousness curve C_2 over the reverse path C_2, of $f \cdot dr$, which can be written as

$$\int_{C_1} f \cdot dr + \int_{-C_2} f \cdot dr = 0$$

Therefore, as a closed loop integral, we can write

$$\oint_C F \cdot dr = 0$$

It means that the line, if we have a potential in a region, and it may be everywhere, then the line integral between any two points is independent of the path. This could be any closed path where our vector field f has a potential, or where it is the gradient of a scalar field, or where it is conservative. And so this can be written as a closed path of C_1 plus the reverse of C_2.

That tells us that at any point in the region where this is valid, the line integral from one point to another is independent of the path; and because of that, a closed-loop line integral, or a closed-line integral, so if we take some other place, if we take any other closed line integral or we take the line integral of the vector field on any closed loop, it will always become 0 because it is path independent.

It can be interpreted as the fundamental nature of information consciousness that remains constantly pervaded and uniform in its distribution over any unit of time and space irrespective of any shape and symmetry of the trajectories of its movement.

More precisely, it can be diagrammatically represented as Green's Theorem.13 as shown below:

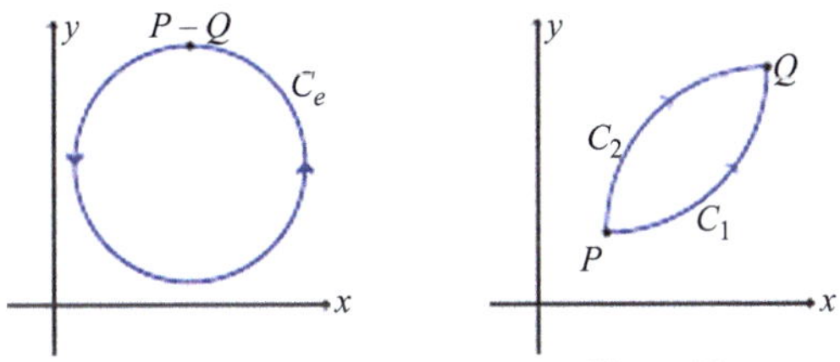

Figure (i) Figure (ii)

Extended to infinity of space (dr) and time (dt):

This form of line integrals can be extended to infinite series of information consciousness cycles spread over space-time:

$$\oint_{C\to\infty} F \cdot dr \cdot dt = 0$$

where the cumulative path independent values of all information consciousness C series loops will still be zero (0) because the sum of the path values of all individual closed-loop integrals will be zero. It can be more clearly demonstrated as follows:

$$\oint_{C_1+C_2+C_3+C_4+C_5+\ldots+C_n\to\infty} F \cdot dr \cdot dt = 0$$

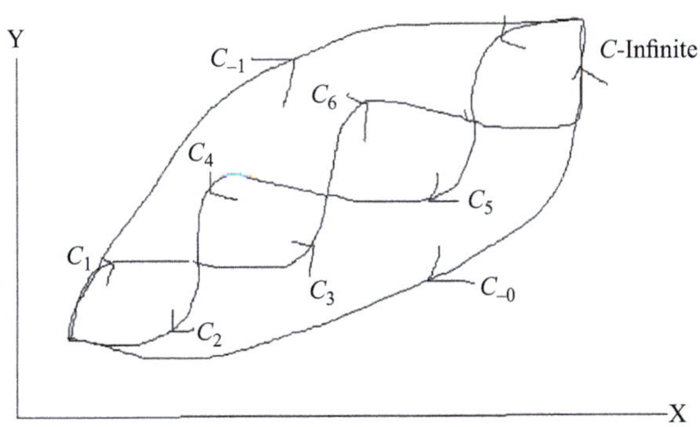

It can also be elaborated as

$$f(C,S,T) = \frac{\Delta C^{0\leftrightarrow 1}}{(S,T)^{\infty}}$$

In the line integral form as

$$f(C,S,T) = \frac{\int_0^1 C}{(S,T)^{\infty}}$$

Or

$$f(C,S,T) = \frac{\int_0^1 C\,dC}{(S\,(C,T)T)^{\infty}} = \textit{Diverges}$$

Steps:

$$\frac{\int_0^1 C\,dC}{(S\,(C,T)T)^{\infty}}$$

$$\int_0^1 C\,dC = 1/2$$

$$\int_0^1 C\,dC$$

Computing the indefinite integral:

$$\int C\,dC = C^2/2 + C$$

$$\int C\,dC$$

Applying the power rule: $\int x^a\,dx = \frac{x^{a+1}}{a+1}, a \neq -1 = \frac{C^{1+1}}{1+1}$.
Simplifying further, it becomes$= C^2/2$.
Adding a constant to the solution, $\frac{C^2}{2} + C$.
Computing the boundaries: $\int_0^1 C\,dC = \frac{1}{2} - 0$.

$$\int_a^b f(x)dx = F(b) - F(a) = \lim_{x\to b-}\left(F(x) - \lim_{x\to a+}(F(x))\right)$$

$\lim_{C\to 0+(C^2/2=0)}$ and $\lim\limits_{C\to 1-(C^2/2=1/2)} = \frac{1}{2} - 0$, simplifying further, $\frac{1}{2}$.

Therefore, $\infty = \textit{Diverges} = \textit{Creational expansion}$ (or Convergence in case of negative value).

So, reversing the above phenomenon will lead to Convergence back to origin of consciousness. The idea is also portrayed by Penrose diagram [16] of time and space as follows in terms of future and past time like Infinity $T = -/+(\infty)$ and positive/negative space like infinity, i.e. $X = -\ or + (\infty)$.

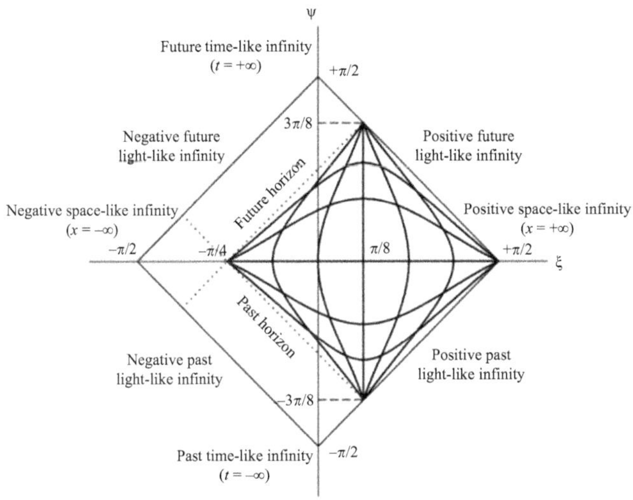

Additionally, information consciousness manifested as potential and kinetic information can also be shown as

$$f(C) = \int_0^\infty E \cdot dt$$

where E = Kinetic + Potential Consciousness.

4.5.4 Consistency with energy conservation law and Schrödinger equation

Consistent with the conservation law, the total energy E of a particle can also be expressed as the sum of kinetic energy T and potential energy V; this sum is also the frequent expression for the Hamiltonian H in classical mechanics:

$$E = T + V = H$$

Explicitly, for a particle in one dimension with position x, mass m and momentum p, and potential energy V which generally varies with position and time t:

$$E = \frac{p^2}{2m} + v(x, t) = H$$

For three dimensions, the position vector r and momentum vector p must be used:

$$E = \frac{p \cdot p}{2m} + v(r,t) = H$$

Schrödinger equation

Taking the discussion to another level of particle exhibiting wave properties, the time-dependent Schrödinger equation predicts that wave functions can form standing waves, called stationary states (also called 'orbitals', as in atomic orbitals or molecular orbitals). These states are particularly important as their individual study later simplifies the task of solving the time-dependent Schrödinger equation for any state. Stationary states can also be described by a simpler form of the Schrödinger equation, the *time-independent Schrödinger equation* (TISE), where E is a constant equal to the total energy of the system. This is only used when the Hamiltonian itself is not dependent on time explicitly. However, even in this case, the total wave function still has a time dependency.

$$\mathrm{H}\,|\Psi\rangle = \mathrm{E}\,|\Psi\rangle$$

In words, the equation states,

> *When the Hamiltonian operator acts on a certain wave function Ψ, and the result is proportional to the same wave function Ψ, then Ψ is a stationary state, and the proportionality constant, E, is the energy of the state Ψ.*

4.5.5 Collective observation and power law

This section explores the concept of Power Law which characterizes the attainment of critical distribution and inflection points after a certain quanta of threshold energy response limit being reached; a phenomenon widely observed in the nature and cosmos, it also explains consciousness dynamics where not an individualistic photonic observation, rather a minimum threshold of collective observation will eventually make a quantum event collapsing into a classical state.

Despite the enormous progress made in explaining and understanding the mysteries of quantum phenomenon especially in last few decades, the question of measurement problem remains unanswered. Several theorists including the stalwarts of quantum physics have remained caged to their hostility of accepting 'Consciousness' as that final frontier which might lead to the answer. It is highly probable to get an answer by extending the idea of an observation made by an individual or say a photon into the higher realm of collective observations made by more than one sentient observer. It has been comprehensively elaborated in Eastern philosophy about the effect produced by collective intentions expressed as an act of observation by people in a cohesive group resulting into realization of an intended consequence or 'Event' either Good or Bad.

The concept of Power Law indicates significantly towards this direction where not an individualistic photonic observation, rather a minimum threshold of collective observation will eventually make an event happen classically.

This particular proposition has immense potential to put to rest speculations and dilemmas that surround and divide the respective domains of classical physics and quantum physics being still devoid of finding out that illusive bridge of theory of everything.

According to this perspective, any macro-level event is a relatively permanent state of classical physical properties of the collapsing wave of super-positioned electrons that shape up into a concrete existence, whereas a micro-level event is simply a temporary transitory phase of electrons placed in a quantum state. In other words, when intentional collective observations in quantum state become big enough to feel significant quantum gravitational pull, 'Events' at macro level do take place under the classical state which is also akin to the transition from potential to kinetic state of energy (Figure 4.10).

4.5.6 Cosmic-scale dynamics of evolution and contraction-phases of universe

The above hypothesis can very well be expanded and mapped further on the cosmic level to explain the physical dynamics of evolution and involution/expansion and contraction/infinity to singularity as a series of 'eons' of big-bangs suggested by some scientists early on.

In this case, the Power Law dynamics can be mathematically illustrated and expressed by taking, for instance, a hypothetical exponent value of 5/6 and −6/5

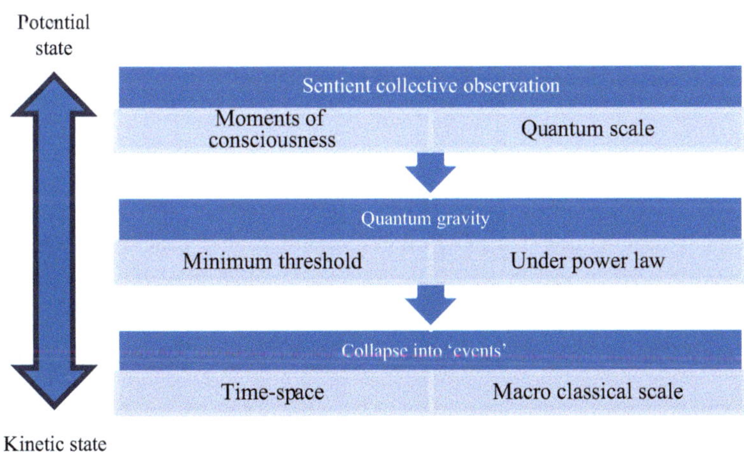

Figure 4.10 Potential to kinetic state of consciousness

both for evolution and involution processes, respectively. Starting with the single exponent, i.e. $\infty^{5/6}$, it would equal to ∞ suggesting towards '*Evolution*' of universe; however, by treating it further up with the reverse power exponent of 5/6, it would lead itself to 0, i.e. 'Contraction' of Universe (moving towards singularity) as inferred mathematically in the following:

$$
\begin{aligned}
&(\infty^{5/6})^{-5/6}\\
&\infty^{\left(\frac{5}{6}-\frac{5}{6}\right)}\\
&= \infty^{-1}\\
&= \lim_{x\to\infty} (1/x) = \lim_{x\to\infty} \left(\frac{1}{x}\right) = 0(\text{Contraction})
\end{aligned}
$$

4.5.7 New horizons

Every stretch and corner in the vastness of this cosmos is required to remain in a perfect balance all the time in order to create and sustain the 'human race' as it is the only form existing among all animate and inanimate forms which has the capacity to re-unite with the reservoir of consciousness and attain its eternal purpose of evolution and existence as it is. There may be innumerable number of Universes all originated from the same source of consciousness and forms like 'humans' may be existing out of our cosmos too.

They may have a similar pattern of evolution and dissolution, but their term of emancipation can come only when their cosmos is perfectly aligned with the source-eternal. Man is made after the image of God truly resonates with the theory of consciousness as it is only the human form of man which has the switches and the capacity to comprehend and correspond with the ultimate reality while following the natural path and laws of physical creation and its evolutionary tracks. Information consciousness: As and when, a human form acquires or gets access to information consciousness (as proposed on the lines of Integrative Information Theory of Gulio Tononi) of the level of its original reservoir, he will instantaneously get free from the lower information deficiency or entropy and qualify to get back to the place aboriginal and remain there forever.

The esoteric teachings about macro- and microcosm in their most extreme form are one and the same as per Buddhism, Bible, Yin/Yang and most scientifically elaborated in the Religion of Saints, i.e. Sant Mat of Indian origin in particular Radhasoami Faith founded by Revered Soamiji Maharaj, Dayalbagh, Agra.

4.5.8 Summary of key points

- Information consciousness is a state represented by $\Delta C^{0\leftrightarrow 1}$.
- Time – Space Dynamics $= (s,t)_{\infty}$.

- All pervasive 'environment' of consciousness existing as a fundamental reality like gravitation, electro-magnetism, mass, charge, fundamental particles, etc.
- State of consciousness moving between and from 0 to 1 (State of 'unity' or null to the state of 'duality' of self-conscious localized entity) and back forth.
- Evolution from 'zero' which is a state of global omnipresent entity to 'one' which is a uniquely positioned local entity.
- C_0 (potential state) to C_1 (kinetic state) denotes transition from potential state of information/consciousness to kinetic state of consciousness over time and space.
- This mechanism of evolution spanning over infinite space and time.
- Local entities (animate or inanimate) emerging from global entity and converging ultimately over infinite space and time.
- Movement from 0 *towards* 1 *is creational* or evolutionary mechanism while movement from 1 *towards* 0 *is destructive* or involutionary mechanism.
- Cosmos came to existence through the creational mechanism and will infinitely evolve and ultimately lead towards dissolution/involution and re-creation again (series of 'eons' – Roger Penrose, Higgs boson theory).
- All forces of nature and cosmos exhibit creational and destructive; though a purposive quality as they fundamentally contain the entity-particles evolving under the above mechanism since the very beginning.
- Consciousness experientially and experimentally measurable through integrative informational capability possesses by all entities animate or inanimate including fundamental particles (Guili Tononi – Integrative Information Theory).
- More integration, more self-consciousness (human brain cells made up of neurons and microtubules capable of self-learning, information processing and responsiveness to the environment).
- Inanimate objects devoid of neural network of information-integrative capability are least conscious.
- Localized consciousness governed by the evolutionary mechanism of the pre-creational initial condition of primitive information contained (creational tendencies, from 0 towards 1) in all the entities and experienced by them through varying patterns of space-time geometry.
- 'Qualia of consciousness' are those many moments of such experiences under the space-time fabric of holistic creation realized subjectively by more sentient entities than other lesser/inanimate ones (Penrose–Hameroff orchestrated reduction theory).
- The above descriptions about consciousness and creation are compatible with all other existing theories and discoveries, i.e. theories of relativity, gravity, electromagnetism, quantum mechanics and so on.

4.6 Practicing high-order consciousness at low-order reality: towards building a corporate model of consciousness system dynamics

Inspired from the downward causation and inspired evolutionary epistemology of consciousness which states that the processes at the lower levels are restrained by and act in conformity with the laws of higher levels of spiritual-psycho-physical realties, one can adventure to construct a dynamic and integrative corporate model of practicing socio-physical-psycho-spiritual consciousness in a day-to-day worldly routine. Bohr, Pauli and Heisenberg, the chief architects of quantum theory, had also made it abundantly clear that sharp dichotomies and contrarieties can always be replaced with far more subtle and sophisticated complementarities between different aspects of the same reality. Similarly, consciousness system dynamics also displays a unique balance of 'integration', cohesion and robustness at the global level and differentiation and multiple realization at the component level.

A corporate model of consciousness presented in the paper emphasizes over the management and real-life adoption of the high-order spiritual consciousness aspects in a low-order psycho-physical-social living by judiciously blending spiritualism with a guarded materialism. This form of Consciousness Management System can be envisaged as a collection of decision components which are co-ordinated together to perform a function or attain a specific goal. These systems interact with their environment across a separating boundary while still maintaining gradation and hierarchy.

The various aspects identified and integrated in our corporate model are related to diverse practices ranging from managing an individual's initial and desired states of spiritual consciousness, social diffusion of consciousness through pure spiritual corporate living practices, better healthcare practices and controlling the supply of counterfeit medicines through community consciousness and sensitization, removing vulnerabilities of poor artisans in traditional textile clusters through innovative capacity building and competitive consciousness, promoting indigenous knowledge for disaster mitigation in tribal communities, to finally, exercising socio-environmental–humanitarian initiative of capacity building in the low-impact eco-communities by ensuring supplies of adequate physical, social, economic, technological, environmental and spiritual resources.

Using the modelling and simulation framework of system dynamics, a holistic influence corporate model has been propounded which highlights the mutual interplay of influencing factors related to the diverse sub-elements of consciousness system as elaborated above. The model primarily investigates the structural dynamics of circular causality across the high-order–low-order spectrum of consciousness practices. It, therefore, captures the key structural relationships that define a complex consciousness management system based on inherent feedback channels projecting causality. Action and control are at the heart of this method, and with the support of empirical facts and data, long-term impacts of various decision scenarios have also been simulated to assess alternative policy of practising a corporate model of high-order spirituality at low-order reality under a holistic perspective. The present research has also identified the key policy intervention points that are critical for this purpose. It was also intended to assess the cascading effect that these interventionist decisions create over the entire consciousness spectrum.

4.6.1 Quantum business management and new venture creation as business oriented manifestation of cognitive-spiritual consciousness

The fundamental and intriguing challenge that the new entrepreneurial ventures of today constantly face is to sustain the supply of a certain quanta of creative and productive resources for its present and future development and expansion. An entrepreneur constantly requires these basic impetuses or capabilities to take calculated risk and nurture a business idea to fruition despite unpredictable and uncertain internal–external environmental shocks that surround a business system. As what the quantum physics has taught us that the fundamental particles and their grosser matter forms are understood to evolve into complex intelligent and self-governing forms of 'Life' and to sustain these forms; they need to constantly derive energy from a subtler higher source. Also, all evolving systems and life, ecology and economic systems do maintain distinct recognizable states and are resilient to external shocks despite ever-changing components and continuous export of entropy at micro scale due to strong feedbacks and causal interdependence under the 'Source-Sink' dynamics of exchange of energy.

Accordingly, the prime intelligible source of all creative business ideas can be attributed to be a finer and subtler 'Creative Force or Entrepreneurial Spirit' of doing business which has inherent potential values and quality to initiate an action-oriented kinetic process of decision-making, somewhat equivalent to the property of quantum gravity living in n-spatial dimensions to motivate the transport of creative energy in very natural discrete packets and push the decision-makers to acquire higher-order state to pursue a creative idea and acquire the needed organizational abilities.

Therefore, the creative entrepreneurial forces, in association with the creative mind and resources, are together involved under an implicate order of charting out a subsequent evolutionary path or course of action and thus becoming successful in materializing the idea of a business venture. Through the contemporary quantum description, it is already proposed that there is no ultimate solid material substance from which matter is constructed, in fact, there is only energy. As per Bohminian quantum mechanics propounded by David Bohm, the configuration of quantum particles evolve via a deterministic motion choreographed by its wave function (called Information wave or probability wave by Max Born and Schrodinger) and that can be called an orchestrated wave-particle dynamics. It can therefore be suggested that there is an underlying ontological level or sub-quantum medium or creative source beneath the epistemological physical processes facilitating the realization of a business idea or venture.

4.6.2 Quantum postulates of new venture

- A company is like an electron in the lowest energy state, and it acquires capital, labour and resources to build new capabilities and hence move up to a higher energy state.
- Individual stakeholders and teams also possess a dynamic entanglement or tension between them in their relationships, mirrored unconsciously across organizations, sometimes recognized as 'culture'.
- In a dynamic business environment, managers need an advanced ability to respond to the shifts and changes within and outside their organizations – riding these developments like a wave.
- Management involves humans, and so there is a lot of unpredictability just like subatomic particles. Quantum business is also about probabilities and randomness.
- Action-oriented kinetic process of Observation/Decision-making is somewhat equivalent to the property of 'Quantum gravity' living in *n*-spatial dimensions.
- Underlying ontological level or sub-quantum medium, i.e. Quantum force field or Creative source beneath the epistemological physical processes in businesses.
- Corporate consciousness/Creative ideas/Intuition as primary trigger for new venture creation or corporate innovations and strategic formulations.
- The configuration of quantum particles evolves via a deterministic motion choreographed by its wave function (called Information wave or probability wave by Max Born and Schrodinger) – orchestrated wave-particle dynamics (ideas to venture).
- The business creation, therefore, is holistic in nature; however, it could be split in terms of partial views about its comprising elements within the notions of implicate order and explicate order displaying certain aspects of the holomovement.
- The observer as the entrepreneur being the reservoir source becomes one 'pole' and the process of business evolution another 'pole', but this is only an approximation; really, they are still parts of the same total process.
- It can finally be reiterated that this creational entrepreneurial spirit with the initial credit of some quanta of tendency has intelligence, information, memory and knowledge as per its ontological evolutionary or transformational potential order of values and quality that can be rendered kinetic under a suitable environmental condition.

4.6.3 Conclusion

The conventional quantum force field of quantum mechanics can be corroborated with the creative idea generation or entrepreneurial spirit from which the classical space-time-oriented product or business venture can be supposedly abstracted or realized. It can be said that the potential business ideas with inherent values and quality choose an optimum course of action or own path and collapse into a physical real product as per the inherent implicate narration of its entrepreneurial spiritual evolution. It has tendency, knowledge, information and above all the inspiration supplied from the engrained and encompassing business spirit to do so, and it always remains under influence to realize that potential into kinetic resources. So, the creational idea behind any business paves way for some new kind of holistic description where we analyse deeper to the level where we try to explore the implicate order by radically changing our perspective of describing the reality of business. It can be hence inferred that all successful business ideas and ventures have genesis in the most fundamental finer entrepreneurial creative force which possesses values and quality properties independently of external certainties and disturbances out in the market. That would indeed be the intrinsic momentum and implicate positioning of a creative idea, something intrinsic to the entrepreneurial business process as a state of 'being'. It is what is. What we see are the appearances of a physical product or business, and we have to explain these appearances in terms of 'what is'. Therefore, the intrinsic values and quality of a successful business venture that we attribute to the external resource-based processes can basically be attributed to the imbibed entrepreneurial creative quantum force imbibed by an entrepreneur. The business creation, therefore, is holistic in nature; however, it could be split in terms of partial views about its comprising elements within the notions of implicate order and explicate order displaying certain aspects of the holomovement. The observer as the entrepreneur being the reservoir source becomes one 'pole' and the process of business evolution another 'pole', but this is only an approximation; really, they are still parts of the same total process. We can call this 'Business spirit or creative force' as entity containing energy and supplying it to the matter (post idea creation) as per the inherent implicate order of evolution. As also opined by quantum-spiritual scientists, it is not physical reality when measured from the sense of Plank's length of 10^{-5} metre, but finer-grained particles beyond Planck's length 10^{-35} metre may exist, although they are not matter anymore. Even string theorists admit its multiple landscapes with higher degrees of freedom both for particle fined than Planck's length as well as distances greater than 10^{10} light years. It can finally be reiterated that this creational entrepreneurial spirit with the initial credit of some quanta of tendency has intelligence, information, memory and knowledge as per its ontological evolutionary or transformational potential order of values and quality that can be rendered kinetic under a suitable environmental condition.

Acknowledgements

The author humbly offers his deepest sense of gratefulness and obeisance to the Most Revered Prof. P.S. Satsangi Sahab, Emeritus Chairman, Advisory Committee on Education, Dayalbagh Educational Institutions for His Eternal Blessings, Grace and Guidance, at every step of his spiritual-information conscious living.

References

[1] Sterman J.D. (1989), Modeling managerial behavior: Misperceptions of feedback in a dynamic decision making experiment. *Management Science* 35(3), pp. 321–339.
[2] Bhushan S. (2009), System dynamics conception of cognitive consciousness: A convergent systemic interface to artificial neural network and genetic algorithm. *Journal for Interdisciplinary Research on Religion and Science – JIRRS (Romania)* 2009(5), pp. 1843–4142 (E-Journal).
[3] Jones D. (2000), States of non-mind. *Nature* 403, p. 263, Science Frontiers #128, March–Apr 2000.
[4] Penrose R. (1989), *The Emperor's New Mind*. Oxford: Oxford University Press.
[5] Penrose R. (1994), *Shadows of the Mind*. UK: Oxford University Press.
[6] Hameroff S.R., and Penrose R. (1996), Orchestrated reduction of quantum coherence in brain microtubules: A model for consciousness. In: *Toward a Science of Consciousness – The First Tucson Discussions and Debates*, S.R. Hameroff, A. Kaszniak and A.C. Scott (eds.). Cambridge, MA: MIT Press. Also published in *Mathematics and Computers in Simulation* 40, pp. 453–480.
[7] Hameroff S.R., and Penrose R. (1996), Conscious events as orchestrated spacetime selections. *Journal of Consciousness Studies* 3(1), pp. 36–53.
[8] Umezawa H. (1993), *Advanced Field Theory: Micro, Macro, and Thermal Physics*. New York: American Institute of Physics.
[9] Stapp H. (1993), *Mind, Matter and Quantum Mechanics*. Heidelberg: Springer Verlag.
[10] Stapp H.P. (2007), *Mindful Universe: Quantum Mechanics and the Participating Observer*. USA: Springer.
[11] Laszlo E. (2007), *Science and the Akashic Field: An Integral Theory of Everything*. Inner Traditions; 2nd edition, ISBN-10: 1594771812, USA.
[12] Campbell D.T. (1990), Levels of organization, downward causation, and the selection-theory approach to evolutionary epistemology. In: *Scientific Methodology in the Study of Mind: Evolutionary Epistemology*, E. Tobach and G. Greenberg (eds.). Hillsdale, NJ: Erlbaum, pp. 1–17.
[13] Campbell D.T. (1974), Downward causation in hierarchically organized biological systems. In: *Studies in the Philosophy of Biology*, F.J. Ayala and T. Dobzhansky (eds.). USA: Macmillan Press, pp. 179–186.

[14] Calyton P., and Davies P. (2008), The Physics of downward causation. *The Re-Emergence of Emergence*. UK: Oxford University Press, p. 36.
[15] Satsangi, P.S. (2011), Cosmology of consciousness: Towards quantum-theoretic systems modelling; spirit-mind-brain interactions. Vision Talk at the Inaugural Workshop of the Centre for Consciousness Studies, DEI, Dayalbagh.
[16] Penrose, R. (2016), *Fashion, Faith and Fantasy*. New Jersey, USA and Oxford, UK: Princeton University Press.

Chapter 5

A sustainable knowledge framework for technological advancements in humanitarian assistance

Yamini Meduri[1]

The current political scenarios across the world are competing like never before and every country has either directly, indirectly or virtually become a threat to every other nation. Gone are those days when nations had friendly neighborhood. Be it the Syrian crisis or the North Korean nuclear weapon testing, crisis is constantly expected, and it is time for us to become resilient if any. With increasing competition globally, business landscape is also operating in a completely volatile, uncertain, complex and ambiguous environment to which a linear model of response might only "blind" the organizations to the reality. Sustaining in such fierce competition and constant changes is a new crisis that organizations across the world are facing today [1] and it is not necessary that any linear model of response could help.

Beyond these, the adverse effects of human behavior with the natural environment activated the phenomenon of climate change like never before. Increasing number of disasters, both natural and man-made, around the world triggered an increasing amount of research in this field. Most of the disasters today have revealed that no one country is fully equipped with relief resources that can be deployed for every magnitude of disasters and that's when the countries are coming together to help the affected. In a developing nation like India where the adult literacy rate is 69.3%[*], it is all the more difficult for local communities to respond to the increasing intensity of disasters. Be it the Uttarakhand Cloudburst or the Visakhapatnam Cyclone or the Chennai Floods, the traditional knowledge of the local communities didn't effectively help in facilitating relief operations.

At the onset of any disaster, relieving the vulnerable from pain at the earliest becomes the priority of every stakeholder involved in disaster relief operations. One of the key performance indicators for any disaster relief operation is the average response time, which is affected by not only the nature and intensity of the disaster but also the availability and quality of relieving resources deployed in

[1]Vignana Jyothi Institute of Management, Bachupally, Hyderabad, Telangana, India
[*]UNESCO Institute for Statistics: https://en.unesco.org/countries/india

the disaster site around the time of the trigger event. As the time ticks through the different phases of the disaster management process, the metric called "speed" gains momentum [2] and becomes the defining factor to save multiple lives. Speed in disaster management is subjective in nature and defining it objectively becomes difficult. Also the point of reference for the definition also alters the definition big time. For a humanitarian organization, every task is generally operated at the maximum speed, and for the affected, the operation seems to be slow until help reaches. In such a volatile and complex scenario, the role of the humanitarian organizations complicates further.

With increasing nature, intensity and number of disasters, it is pertinent for emergency relief organizations to be ready and the recent advancements in technology space come in handy. However, it is observed that the critical 72 h immediately after the trigger event, technology is the least used in managing the impact of disasters [3].

5.1 Understanding the context

Disaster research has increased exponentially from the 1990s and the reasons is well supported by the data of EM-DAT† which show a significant increase in the number of disasters that occurred across the world [4] and the total number of affected multiplied even more. With less than 30 disasters in 1990, 2016 reported over 342 disaster events across the country with the people affected increasing from 100 to 700 million over time. This enabled the research community to study the processes, technologies and possible solutions even more with only 846 research records to over 38,729 records in 2018 (Proquest). With increasing focus on disaster management by humanitarian organizations, local governments, international associations and academia, it is crucial to understand what is it that the people involved are required to know. Also, it is pertinent to remember that disasters can only be managed—their impact can only be reduced but not prevented totally. Hence the current chapter intends to study the humanitarian logistics (HL) dimension of disaster management so that the vulnerability of the affected can be reduced.

Ever since the Indian Ocean Tsunami in 2004, which was the biggest not only in terms of the havoc it has created and the damage it has done but also to the level of relief operations that were carried out, humanitarian organizations started to redefine their strategies and operations. The work in [5] defines HL as

> the process of planning, implementing and controlling the efficient, cost-effective flow and storage of goods and materials, as well as related information, from the point of origin to the point of consumption for alleviating the suffering of vulnerable people. The function encompasses a range of activities, including preparedness, planning, procurement, transport, warehousing, tracking and tracing, customs clearance, etc.

†EM-DAT: The Emergency Events Database—Universite catholique de Louvain (UCL)—CRED, D. Guha-Sapir—www.emdat.be, Brussels, Belgium.

To keep it simple, [6] defines it as a process "to procure, transport, receive and distribute supplies to the site of relief" and [7] defines logistics for humanitarians as "the processes and systems involved in mobilizing people, resources, skills and knowledge to help vulnerable people affected by disaster."

5.2 Need for knowledge framework

Knowledge management is a carefully created term that involves problems, data, information, strategy, solutions concepts, etc. It is often misinterpreted with the capability of employees but the "knowledge" in knowledge management spreads to define much more than just the employee's capabilities. A knowledge framework is expected to provide better quality solutions for complex business problems by enriching data availability within the system, information procurement and processing within an organization. Though knowledge management framework doesn't directly involve or impact the decision-making processes, its role is becoming important with passing time for organizations [8].

Developing a knowledge system for any kind of organization is a multilevel and in-depth analysis project which involves not only the multiple stakeholders formally inducted into the system but also the data from multiple sources. Reliability and validity of such data makes a huge difference in the disaster relief operations. At the onset of a disaster, multiple actors with multiple capabilities start contributing from multiple points on field. At a disaster site, the last thing that we want is redundancy and it is important that this is addressed. Also, more data inflow with more actors involved can increase the data processing time leading to losing the critical 72 h in understanding the situation. According to [9], the characteristics of such data are multifaceted, which include the following:

1. large number of producers and consumers of data;
2. time sensitivity of data;
3. trustworthiness of data sources;
4. combination of static and streaming data; and
5. heterogeneous formats of data.

One thing that is certain in the most uncertain environment is that there are huge data that are generated and consumed at the disaster event and all such data need to be managed well. The process of data management includes subprocesses such as data identification, data processing, data recovery, data integration and data protection. All these subprocesses operate from multiple nodes and equally participate in the disaster management processes. The information and communication technologies (ICTs) suggested in each of these subprocesses are presented in Table 5.1.

Disasters of different magnitudes generate and use different levels of data, and central nodal point to collect, process, manage and disburse data within a minimum time frame makes it more important and relevant for disaster relief organizations.

Table 5.1 Data-related activities and suggested technologies

Data-related activity	Technology suggested	Literature evidence
Data identification	Social media, GIS data, satellite data	[10–13]
Data processing	Sentimental analysis, data mining	[9,12]
Data recovery	Cloud technology	[14]
Data integration	Semantic web technologies	[9]
Data protection	Cloud technology	[15]

Multiple disasters and the way they were handled and managed in India suggest that a lot of knowledge created is often lost in the test of time. This leads us to the need for developing a sustainable knowledge framework that develops, stores and delivers such knowledge which not only makes disaster management more reliable but also can reduce the operational cost of handling disasters. The framework suggested in the current chapter is expected to create a bridge between the technological Know-How and field operations involving multiple actors during the disaster relief operations.

5.3 Methodology

The current chapter uses qualitative methodology of research which involves primary data from personal interviews conducted, observations during the disaster event, participation in the data generation and consumption process. In one of the disaster cases discussed in the chapter, the author of the chapter participated in the data generation process, and her experience framed the necessity for suggesting an integrated model. The secondary data used for the chapter are collected from the research articles and reports obtained from Proquest, Google Scholar, Emerald Insight, books and journals, etc. The chapter also presents case studies of two disasters that occurred in India—Hudhud Cyclone, 2014, and Kerala Floods, 2018—and how ICT was used with different nodal points in the disaster relief and rehabilitation operations. The case studies were chosen based on the differences in how technology was used in the different situations by different stakeholders. The ground level information was collected through interviews—30 volunteers and 24 officials in different capacity during Hudhud Cyclone; 25 ground volunteers and 19 officials in different capacity during Kerala Floods. The analysis of how these two disasters were handled can help us understand the need for developing a sustainable and integrated knowledge management system, the primary objective of the current chapter. The information is carefully curated to meet the objectives of the chapter and is further processed qualitatively to design a knowledge framework that may support disaster relief operations.

The further sections of the chapter briefly explain the disaster profile of the nation in general and the states in particular. The nature and intensity of the disaster, role of different stakeholders, use of ICT in the case disasters, etc. will be explained. Based on the review of literature and the case analysis, the knowledge framework will be proposed along with the assumptions and implications for disaster management in India.

5.4 Disaster profile in India

India as a nation is prone to disasters of different types wavering from sudden onset disasters to slow onset disasters. According to the National Disaster Management Authority of India, 75.8% of the 7,516-km long coastline is inclined to floods and cyclones, 68% of the agricultural land is likely to be affected by droughts while 12% is prone to river erosion. Also, more than 58.6% of the land is susceptible to earthquakes of different intensities. Other man-made disasters only add on to the vulnerability profile of the Indian subcontinent.

A total of 698 natural disasters stuck the Indian subcontinent from 1900 to 2018 (EM-DAT) with around 3,97,048 average deaths and 10,43,83,653 people affected on an average. With increase in the number of disasters over the last few years, it can be observed that the number of deaths has come down but the number of people affected by the disasters and the total damage has increased exponentially.

5.4.1 Disaster profile of the state of Andhra Pradesh

Andhra Pradesh (divided) has 974-km long coastline, most of which is prone to different types of disasters ranging from cyclones, floods and droughts at times. Nine out of 13 districts of the state are declared to be prone to cyclones and floods whereas the rest are prone to drought. The undivided state had experienced 32 cyclones after a severe cyclone in 1977 with an average estimated loss of INR 1,395.44 Cr. The magnitude of the Hudhud Cyclone was way beyond the one in 1977, making an estimated loss of INR 21,640 Cr contributing to almost 48% of the total damage caused by the cyclones (EM-DAT).

5.4.2 Disaster profile of the state of Kerala

Kerala is situated between the Western Ghats and the Arabian Sea. The coast length is 580 km and the width of the state varies from 34 to 120 km. The topography consists of hot and wet coastal plain that gradually rises in elevation to the high hills and mountains of the Western Ghats. The state is prone to several natural calamities like landslides, floods, lightning, drought, coastal erosion, earthquakes, tsunami and windfall. The highlands of the state experience landslides. The characteristic pattern of this phenomenon is the swift and sudden downslope movement of highly water-saturated overburden containing a varied assemblage of debris material ranging in size from soil particles to boulders, destroying and carrying with it everything that is lying in its path. All except 1 of the 14 districts in the state are prone to landslides. Wayanad and Kozhikode districts are prone to deep-seated landslides whereas Idukki and Kottayam are prone to shallow landslides.

5.5 Disasters in discussion

The current chapter discusses two major disasters in the states of Andhra Pradesh and Kerala in the years 2014 and 2018, respectively. A brief of these disasters is presented in the following subsections.

5.5.1 Hudhud Cyclone

According to the Indian Meteorological Department (IMD), a cyclone could be categorized into five categories based on wind speed: low pressure (<31 km/h), depression (31–49 km/h), deep depression (49–61 km/h), cyclonic storm (61–88 km/h), severe cyclonic storm (88–117 km/h) and super cyclone (>221 km/h). The observed cyclone, Hudhud, was way beyond the "severe cyclonic storm" and hence was declared a "very severe cyclonic storm." IMD rightly forecasted that Hudhud would be an urban disaster with a landfall at the noon of October 12, 2014, over the city of Visakhapatnam with winds of 205 km/h in two spells. The intensity of the velocity of the winds was so high that it uprooted huge trees, broke telephone lines and electricity towers, damaged electric poles, damaged houses, etc. all along the city. The devastation was huge in the second spell compared to the first.

According to the report of the Ministry of Home Affairs, 3.3 lakh-hectare cropped area was affected, 40,379 houses were damaged and 4,777 cattle heads and 61 human lives were lost. "... the decision about switching off the power supply largely affected the number of human lives lost...," highlighted the District Collector. About 90,000 electric poles were damaged, High Tension lines swiped off, 500 communication towers damaged, highways damaged and all continued to be affected until the midnight of October 13, 2014. Once the velocity of the winds slowed down, the different resources started relief operations. Also, 1,800 private industrial units were damaged; whereas in the state-owned industrial park, the number is around 1,500 units.

5.5.2 Kerala Floods

The state of Kerala generally does not experience severe floods like the Gangetic plains, but recently, the incidence of floods in the state is becoming more frequent and severe. The latest severe one was the August 2018 floods. The last time when Kerala saw bad floods was in 1924. This was called the great flood of 99 (because it occurred in the year 1099 as per the Malayalam Calendar). The main reason for the extreme floods in the state in 2018 is the high-intensity rainfall. There are other reasons that are man-made as well. These include wrong land-use practices and mismanagement of water resources and forests, reclamation of wetlands and waterbodies, change in the land-use pattern, construction of networks of roads, establishment of more residential settlements, deforestation in the catchment areas, unchecked quarrying, etc. Increased habitation in the flood plains has resulted in increased flood damages. Kerala is home to 53 large dams with a collective capacity of nearly 7 trillion liters. The continuous rains in August forced the authorities to release water into the already flooded areas. Gradual phased release of water could have avoided the level of disaster. Idukki dam opened all shutters for the first time after 26 years. Filling up reservoirs before the monsoons is an open invitation to such disasters.

5.6 Similarities and differences in handling disasters

There is a lot of similarities yet differences in the way both the disasters were handled, onset 4 years apart. However, operations can be classified into four major categories, namely Leadership, Disaster Response Force, Local Communities and Technology. Each of these categories had contrasting differences in the role in both the disasters yet both the disasters were managed. The following sections of the chapter explain briefly[‡] each of these categories and technology in detail in managing the two different disasters. The following sections will form a base for the proposed knowledge framework for the improvement of the disaster relief and rehabilitation processes.

5.6.1 Leadership

The leadership of the states played an important role in how the state administration was prepared for the disaster event. The then Chief Ministers of the states have adopted diverse styles in managing the disasters in their respective states. Though the economic, political and cultural environment of the states was alarming at the time of the onset, the leadership stood strong to lead the relief and rehabilitation efforts.

The state leadership of Andhra Pradesh was proactive in responding to the signals from the state-of-the-art technology at IMD which perfectly predicted that it is an urban disaster providing a good 5-day preparation time. The quick first-level decisions were evacuation of coastal communities, cutting off power well in advance to reduce fatal emergencies, setting up of relief centers, mobilizing resources from state and center, etc. Hundreds of thousands of people have been moved to refuge in 223 relief camps. However, lack of previous experience of the magnitude of the disaster made it more difficult to sensitize the local population and support evacuation. "... at one point of time, we had to literally drag them out as they were not ready to leave their homes," said a police officer during the interview. In spite of this, 3.5 lakh people were shifted to relief centers where access to basic amenities like water, food, etc. was made well before the landfall. Enough

[‡]Detailed analyses of the disaster management of the two disasters are presented in the following:
Hudhud Cyclone: Presented a paper titled "Multi-stakeholder participation in the disaster recovery: A case study" at Humanitarian Technology 2016: Science, Systems & Global Impact, by ISCRAM in June 2016 on full grant from SERB (Science & Engineering Research Board), Dept. of Science and Technology, Govt. of India under the Young Scientist Category.
Kerala Floods: Co-presented a paper titled "When humans saved the God's own country: A case study of community participation in disaster relief" at 4th World Congress on Disaster Management organized by DMICS in Mumbai, January 29 to Feb 1, 2019, along with Ms. Jayashree V.

supplies of potatoes and onions were procured from neighboring states and kept ready. The telecom operators set up call centers in states of Andhra Pradesh and Odisha and communicated to every subscriber about the details through a text message. The South Central Railway cancelled trains in advance to reduce the damage and vulnerability of people. The level of preparation did help in relief, recovery and restoration of the cyclone-hit area.

"... our prime motto of the relief operation was 'Zero Casualty' and we were ready to face the filed round the clock to being back normalcy in the minimum possible time," elicited the Chief Minister, who camped in a van for 8 days overseeing the operations. This ensured the people of the city to be brave and face the operations.

The flood situation in Kerala was handled with the tacit knowledge of the local communities. With the situation becoming almost a flash flood, the on-the-spot evacuation plan and the natural tacit knowledge of the local communities to deal with floods added to support the state administration. "In most cases, the damage was higher because the dams were open in the night without a proper evacuation plan," opined a village panchayat officer during the interview. The high political situation of the state did not show the solidarity of standing strong with the state. However, the state administration stood strong with the local communities and public administration offices in relief and rehabilitation efforts and brought in normalcy within 2 weeks. Most of the decentralization of decision-making worked in faster decision-making and disbursement of relief efforts. With the support from the local administration authorities, coordinating the relief operations was possible. With most of the communication channels interrupted due to the inundated floods, the decentralized authority made it possible for Kerala to get back to normalcy within 2-week time.

Be it Andhra Pradesh or Kerala, the leadership of the state and the understanding of the vulnerability of the disasters turned out to be an important parameter that determines the way the calamities are handled. The two instances are an important pointer that elucidates the need for the involvement of state leaders in entire disaster management process determines the efficiency of the management process.

5.6.2 Disaster response force

Both the disasters, the Hudhud Cyclone and Kerala Floods, have seen a well-coordinated effort of the disaster response forces all over the nation. Multiple organizations have come forward to provide the necessary relief and rescue operations for the people affected due to the disasters.

During the Hudhud Cyclone in the state of Andhra Pradesh, the state police force deployed specialized platoons like 40 Octopy forces, 180 Greyhounds, 36 APSP battalions for the relief operations. The National Disaster Response Force (NDRF) deployed 12 specialized teams for cutting the fallen trees and clearing off the highways for regulating vehicular traffic with relief materials other than the regular 45 teams that participated in the direct relief operations.

With more than 45 public departments actively involved in the emergency relief activities and with the communication channels cut due to the cyclone, the access for right information was far difficult increasing the vulnerability multifold. "...the District Collector's office & home, office of IMD, Police Stations all were damaged. For a while, we didn't know how to react or where to go first," recalls a police officer who was on duty for straight 6 days at the time of disaster. Power supply companies deployed over 6,000 employees to restore the power supply channels and the hundreds of non-governmental organizations (NGOs) came in to assist in the relief operations.

In the state of Kerala during the floods of 2018, the rescue teams from every possible organization joined to support the efforts of the state administration in providing the necessary relief operations for the affected. The state police worked with their complete force in the relief and rescue operations, 26,000 personnel deployed at any given point of time, who were joined by 58 NDRF teams deployed from different parts of the country in the state. The Indian Air Force deployed close to 3,000 service personnel for the relief operations with 29 helicopters involved in not only lifting and dropping the relief material but also participating in 583 air-lifting operations. "... airlifting a pregnant lady due for delivery was the ultimate achievement," recalled the officers in one interview. The Indian Air Force, also for the first time, set up mobile hospitals serving the affected. Looking at the intensity of the disaster and magnitude of vulnerability, Indian Navy from the South Naval Command deployed 92 fully specialized rescue teams equipped with the necessary technology to save the affected people of Kerala. "... the data centers from student volunteers helped us with accurate locations and made our work more easier," told one commanding officer. The motorboats and the abundance of natural water routes made it possible for Indian Navy to provide all the necessary support in the time of crisis. The efforts were also supported by the Indian Army which deployed over 27 teams to help the state in the rescue operations. A central control room operated by National Disaster Management Authority (NDMA) helped the state understand the intensity of the situation and get real-time data to channelize the efforts.

5.6.3 Local communities

In both the events, the local communities played a crucial role in the entire disaster management process. Despite being hit by the disaster themselves, the solidarity shown by the local communities in handling the vulnerability of the disasters and helping the affected had to be appreciated. The local communities' part of the disaster relief operations can be considered as the most important and prominent aspect of the knowledge management system that is developed through this chapter.

The social capital structure [16] not only enabled trust and belief in the relief operations but also brought forward the local communities to facilitate a speedy recovery. Students, volunteers from various NGOs, local communities and many more participated in clearing debris, removal of fallen trees, setting up communication towers, etc. Most of all, the multi-stakeholder participation facilitated in recovering back to normalcy in record 8 days after a massively destructive "very severe cyclone" hit the city of Visakhapatnam.

The local households which recovered faster than the rest participated in community cooking drives where foods were cooked and distributed to the affected areas. The business community had come forward voluntarily to participate in the reconstruction of the affected areas free of cost. The telecom companies put in all efforts to restore the communication infrastructure that was damaged due to the cyclonic winds in the least possible time. The representatives of the insurance companies worked with toll-free centers to clear the insurance claims in a fastest time possible. The magnitude of the disaster attracted notable donations in cash and kind from all sectors of the society routed through state administration and several NGOs.

Students participated in the disaster relief operations in altogether a different scale. With the 5 days in hand for preparing for the onset of the disaster, the students of the city collaborated in identification, curation and management of data that can be of use immediately after the onset of the disaster. The Hudhud Cyclone marked as one of the first disasters in India where the technology created by the local communities was used for the relief and rescue operations. The event showed a huge level of service orientation in the local communities which multiplied across the country during the disasters that followed.

Taking cue from the Hudhud Cyclone, the local community participation during the Kerala Floods defined a new level. Multiple stakeholders participated in the rescue and relief operations. Taxi aggregators like Ola and Uber provided free transport for the affected to the relief shelters. Food delivery apps like Swiggy and Zomato allowed its customers to donate food through their apps and the organizations took care of supplying it to the affected directly. Multiple restaurants also joined hands by making use of infrastructure available to prepare food while the food delivery apps took care of the delivery part. While companies like Paytm enabled its customers to donate cash, Amazon and BigBasket on the other hand helped its customers to donate food and relief materials from their inventory. On the one hand, Facebook made the disaster go viral and it not only partnered with Goonj, an NGO that is working on the relief operations in the state but also facilitated donations by creating a "Crisis Response Page."

The efforts of the student community raised the bar to newer heights discussed next in detail in Section 5.6.4. "... The data provided by the student community which had setup on-the-spot call centers supported the relief operations. They knew where to go and we just went there and did our bit," recalled a policeman who was at work for 10 days straight and rescued thousands of affected citizens just following the directions of the data centers.

While the technology enabled stakeholders participated through their bit, the Kerala Floods experienced higher levels of humanity from the fishermen community of the affected and non-affected districts. Known for their skills of rowing through the harshest waters using the specially designed boats that row through the toughest undercurrents, the fishermen community rescued thousands of affected and rowed them to the relief shelters. "... I just did what I know. It's nothing much," said a fisherman who participated in the relief operations with all modesty.

In both the disasters, communities played a very important role in the rescue and relief operations. In every level possible, the local communities contributed to the relief operations at the disaster-struck regions when they themselves were affected too. This service orientation in the communities is what the disaster management banks on.

5.6.4 Technology

Technological innovation during disastrous events is seeing a tremendous ascent in design and execution. In earlier disasters, the technological interventions were designed and implemented with the government as the focal point. As the magnitude of disasters increased and the technological disruptions crowded the lifestyle of individuals, the use of technology in the rescue and relief operations also changed its course from government to communities. The two cases in discussion clearly show the transition from government to communities in developing and using technology for disaster rescue and relief operations.

The Hudhud Cyclone has been an excellent example for how much the government can make use of technology in succeeding in its efforts to relieve people from the vulnerabilities of the disaster. The Government of Andhra Pradesh (GoAP) made the right use of technology in the different phases of disaster management and there were many first timers for a lot of stakeholders who were involved in the entire process. The telecom operators set up call centers in states of Andhra Pradesh and Odisha and communicated to every subscriber about the details through a text message. The text message included details of the relief shelters, toll-free numbers, intensity of cyclone, warning alerts, etc. This enabled the citizens of the state to be alert at the onset of the disaster. Since it was known that the communication channels are going to be affected by the landfall, the text messages kept the families far away from the onset site to be aware about what was happening in Visakhapatnam. The text messages relatively reduced the panic that generally escalates in any crisis situation.

One of the natural advantages of the cyclone was that the landfall happened in the midnoon of the day, but the state prepared the citizens to face the landfall for any time of the day. This enabled the state to reduce the loss of life drastically as they were warned to stay at home and not move out as the intensity of the winds was not known.

The Satellite Imagery and Space Technology of IMD was closely monitored to understand the situation in the closest quarters so that the government can be rightly prepared for the disaster. The 5-day time in hand helped the government to put in place the right evacuation plan to reduce the casualties. The state leadership also believed on crowdsourcing as a perfect solution to understand the real damage in real time. With 5 days in hand before the onset of the disaster, the GoAP in collaboration with National Remote Sensing Centre launched a web portal that can be used for collecting real pictures of the damage from the different localities of the onset site. This was expected to help faster decision-making in the crucial first 72 h of the relief operation. The government used Geographic Information System (GIS), Global Positioning System (GPS) and remote sensing technologies to create an interactive map of the havoc created by the harsh cyclonic winds. The web portal was later developed into an app that works on an Android to enable the citizens use it more effectively and for the government to get more relevant and real-time data.

The government ran a social media campaign to collect data to match it with the data collected with the images of the web portal. The campaign not only resulted in data but also created a volunteer network that is readily available for post-landfall rescue and relief operations. Owing to the social media campaign, through WhatsApp, Facebook and Twitter, a huge data was received by the portal and the app. A real-time data mining tool and big data analytics was rampantly used to understand the ground realities of the situation. The analytics helped in understanding the right locations to place the inventories of the relief materials, target population, identify relief shelters, etc. This enabled the government to keep the casualties low as the infrastructural damage was inevitable.

After the winds and rain subsided, the same data were used to prioritize the deployment of men and machinery to clear the debris created by the havoc. The analytics-enabled program environment supported the government in faster and accurate decision-making. Beyond the pictorial data, the portal (and app) attracted data about multilevel increase of prices of utilities, which helped the government to intervene immediately and control the situation to normal pricing.

During the Hudhud Cyclone, the understanding of technology and capitalizing on the ability to use it in such a nerve-breaking environment was possible with the intervention of the government and being the focal point of key decisions. The support of the local administration to the state authorities played a prominent role.

After Hudhud, it was expected that the use of technology in managing disasters to increase as the knowledge base was already created and it only needs to be upgraded. However, the knowledge transfer was not evident because of which the disasters after Hudhud did not see technology being used in the right direction for the rescue and relief operations.

The case of Kerala Floods and use of technology has shown the country, and may be the world, how community-driven technology interventions can help in faster rescue and relief operations. The disadvantage of the Kerala Floods incident is not only that the state did not make use of the information from IMD to plan for sufficient evacuation but also that the dams released water mostly during the night which caught people in the darkest situation. The inundated floods made people move to the rooftops not knowing how to save themselves. As expected, the communication channels were affected by the bad weather conditions and panic built up as the time passed by. In order to support the government in the relief operations, the community participation was triggered through technological interventions. Further to the efforts of IMD, organizations like GeoSpoc and Satsure worked on decoding the data from satellite imagery and presented the information to the Government of Kerala to channelize their efforts.

Crowdsourcing and data mining were used as important tools to make use of the data that were flooding from different data sources such as Facebook, Twitter, WhatsApp, Slack, text messages, Google codes, etc. Big data analytics helped in understanding the patterns of requests and provide relief to the affected. In addition to the satellite imagery and geospatial data, the crowdsourced data helped in understanding the accuracy of the location requiring help.

Kickstarted by the initiative of Google, it's introduced specific services such as Google Person Finder, Plus Codes and Map Pins which enabled an offline user to share their location with the family and/or the rescue teams. This enabled many individuals, groups and organizations to send help. Twitter created a specific hashtag for the disaster, #KeralaFloods, which captured a lot of data across platforms for finding people, providing help, mobilizing funds, deploying resources, distributing relief material, etc.

A team of students from IIT Guwahati worked to develop an open source application, Keralarescue.in, which started collecting data about the rescue and relief requests and identifying the GPS locations of the stranded citizens. The team coordinated with the central nodal point of NDRF to provide help to the requests that the application collected. This helped the trained relief personnel deployed on the field to prioritize their resources. As the number of requests increased, 1,500 professionals from different parts of the world joined to develop the application and handle the requests voluntarily. Being an open source application, dynamic and real-time optimization was possible. With a lot of data flowing in from multiple sources, a team of student volunteers from across the country stepped into verifying the data before passing it on the on-field workers. Soon, the team was joined by institutions like IISc Bengaluru and IIT Hyderabad to maximize the impact. A cloud telephony-based solution was created to handle the traffic to ensure all the requests are addressed. The portal registered close to 50,000 volunteers who addressed 45,000 requests that were posted in the portal.

Another team worked closely with ground volunteers stationed at the unaffected places in Kerala in mobilizing funds to distribute groceries and daily essentials in the affected parts of the state. A group of volunteers from different states came together to support the affected. Each day fund raising campaigns were run through WhatsApp groups and the fund raised was sent to the ground volunteers who would buy, segregate and distribute the relief material to the affected. By the end of each day, the bills and photographs of the groundwork were sent to the donors to build transparency in the entire system. Close to INR 2,00,000 were mobilized in a span of 4 days, which helped the families of two districts with the immediate daily utilities.

A travel blogger set up a contact with the relief personnel working in the state and supported in providing information to the response forces about medical emergencies in specific and relief requests in general. On average, 300 distress calls a day were addressed by this one-man army.

A Kerala student group, 11,000 of them, from IIT Hyderabad joined together in locating victims and mapping flooded areas. The information collected by mining the data from Facebook was used by the team to contact the stranded victims and assured them of help. The contact was established through a hot line that was launched using the IVR technology which verified most of the data. Once the data were verified and the contact was established, the team posted an update to the data source. This not only reduced the panic but also increased the flow of data as the visibility of the work increased. The team went an extra mile to contact the victims once again to understand if the aid was received or not. This enabled a full support to the victims.

A few of the many initiatives to provide relief for the vulnerable during the Kerala Floods show the impact of community-driven technology interventions can help in maximizing the relief efforts of the ground personnel. The technological interventions not only supported the government, or the disaster response forces, in faster and right decision-making but also created a platform of assurance for the victims and their distant families. These initiatives had shown the disaster management communities as to how much knowledge is available for harnessing to improve the efficiency of humanitarian relief operations.

5.7 Lessons learnt from Hudhud Cyclone and Kerala Floods

As discussed, multiple stakeholders participated in different capabilities during the Hudhud Cyclone of Andhra Pradesh and inundated floods of Kerala. Brief summary and key takeaways from the disaster are illustrated in Table 5.2.

It has been observed, and found to be true, that a lot of knowledge created during a disaster is often lost in the transition and is not available for the next onset. Hence, it has become essential that a central nodal agency collects, stores, monitors and disburses such knowledge created during a disaster. The proposed knowledge framework is expected to create a base for thinking in this direction.

5.8 Knowledge framework

The form of knowledge framework that is considered for the purpose of the chapter is to design a supply chain network for the knowledge created during the rescue and relief operations. For the purpose of this framework, it is assumed that the data produced and consumed by different user nodes is stored in a cloud location. This will enable us not only to gather a huge amount of data that is processed but also provide information readily available for immediate usage. The proposed knowledge framework is presented in Figure 5.1.

The four parameters, as suggested in the earlier sections of this chapter, are divided in to subsections in order to showcase the magnitude of these parameters. For example, the local communities are formed by local community, crowdsourcing and technical expertise parts of the model (Figure 5.1) while technology also includes technical expertise, social networks and crowdsourcing (based on the data sourced) parts of the model. The leadership is the culmination of government, technical expertise and crowdsourcing based on the roles taken by these

Table 5.2 Key takeaways from the cases in discussion

Parameter	Hudhud Cyclone	Kerala Floods	Key takeaways
Leadership	On-the-go decision-making	Triggered by the political situations	Leaders' understanding of technology can make a huge impact
	Decentralization of authority	Ambiguity in decision-making	Decentralization of decision-making can save time.
	Effective utilization of time in hand	Inappropriate use of time in hand	Presence should ease the panic among the affected
	Harnessed the potential of technology	Technology not completely harnessed	
Disaster response force	Multiple teams from different organizations	Multiple teams from different organizations	Response forces should be ready at every instance
	Participated in all phases of disaster management	Participated in rescue and relief operations only	Engaging the response forces in every force can reduce damage
Local communities	Huge participation witnessed from multiple communities	Communities became the focal point in the relief operations	Huge potential in the local communities to be harnessed
	Understanding and adhering to warning systems	Supported the response forces with real-time data	Tacit knowledge of local communities helps relief operations, especially with topography of the place
	Standing up for affected local communities		
Technology	The government is the focal point in using technology	Community is the focal point in using technology	A combination of government and community-driven interventions
	Crowdsourcing real-time data	Data mining through volunteer network	A lot of knowledge lost in transition
	Satellite imagery and real-time data mapping	Humane effort to connect the disconnected	

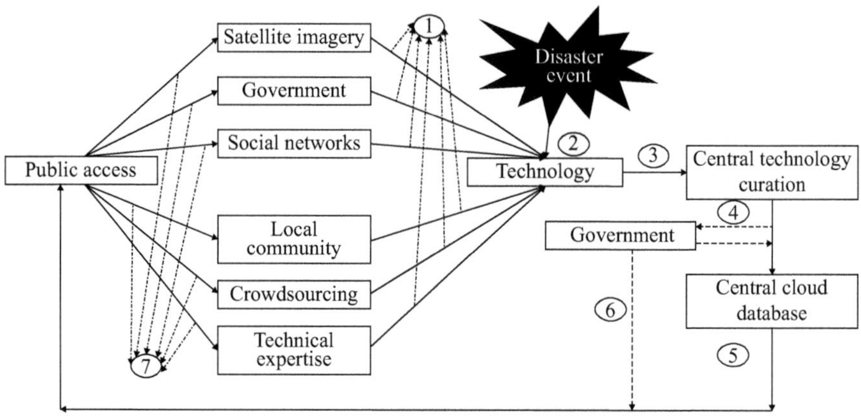

1 Technology solution created
2 Technology utilized for a disaster event
3 Technology posted for central curation
4 Curated technology as per Government Regulation and stored in cloud location
5 Technology available for public access
6 Government checks the identity before public access
7 Technology available for public to use or improve

Figure 5.1 Knowledge framework for disaster management

components proactively during a disaster event, and the disaster response force is included in the government as they are often directed by the government-led organizations. Hence, the four parameters listed in the earlier sections of this chapter may not be directly visible in the model (Figure 5.1).

The knowledge framework that is proposed in this chapter attempts to design a possible network of technology that is designed from the scratch at the time of every disaster and is often lost after the relief operations ends.

The technology that is generally designed and executed for the disaster event using satellite imagery, government interventions, social media campaign, community participation, crowdsourcing data and technology expertise is often lost to the test of time. The framework proposed insists to collect and curate every such technology designed during a disaster based on the type of disaster, problem area addressed, solution generated, etc. and to store it in a cloud location. This cloud location will be managed and monitored by the government (both State and Center).

Instead of developing the technology each time from the scratch and disposing it after the impact of the disaster subsides, as it is the case now, it is important that we build from where we stopped. The proposed framework suggests using a central nodal point where the data, code, technology, prototypes, idea charts, idea summaries, challenges faced, lessons learnt, possible errors, etc. can be stored for public use. This nodal point will be created, monitored and managed by the Central Government of the country. The nodal point will also have to develop an expertise that can check and curate information that flows into the repository. The expertise may be open-sourced based on the competencies, research interest and service orientation. The experts are responsible and accountable for the different activities listed in the Table 5.1.

Further to the curation of data, the relevant information would be then proceeded to be stored in a cloud location. The cloud location should be accessible to different stakeholders who can upload and download information. However, the accessibility will have to pass through a secured line to ensure that the data or technology is not misused. To enable this, the nodal center will devise a validation mechanism.

This framework can ensure a proper knowledge collection, management and dissemination to the right users to ensure that the knowledge created does not go waste. It is important for countries, which are more disaster prone, to invest in such technologies so as to capitalize on the expertise available with the public and private organizations, local communities and researchers. The primary advantage of having a central accessible nodal point is to start from where it was stopped and not to start from the scratch. Also, the framework doesn't disturb the interventions that are bound to be designed at the onset of the disaster but adds value and reduces time in designing the solution concept at the onset of the disaster. With the first 72 h being very crucial for determining the efficiency of a humanitarian relief operation, the knowledge framework helps in reducing the time spent on developing a technology.

5.9 Conclusion

Increasing disasters and the magnitude of impact made it necessary for the multiple stakeholders to respond at the onset of any disaster. With the advancement of technology and its reach to the communities, technology has become an integral part of the lives of people, and in the time of disasters, such knowledge is coming in handy. With advanced information systems such as the Internet of things, crowdsourcing, cloud computing, artificial intelligence, big data analytics, etc. relief operations at the onset of a disaster are rightly supported with effective decision-making systems.

However, over a few disaster instances in India, it was observed that the technological solutions developed during disaster instances are becoming inaccessible and, hence, obsolete after recovery and rehabilitation phases of the disaster management process. Every next event the knowledge creation starts all over from the scratch again. To minimize this, it is vital that the knowledge created during one disaster is available for use (and/or reference) during the next disaster event, if need be. At a time of utmost complexity and ambiguity, searching for such knowledge without a proper reliable source becomes a dreadful task making the complexities multiply. Hence, it is important that a supply chain of knowledge created during the rescue and relief operations of a disaster is available for research and/or development of a better solution for disaster management. Hence, the current chapter proposes a sustainable knowledge framework that not only captures the data related to technologies but also monitors and manages the inflow and outflow of such data.

The proposed knowledge system will identify, curate and manage the crowdsourced data and technologies developed in a cloud location which will be accessible by any stakeholders who can make use of such information. It is an integrated system that shall capture knowledge from data generated by the satellite imagery, crowdsourcing through social media campaigns, big data, machine learning techniques etc. and makes it be available during the times of crisis. A system at the central nodal level of any country is expected to save the precious 72 h after the landfall of a disaster.

The knowledge system suggested in the present chapter is only at the ideation phase, and the implementation phase requires the consent and resources from the multiple stakeholders who involve in the disaster management process. The use (and/or misuse) of data is seen as one of the major threats to the system which can be addressed using a right mix of encryption and decryption of data. When the security features are addressed and the system is monitored by a central focal point, like NDMA in India, the primary objective of relieving the vulnerable in the minimum possible time can become true.

References

[1] Kutz MR, and Bamford-Wade A. Understanding contextual intelligence: A critical competency for today's leaders. Emergence: Complexity & Organization. 2013;15(3):55–80.

[2] Meduri Y. Humanitarian logistics: Challenges for human resource management. Universal Journal of Industrial and Business Management. 2014; 2(6):135–41.

[3] Heide EA. Common misconceptions about disasters: Panic, the disaster syndrome, and looting. The first 72 hours: A community approach to disaster preparedness. iUniverse. 2004;337.

[4] Guha-Sapir D, Vos F, Below R, and Ponserre S. Annual disaster statistical review 2011: The numbers and trends. Centre for Research on the Epidemiology of Disasters (CRED); 2012.

[5] Thomas AS, and Kopczak LR. From logistics to supply chain management: The path forward in the humanitarian sector. Fritz Institute. 2005;15:1–5.

[6] Russell TE. The humanitarian relief supply chain: analysis of the 2004 South East Asia earthquake and tsunami (Doctoral dissertation, Massachusetts Institute of Technology).

[7] Van Wassenhove LN. Humanitarian aid logistics: Supply chain management in high gear. Journal of the Operational research Society. 2006;57(5):475–89.

[8] Anonymous (1). Defining knowledge management: Knowledge management or not knowledge management? That is the question. Strategic Direction. 2005;21(10):22–24.

[9] Hristidis V, Chen SC, Li T, Luis S, and Deng Y. Survey of data management and analysis in disaster situations. Journal of Systems and Software. 2010; 83(10):1701–14.

[10] Sangameswar MV, Rao MN, and Satyanarayana S. An algorithm for identification of natural disaster affected area. Journal of Big Data. 2017;4(1):39.

[11] Voigt S, Kemper T, Riedlinger T, Kiefl R, Scholte K, and Mehl H. Satellite image analysis for disaster and crisis-management support. IEEE Transactions on Geoscience and Remote Sensing. 2007;45(6):1520–28.
[12] Montoya L. Geo-data acquisition through mobile GIS and digital video: An urban disaster management perspective. Environmental Modelling & Software. 2003;18(10):869–76.
[13] Zhang D, Zhou L, and Nunamaker Jr JF. A knowledge management framework for the support of decision making in humanitarian assistance/disaster relief. Knowledge and Information Systems. 2002;4(3):370–85.
[14] Challagidad PS, Dalawai AS, and Birje MN. Efficient and reliable data recovery technique in cloud computing. Internet of Things and Cloud Computing. 2017;5(1):13–18.
[15] Wei C, inventor; YOTTAA Inc, assignee. System and method for performance acceleration, data protection, disaster recovery and on-demand scaling of computer applications. United States patent application US 12/717,297. 2010 Sep 9.
[16] Nakagawa Y, and Shaw R. Social capital: A missing link to disaster recovery. International Journal of Mass Emergencies and Disasters. 2004;22(1): 5–34.

Chapter 6

ICT for early assessing the disaster amplitude, for relief planning, and for resilience improvement

Horia-Nicolai Teodorescu[1,2] *and Mironela Pirnau*[3]

There is a well-established relationship between information provided by authorities during emergency situations and community resilience [1,2], as well as between the relief planning, the time of response and the outcome of the relief intervention. All the premises in these involvements are essentially based on the correct and fast gathering of the data on the amplitude of the disasters and on their extent and immediate consequences [2–7]. The role of the information and communication technologies (ICTs) in emergency situations consists in data collection, data processing, data aggregation, knowledge discovery on various disaster situations, risks assessment and prediction of the disaster effects, decision-making, relief control, and guiding the people affected by disasters, among others. In many disaster cases, data can be collected on social media (SM) and particularly on social networks (SNs); SN can also be used for relief control and for guiding peoples affected by disasters. This chapter addresses the optimization of the early assessment of the disaster amplitude and features based on data collected from SN, with the aim of improving relief planning and for increasing resilience to disasters based on the analysis of the patterns of the responses on SM.

According to the work group LODGD (Linked Open Data for Global Disaster Risk Research) [8] for the best practices of data managing regarding disasters, there are two important stages: the data definition (lower level) and data aggregation (higher level). At the lower level, one finds knowledge about the disaster taxonomy and the data dependence upon the disaster events. The identification of the relation between the specific disaster events and the data needed for the discovery of the semantic description is a major concern at this level. The upper level includes specific technical methods used for the data aggregation and for the derivation of knowledge on the disaster. We address both levels in an integrated approach.

[1]Institute of Computer Science, Romanian Academy, Iasi, Romania
[2]'Gheorghe Asachi' Technical University of Iasi, Iasi, Romania
[3]'Titu Maiorescu' University, Faculty of Informatics, Bucharest, Romania

The chapter is organized essentially on four parts. In the first and second sections, extensive evidence is gathered from the literature and from authors' own data collections showing that the major events elicit fast responses on the SM, but with varying slopes and laws of increase in the number of messages per unit time. The gathered evidence indicates that the data collected during the early stages post-disasters can be used to infer the numerical and geographical extent of the disaster and of its consequences.

In the third section, after reviewing the models proposed in the literature for the SNs activity flows, a thorough discussion of the models is presented and a set of models is selected for a specified set of disasters. The next part of the chapter (Sections 4 and 5) shows how to improve information extraction and predictions on the disaster effects, based on the responses on the SM and making use of the response models. This final part of the chapter proposes further improvements to the relief planning based on the derived information and predictions, in conjunction with models of resilience tailored to the type of disaster and to the local conditions.

6.1 Introduction: ICT in assessing disasters' amplitude and for disaster relief – a literature review

Learning in due time about the incidents that happen during and after an emergency may contribute both to reducing the number of injuries and to adapting the management of the disaster-generated impact. While traditionally the media sources such as television, radio and press are responsible for the diffusion of information regarding an emergency situation, nowadays the new ICTs play a key role during an emergency situation. ICT methods and systems serve for producing information by those involved in emergencies, for processing that information and fusing it with information from other sources, and for returning information and instructions to those involved in the emergency. In this respect, [9] rightly stresses that '*the social networks clearly blur the traditional limits between the producers and the consumers of information*'.

ICT means proved to be an efficient tool in disaster management, mainly for the informing, helping, rescuing and monitoring procedures, as well as concerning the limitations of the economic effects that may occur after a disaster situation. ICTs for disaster management typically refer to digital resources and tools grouped in a communication cluster. Their purpose is to offer a real-time useful feedback designed for the needs of most people. Many studies were focused mainly on the ICT importance in managing disasters and on the manner that ICT can help in disseminating earthquake alerts [10,11].

As ICTs allow the access to real-time information, the usage of these technologies enables the possibility of defining decisions and making decisions in real-time manner. According to the researched conducted by [12], '*the emergency plan needs to be a living document that is periodically adapted to changing circumstances and that provides a guide to the protocols, procedures, and division of responsibilities in emergency response*'.

The ubiquitous wireless communications and applications available on smartphones enabled the analysis of the seismic damages and other emergency situations [13]. The algorithms used to locate the injured people and of the survivors represent the key element regarding the communication for emergency research communication. The techniques used to locate casualties were mentioned in various papers. The studies [14,15] argue that '*Earthquake prediction methods and disaster management strategies using ICT and Wireless Sensor Networks*'. The usage of wireless sensor networks and ICT for disaster-sensitive areas proved to be beneficial for reducing the effects of devastating situations and for protecting people [16].

When natural disasters take place in the areas where it is hard to implement a periodical monitoring, ICT usage may enable the use of devices that detect the natural risks before a natural disaster occurs and without the human presence. However, the implementation and deployment of ICT systems depend on the governmental priorities and may be limited by the available resources [17].

The Internet and online communication web platforms development has the aim of facilitating and generating a better communication between people in the periods before and after the disaster by creating information and communicating platforms that are based on people's collaboration on SNs. In crisis situations, communication sources such as SNs, both correct and false information may become viral, rapidly spreading through different SNs and platforms (Hurricane, Sandy). In such cases, the governmental organizations and first responders should facilitate the diffusion of truthful information that must reach the people effectively. Simultaneously, the first responders should solve the problems that may occur due to false or discrepant pieces of information, and report such information [18].

There are numerous SM platforms used by the crisis and intervention agents in case of emergency, such as those summarized in Appendix A.1. Some of these platforms were deployed by national or state agencies, but many have been developed by companies (e.g., Google Inc.), inter-governmental associations (the United Nations or UN) or inter-agency organisms. Also, there are several associations and organizations that support the research projects that study disasters and the containing of their effects. In the USA, the Incorporated Research Institutions for Seismology (https://www.iris.edu) is an association that includes over 120 universities from the USA and that offers several online instruments which allow the users to learn about global and regional seismicity. Also, in the USA, the American Association for Advancement Science (AAAS), National Science Foundation and several state-level organisms sponsor projects in the domain. A grant managed by AAAS (Grant Award Number 1560948 NSF, 1 May 2016 – 30 June 2019), addressed rules and best practices for the ethical usage of geographic information and of the relevant remote images in case of crisis situations.

The SM resources and the ICT tools can be used by the people to alert the authorities in case of emergency, thus offering the possibility of quick and convenient assistance and '*it also raises the transparency of requests – if a request is made over a platform like Twitter, it has a public audience to hold the emergency responder account-able*' [19]. Appendix A.1 presents some examples of information applications that may help to better manage the disasters and the crisis situations. For a list of mobile applications (Android and iOS) developed to warn and manage the disaster and crisis situations, see Appendix A.2.

Substantial efforts and resources have been used so far in developing tools for collecting and processing information from SNs and for helping rescue and relief operations, yet numerous limits still hamper the use of these tools. For example, it has been noted that, because in disaster situations it might be more difficult for foreign citizens to understand the local announcements and the local instructions due to the language barrier, one should use pictorial information for helping evacuations [20]. The paper [20] offers an overview of other issues related to the essential role of information technology and of ICT regarding the diffusion of information, the rescue requests and the emergency communication provider. Also, [20] posits that SN use cannot solve all problems and '*suggests disaster preparedness and response guidelines for tourists when traveling overseas*'.

6.2 Empirical characterization of SN responses to major events and disasters

6.2.1 Introductory issues

Several natural questions arise in relation to the development of media platforms and ICT technology: How SM and SN can help to better monitor and evaluate the losses after disasters and predict the extent of the effects of the disaster? How the monitoring of SM/SN can provide a continuous flow of information for disaster management? And how the information collected on SM/SN should be processed to effectively avoid the occurrence of victims in case of repeated or chained incidents?

There are multiple researches that highlight the idea that the SM represent a promising tool for managing the disaster response. Before, during and after disasters, the population may distribute texts and data related to disasters by using the SM platforms. Many academic studies indicate that the attention should be directed towards the content of SN regarding disasters. Still, up to the present moment, there are only a few convincing and proven applications [21,22].

Among the benefits of using the information on SN to improve disaster management and response are the growth of the awareness level regarding such situations, the promotion of the emergency information flow, the prediction of disasters and the coordination of the salvation efforts. The study [23] develops a framework for identifying the emergency situations and recent catastrophes, based on the data collected from SNs, using keywords for a regional search. Knowing the most frequently used words in the posts regarding a certain class of events allows one the identification of logic conditions for searching on SN events related to emergency situations. In order to analyse data collected from SM in view of identifying the emergency situations, it is essential to establish the vocabulary used for a regional search, as it is shown in the researches [24–27].

Many searching procedures on the Internet network are performed with the help of search engines that contain a web crawler (also known as web robot or spider). This is used to examine and download web pages with the purpose of facilitating the process of collecting data efficiently. Current search engines are able to verify if the pages of a website previously indexed were changed and to report the changes emerged in the central system. This is useful for an earthquake situation or disaster occurrence. IT tools offer the possibility to monitor disasters and are useful in the process of making decisions in emergency conditions.

The aim of the monitoring process is to analyse and gather information with great occurrence frequency leading to devastating situations, for the quick decrease of the disaster effects [28]. The paper [29] analyses the dynamics of SNs in relation to the external events such as emergencies in the ICT context regarding disasters.

The study [30] focuses on the hazards that tend to affect a large area such as Romania, a country in Europe with major seismic records. The analysed data were extracted by using a web crawler. This study can be extended to searching for posts within all the web pages on condition that they fulfil the logic condition suggested by the researches [24,31].

6.2.2 Using Twitter

There are numerous research studies that monitor different types of disasters based of the data extracted from platforms such as Twitter and Facebook [32–34]. Many researches devoted to data analysis for disaster situations employed the Twitter platform as the base of their study because it facilitates the use of several methods for the extraction of data. The two methods typically used to extract tweets are searching in www.twitter.com and/or using Search API. For the Search API method, the user should obtain credentials for access through a developer account and then to connect to it (https://developer.twitter.com/en/apps).

The API Search represents a part of the REST API Twitter, and its role is to allow the extraction of a sample of tweets that are relatively recent and published within the last 7 days. This method is connected to the relevance of the searched words [35].

The Search API does not reveal all the posts and there are limitations in its use. When building the interrogations, it is recommended to use the maximum of ten words and other operators (see https://dev.twitter.com/rest/public/search). The REST Twitter APIs represent certain limitations regarding the acquisition period. For example, the requests are limited to a number of 350 interrogations for each 15 min for a user account, or a total of 3,200 total number of tweets for each REST interrogation.

The research reported in [6] was based on the REST API Twitter method in the tweets analysis collected; [6] suggests building of an information system that identifies the posts that contain the texts related to a disaster situation by using the said Streaming API Twitter method. The Streaming API Twitter method allows the extraction of tweets that match with the indicated criteria by the final user (keywords, name of users and locations). According to this method, after the connection was established, the tweets are delivered regardless of the limitations imposed by the Rest API method. It is indicated that the streaming end point connection employs the post method and avoids the long URLs that may be obtained using the 'get' method.

The implicit access by using the post method makes possible reaching up to 400 track words. For a higher access to tweets, it is recommended to contact GNIP by using the following link: https://gnip.com/sources/twitter/. In real-life practice, if the API tweets extraction method is used, there are situations when the connection may be closed. This happens in the following cases: when several applications are ran simultaneously, in this situation the last activated application will remain active; when the application is closed during the reception of the tweets sequence; when the computer that receives the tweets reads the data too slowly; and when the Twitter's configuration is changed. The Twitter Streaming API is free to use but there are limitations regarding the quantity of tweets that can be extracted in a certain time interval. If the required limit imposed by Twitter is overreached, the access is limited for some period of time.

There are several ways of collecting and analysing the data that can be found in SM in relation to disasters. From this standpoint, the studies fall into the following categories:

- researches that focus on the data analysis before an emergency situation;
- researches that analyse the data that run in SM, exactly during an emergency situation;
- researches that focus on the analysis of the flow of data and of information that propagates after an incident in an emergency situation.

Through the content analysis of the SNs and using the statistics of posts regarding a certain event, one can often create a detailed description of the situation, regardless the methods used for the data extraction from the SNs [24,28]. Beyond text messages, acoustic, image and video files bring valuable, sometimes key information for identifying the dangerous events. Studies demonstrated the use of the characteristics extracted from the sounds [36] and characteristics of the images, such as major asymmetries [37], and the relevance of information extracted from the change in movements in a scene [38,39]. It was suggested in [37] that a large number of asymmetries in man-made structures may be an indicator of structural degradations and thus of the effects of such disasters as earthquakes or explosions.

While SNs represent a rich source of information, other sources may be needed for improving the characterization of the disasters and their effects, such as satellite images, aerial pictures, including from drones, sensor networks and working force distributed locally. It was noticed in [6] that the improvement of the search of information regarding the disasters presented in SN must consider the available bandwidth of communication in the first hours after the earthquake, and also consider the potential limitations of the proper functioning of the data collecting centre, due to the disaster effects.

The paper [6] offers a clear view for the efficient collection of information from SN in case of earthquake and suggests the usage of a certain system that may be useful and helpful for the improvement of the searching procedure. The main characteristic of the system is its capacity of processing the collected posts in parallel with the collecting process, in order to accelerate the detection of the relevant information.

For the performed testing procedures based on the papers [6,40], we collected tweets for the period 7–11 December 2018, targeting the days immediately after the earthquakes in New Caledonia. There are numerous locations where specific types of disasters, e.g., earthquakes, are more probable than on the rest of the planet. Alaska, New Caledonia, Dominican Republic, Indonesia and Punta Cana are areas where earthquakes with magnitude over 6 occurred from 30 Nov. to 7 Dec. 2018.

The targeted words for the collected tweets were 'earthquake', 'seism' and 'cutremur' ('earthquake', in Romanian). Due to the fact that the Twitter messages contain posts that may reach 280 characters (starting with year 2017), for the tweets contents analysis a pre-processing operation is requested, before the actual analysis. Because many messages on the Twitter platform contain unnecessary elements, one needs to clean the tweets by applying ordinary expressions in order to obtain normalized sets of data [24,27,35]. The aim is to trace the evolution of messages with small contents that only define certain events [24,41]. The improvement of tracing the evolution of messages triggered by certain events has been extensively studied in [6,24–27,29]. This is exemplified in Table 6.1, which contains information on the posts in the data sets used in this chapter, collected by the authors. If the time zone function for the targeted collected tweets is analysed, one finds that there is a certain margin of error. This may occur because a great number of posts did not have the time zone completed, as emphasized in [6].

6.2.3 Keyword selection

According to [27,31], the lists of the keywords and of the associated occurrence probabilities need special attention. Choosing the right keywords for search as indicators of the extent and type of effects of the disasters is essential for the efficacy of the disaster management and specifically for rescue operations. The statistics of the words collected from the messages is needed for selecting the main

Table 6.1 Unique posts presented together with the places where the events occurred

Posts 2018	Alaska	New Caledonia	Dominican Republic	Indonesia	Punta Cana
7 December	425	289	174	148	142
8 December	389	157	95	124	145
9 December	127	124	98	58	57
10 December	245	89	87	115	79
11 December	198	97	45	22	69

keywords allowing a successful search operation [31]. As a matter of example, we show how much the probability of word probabilities and word co-occurrence probabilities may affect the ease of determining when disasters occur, finding the type of disaster, and determining the effects. For the targeted period of collected tweets, the frequency of the following terms is analysed: 'disaster', 'fire', 'dead', 'killed' and 'victim'. Their distribution is shown in Figure 6.1. The words 'disaster' and 'fire' had higher frequency compared with the frequency of the words 'victim', 'dead' and 'killed'. Therefore, the search for the major effects is not the fastest way for detecting a disaster occurrence.

Whenever synonyms are present in a language, or several languages are used in the area of interest, all the equivalent words should be used. By using the searching procedures such as 'cutremur' or 'seism' and 'grade' or 'Richter' or 'magnitudinea' or 'Vrancea' proposed in [31], it is useful to compare the frequency of the keywords 'earthquake', 'seism' and 'cutremur' (the last two words are in Romanian) as well as the frequency of the pair of words, as presented in Table 6.2 and Figure 6.2.

There are fewer tweets that contain specific pairs of words, but according to [40], the identification of the frequencies of word associations contributes to the relevance of new rules of associations between words and to the forming of a complete picture of the disaster magnitude, central point and extent. Indeed, as a matter of example, an earthquake is sensed and announced by messages from a large area, but relief may be needed only in restricted zone close to the epicentre, where the messages also signal dramatic effects by specific word co-occurrences. Searching co-occurrences may also help removing some of the 'noise' in the SN traffic. For example, the occurrence in the same message of the words 'dead' and 'victim' probably signals that the message is not referring to a disaster; see the corresponding rule in Table 6.3. By using the technique described in [40], the main association of words and their frequency of occurrence were identified and the rules of association of the words in tweets, in case of disasters, for the collected data in the study [6] were derived. Table 6.3 indicates that only the first three rules of

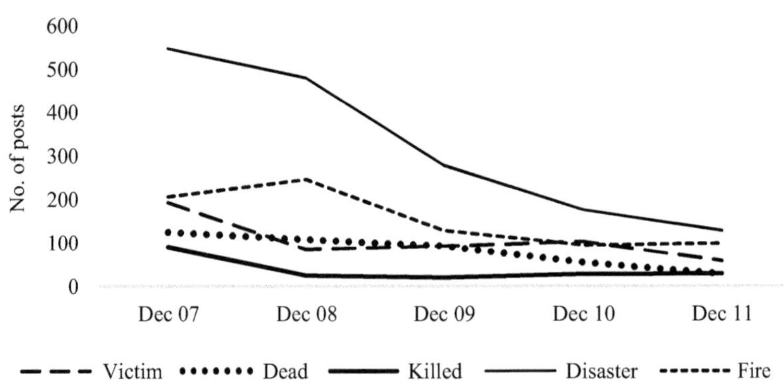

Figure 6.1 Frequency of the words 'disaster', 'fire', 'dead', 'killed' and 'victim'

Table 6.2 Frequency of the words 'earthquake', 'seism', 'cutremur' and of the pairs of words

Days	Earth-quake	Seism	*Cutremur*	Earthquake–seism	*Cutremur*–seism	Earth-quake–*cutremur*
7 December	178	158	11	150	7	3
8 December	157	142	9	128	6	4
9 December	147	125	0	111	0	0
10 December	67	54	4	45	1	3
11 December	98	75	3	21	1	2

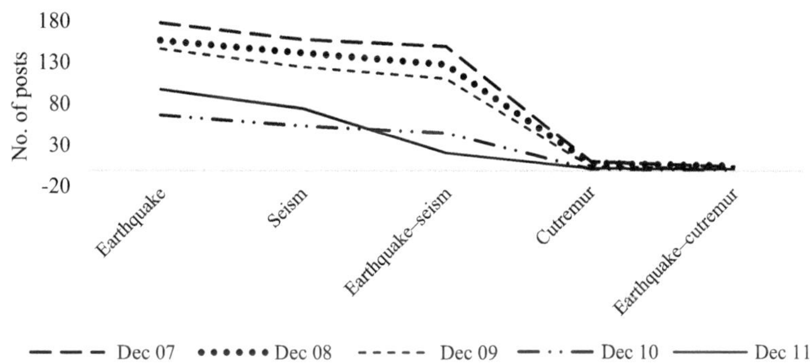

Figure 6.2 Frequency of words and of pairs of words

Table 6.3 Identification of the association rules of words in tweets in case of disaster

Rule	Antecedent	Consequent	Support	Confidence
If dead → earthquake	dead	earthquake	0.177	0.834
If victim → earthquake	victim	earthquake	0.329	0.808
If disaster → earthquake	disaster	earthquake	0.000	0.759
If earthquake → victim	earthquake	victim	0.329	0.435
If earthquake → dead	earthquake	dead	0.177	0.245
If dead → victim	dead	victim	−0.030	0.031
If fire → disaster	fire	disaster	−0.030	0.019
If earthquake →disaster	earthquake	disaster	0.000	0.016
If seism → dead	seism	dead	−0.030	−0.002
If victim → fire	victim	fire	−0.030	−0.003

association have the confidence higher than 0.5. The state of spirit, the opinion and the sentiment of the people must be considered for disaster situations because all these may affect the outcome of the disaster management and mitigation. Knowing the people's opinion may lead to the possibility of quantifying their reactions towards the danger to which they were exposed. The research of the sentiment analysis and of the emotional domain has continuously developed, so the World Wide Web Consortium (W3C) tries to define a certain language that may indicate the emotional level. EmotionML aims to create a standardized description of the emotional knowledge [42]. In order to identify the level of specific attitudes and states related to a certain situation, many research papers demonstrate that it is useful to use the annotated dictionaries that identify the attitudes of the people who post messages after an unfavourable situation [7,25,43].

The real-time identifications of the people's opinion regarding catastrophic situations are based on specific algorithms and represent a delicate task for the text classification process. Using the results in [44], the occurrence frequency of each sentiment score was determined and the latter was calculated using the Stanford CoreNLP, as presented in Table 6.4.

Notice that over 96.5 per cent of the tweets are negative and very negative, while only about 1.92 per cent collected tweets that relate to the earthquake are positive. The results are similar to those reported in [45].

These results must be interpreted carefully because it is a well-known fact that many of the messages may be simple 'noise'; that is, even if they contain keywords used for the search procedure, the semantics and the context, in which they have been used, differ, for example 'terrible' may refer to harsh conditions or to a very positive feeling [31]; see also 'Models' section in this chapter.

6.3 Models

For developing efficient tools for monitoring, analysis and management of disaster effects and for the use of related ITC means, various sets of models are required. These include models for the secondary effects generation and propagation, models of resilience and models of information propagation in disaster areas and beyond. Good models are essential not only for their explanatory power, but for the prediction of disaster outcomes and for prevention purposes too.

Table 6.4 The sentiment score

Sentiment score	Number of tweets	Sentiment	The sentiment score
0	18,993	Very negative	0.0669
1	35,703	Negative	0.9236
2	876	Neutral	0.0206
3	964	Positive	0.0306
4	129	Very positive	0.0109

In this chapter, we do not discuss models for prediction of (natural) disasters; such models pertain to the disciplines studying the causes of the disasters, such as geomorphology for earthquakes, meteorology for tornadoes and oceanography for tsunamis. Also, we are not concerned with models related to the statistics of disasters as in the insurance industry. We are concerned only with the information generation and propagation in SN, where the information relates to disasters. In many respects, these models are sociological response models. However, a complete picture, from root causes to the detailed, multi-dimensional effects of the disasters, should make use of and integrate all these types of models.

Models of information related to disasters in SM/SN fall in several categories, including (i) spatial models: models for the geographic distribution of 'friends' (i.e., probable distribution of news) in an SN; (ii) temporal models: models for the temporal variation of the total number of messages related to an event, after the event is produced; when the event is not almost instantaneous, as in an earthquake, but develops in time (such as for bush fires, where the models are more intricate); and (iii) models of spatio-temporal propagation of social and medical effects of disasters, such as panic and epidemics. The last ones are epidemiologic-like models and may fall in the category of diffusion models.

A potential use of spatial models for the generation of messages after a disaster is the determination of the number of affected people and the extent of the area covered by the disaster. Assume a constant density of the population per surface unit, $\rho(\delta)$, and a probability p_s of response to disaster on a specified SN by sending a message. Then, the affected surface S and the number $N(t \approx 0)$ of disaster-related messages in the first moments after the disaster (before responses to the original messages become numerous) are related by

$$N(t \approx 0) = N(\Delta t) \approx \rho(\delta) p_s S \tag{6.1}$$

Thus, knowing $\rho(\delta)$ and p_s and determining the traffic on the SN related to the disaster, the affected area S can be roughly estimated and the resources for the initial rescue operation planned. However, this estimation is quite basic, because of the implicit assumption that subjects affected by the disaster send very fast messages, before other related messages emerge from unaffected people. Also, the above estimation does not take into account chain effects that increase the probable period of time when people are affected and send messages asking for help. An improved estimation of the number of people affected and of the extent of the region of disaster, based on the dynamics of the number messages sent, may be obtained using the more elaborate models as given next. The next models recognize that not only the people who send messages are required on SN, but those to talk to also; that is, these models analyse the group of 'friends' and their geographic distribution. Taking into account the geographic distribution of friends allows one to determine the spatial distribution of the messages sent as a consequence of the original messages from the disaster area.

6.3.1 Models for connections in an SN – distributions [10,11,46–48]

The efficiency of SNs in disaster relief is partly dependent on the geographic (surface) distribution of people in the network, in the region of interest and close to the region where the disaster occurred. The connectivity in an SN is affected by the distance. Kleinberg *et al.* (see [10,11,49,50]) and Liben-Nowell *et al.* ([48,51]) have proposed and discussed models in the following form, previously also exposed in [47]. For a distance δ_{ij} between the subjects, i,j, the probability of the link is

$$p_{ij} = 1/\delta_{ij} \tag{6.2}$$

An improved empirical law was proposed as a negative power law with negative coefficient θ of the model, ranging in the empirical analysis cited above between 0.5 and 1.2 (similar to Zipf law):

$$p_{ij} = \delta_{ij}^{-\theta} + \eta \tag{6.3}$$

where η is a constant, meaning a random, background, 'white noise'-like distribution of probability, not depending on the distance. The term η becomes important for distances of about 1,000 km ([52], see also [47,49]. On the other hand, we noticed that the noise term contradicts the normalization condition for probabilities, as $\int_0^\infty p_{ij}(\delta)d\delta > \int_0^\infty \eta \, d\delta = \infty$. Based on the simple models (by Kleinberg *et al.* and Liben-Nowell *et al.* [10,11,48–52]) of 'making friends', that is, connectivity in an SN, in [22] we presented reasons for the distribution of 'friends' with a product of power and exponential for large distances. Namely, we suggested to use a factor $e^{-\delta/D}$, where D is a normalization parameter. The suggested model is

$$p(\delta) = A e^{-\frac{\delta}{D}} \left(\frac{1}{\delta^\theta} + \eta \right) \tag{6.4}$$

The exponential factor is close to 1 for small δ values; therefore, for small distances used in the validation of the previous model, the suggested one produces the previous models. The same suggestion was made for the rank-based models of connectivity (between u and v), where the exponent ζ is also close to 1:

$$p(u,v) = A e^{-\frac{\delta}{D}} \left(\frac{1}{r_u^\zeta(v)} + \eta \right) \tag{6.5}$$

where $p(u,v)$ is the probability of a friend at location v for someone at u; see [22].

We noticed in [47] that the use of the exponential factor also limits the effect of the noise η and of any other term that may be used in the expression of the original laws, which is consistent with the data. The value of the constant A results from the normalization condition, which the above expression should satisfy.

Further, we noticed in [47] that an expression of the probability of having friends at a specified distance that takes into account only the distance, but neglects the population density at the specified distance, is counter-intuitive and against other epidemic and social models. Therefore, we proposed the corrected expression of essentially one-dimensional distribution with variable δ:

$$p(\delta) = A\rho(\delta)\mathrm{e}^{-\frac{\delta}{D}}\left(\frac{1}{\delta^{\theta}} + \eta\right) \tag{6.6}$$

Equation (6.6) assumes that the population density $\rho(\delta)$ depends only on the distance δ from the subject. Further refinements are easy to derive and are not presented here. A significant limit of the models discussed above is that they consider static populations; that is, they do not consider the case of travelling subjects who, during the travel, modify by a large value the distances to their 'friends'. This is the case of tourists to a region struck by a disaster.

For non-travelling population at the disaster site, the above model allows us to derive the probability of a response to initial messages from a specified distance from the disaster place.

A detailed analysis of the model and of the constraints was presented in [22], where the detailed expression of the probability, satisfying all constraints (imposed by normalization and finitude of the values), was given as

$$p(v) = A\mathrm{e}^{-\frac{\delta}{D}}\left(\frac{\alpha}{\lambda + \delta(v)} + \frac{\beta}{P(u,\delta(v))}P(v) + \frac{\gamma}{P\left(u,\frac{\delta(v)}{2}\right)} + \eta\right) \tag{6.7}$$

In (6.7), the additional notations are as follows:

α, β, η and λ are model constants;

$P(u,\delta(v))$ is the population in the circular region of radius $\delta(v)$ centred in the point of u, $P\left(u,\frac{\delta(v)}{2}\right)$ has a similar meaning, $P(v)$ is the population density (population on the unit surface) at v, where v denotes a friend of a correspondent u placed at the origin of the plane, that is, $\delta(v)$ replaces δ in the previous equations.

Integrating (6.7) for all correspondents in the disaster area (if the area is known), placed at positions (u_x, u_y) in the plane and then for all $\delta(v)$, one obtains the total number of potential participants to the traffic on the SN, where the participants are placed in the disaster area. These people should be a target for ICT systems for on-site data collection and also for providing support – when not needing themselves support – to the rescue operations. Further recent analyses on connectivity in networks can be found in [53]; refinements of the above models may benefit of these developments.

Assume next that we wish to determine the number of messages related to the disaster after some short time since the disaster, that is, after the first wave of 'friends' send messages related to the original ones.

6.3.2 *Models of temporal evolution of the SN traffic [22,28,41]*

We first use the above derivations to determine the number of messages $N(2\Delta t)$ produced in a second wave of messages, in a time interval, Δt, after the disaster 'first wave' of messages has been sent. Denoting the average number of friends of a person by n_f, $N(2\Delta t) \approx N(\Delta t)n_f p_s$. Based on the same reasoning, the number of messages sent after k time intervals can be estimated as follows:

$$N(k\Delta t) \approx (N(\Delta t)p_s)n_f^k \qquad (6.8)$$

This is a power time-dependent model for the dynamics of the flow of messages related to a disaster; it was first suggested in [41]. The model is easily generalizable to a differential model, also laid down in [41]. These models, which are valid for instantaneous disasters, such as earthquakes, airplane crashes, terrorism acts and explosions, were found to predict very well the initial dynamics of the disaster-related messages on several SNs and for several such disasters. Examples from real life were exposed in [28,41].

The exponential model of increase in the number of messages during the first interval after a brusque disaster (occurring at time moment 0 is derived from the differential model and is [41]

$$N(t) \approx A(e^{\alpha t} - 1) \qquad (6.9)$$

where A and α are constants (but not the same as in the previous equations). To appreciate the dynamics of a typical SN response, Figure 6.3 shows the variation of the number of tweets during a period of 30 days, in the winter of 2015, where the tweets are referring to harsh weather conditions (blizzard) in the north-east of the US; while this cannot be considered a true disaster, several people died as a result of the repeated blizzards in that period. Notice the exponential increase in the number of tweets on the 25th, 26th and 27th days of that period. A smaller interval (hour instead of day) for counting the tweets is required to better match the exponential law.

6.4 Improving information extraction and predictions

As it was stressed in [41], the values of A and α can be estimated after the first moments of data collection on the SN. Using the exponential model for the increase

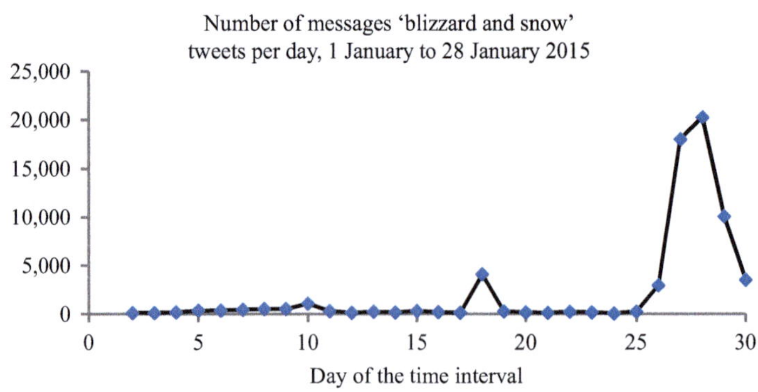

Figure 6.3 Dynamics of tweets during a major weather event

in traffic, as above, from determinations of $N(t)$ at several moments, $t_3 > t_2 > t_1$, during the onset of the response, α is computed [41] from:

$$\frac{N(t_1)}{N(t_2)} = \frac{\mathrm{e}^{\alpha t_1} - 1}{\mathrm{e}^{\alpha t_2} - 1} \tag{6.10}$$

$$\frac{N(t_1)}{N(t_2)}(\mathrm{e}^{\alpha t_2} - 1) = \mathrm{e}^{\alpha t_1} - 1 \tag{6.11}$$

Then, one obtains [41]:

$$A = \frac{n(t_3)}{\mathrm{e}^{\alpha t_3} - 1} \tag{6.12}$$

Knowing, from previous events of the same type, the time interval of increasing the number of messages (for the same SN and population), one can estimate *in advance* the total number of messages and thus the number of affected people. This is invaluable information because the amplitude of rescue operation and the required resources can be planned before the event reaches its peak on the SN. As the time before peaking can be several hours or days according to our empirical study on several events, one gains an invaluable time to plan and deploy commensurate relief means and personnel. The above prove the worth of a good model, applied with automatic ICT means and used in conjunction with automatic relief planning operations. As the first hours are critical for the rescue, one can confidently say that applying such models can help saving lives – maybe not tens, but thousands in a major disaster.

6.5 ICT-based improvements for enhanced disaster resilience

One of the issues largely ignored in studies on ICT support for disaster planners and relief teams is the evaluation of the interface usability [54] and the design of highly intuitive user interface, an issue that requires extensive testing [51,55–58].

The importance of interfaces for efficiency of conveying information under disaster conditions, for decision-makers, planners, relief teams, etc. cannot be enough stressed. The general guidance for usability [59] may be not enough for this type of application, because relief team may work under harsh conditions (e.g., no light, wet conditions) and time pressure. A fast grasping of the information and help in correct interpretation of the data are of matter; difficult to interpret or navigate interfaces may delay or jeopardize the usefulness of the communication tools under stressful situations during emergencies.

Availability and easy finding of alternate communication means, when the typical ones (based on cellular phones) fail under disaster conditions, is crucial [60]. Person-to-person telephony, software-defined networks, software-defined and cognitive radio networking [61], opportunistic networks [62], UAV-assisted emergency communications [63] and use of amateur radio (*radio hum*) networks have been tested for emergency communications. The interoperability of the communication systems is an unsolved problem, which hampers communications during disaster situations. The UN has prepared guidelines for radio communications in emergency situations [64]. Several reviews are available on the mobile applications proposed for disaster situations, for example [29,65,66]. However, despite large research and logistic efforts to improve the reliability of the communication systems during disasters, the problem is yet unsatisfactorily solved. The current solution remains the deployment of mobile base stations in the areas where the static ones have been destroyed, but this solution takes at least several hours, a precious time in disaster situations.

6.6 Discussion and conclusions

Some recent events have shown that SNs cannot be used effectively or at all in many cases of disasters, either immediately after or for short periods after the disaster occurred. Examples include the shutdown of media by the government after Sri Lanka attacks (April 2019), with virtually all platforms' (Facebook, WhatsApp, Instagram, YouTube and Snapchat, but not Twitter) shut down, the loss of connectivity after Nepal earthquake (2015), and large-scale disasters such as Katrina and the tsunamis in Japan and in Indonesia. In case of the last two, the connectivity was a major problem, but not the only one. Especially Catrina and Nepal earthquake proved that the lack of personnel can be a major issue. These facts recall us that ICT systems have a high complexity, with many links that can fail in disasters, and with limited resilience under unexpected conditions.

On the other hand, there are long-established methods that can and have been used for communication in emergencies, mainly 'hum' radio, radio-amateurs informal networks. These networks are encouraged by national agencies, such as US Federal Communications Commission (FCC) in the US, and they are supported by professional associations and even universities; see [67].

In addition, there are currently several proposals for making use of new means, such as ad-hoc people-to-people direct telephony networks.

The models exposed help answer questions such as how fast do SNs respond, how to predict the dynamics of the response of SN in disasters based on initial moments response, and when the peak of the SN traffic related to an event will probably occur?

There are still many unanswered questions and avenues for research. Little is still known on how the type of event influences the vocabulary used, on the reliability of the information on the event the SNs and what influence on the response on SN to disaster the inner structure of the network does play.

It is unclear how the new means of communication, including 5G recently introduced and the expected 6G communications, will affect the use of SN, especially the reliability of communications and in particular SN use under disaster conditions.

Acknowledgements

HNT thanks Dr. Paula Crucianu for finding a few minor errors in a preliminary version of the chapter.

Authors' contributions

HNT conceived the structure of the chapter, wrote Parts 3–6, and contributed to Parts 1–2. MP was the main contributor to Parts 1–2 and the appendix. Both authors agreed with the final version of this survey. Section 3 in this review is heavily based and has parts, including all equations from HNT's previous studies, which, in turn, were built based on the cited literature. The appendix is a compilation of data from other sources, with no other merit except the information systematization and summarization.

Appendix

A.1 Compiled examples of web platforms and applications for disasters

- *Advanced National Seismic System* (ANSS) (https://earthquake.usgs.gov/monitoring/anss) includes a network of regional partners and American universities that collect and analyse data regarding significant earthquakes. ANSS provides information for seismic events, including the effects on the buildings and other constructions, by using modern monitoring methods and technologies. The information provided by ANSS contributes to the generation of national seismic risk maps. It helps the neighbourhoods vulnerable to seismic risks to develop more solid and secure structures.
- *Centers for Disease Control (CDC) and Prevention* (https://www.cdc.gov/) represents one of the main parts of the Department of Health and Human Services. It aims to increase the health security in the UN and to protect America from the dangers regarding health, security and safety. CDC uses the SM to offer the users access to the credible information on diseases (https://www.cdc.gov/socialmedia/).
- *California Integrated Seismic Network* (https://www.cisn.org/index.html) monitoring, research and distribution of the information. It aims to reduce the future impacts of earthquakes in due time. Main members are California Geological Survey, Caltech Seismological Laboratory, UC Berkeley Seismological Laboratory, the United States Geological Survey (USGS) Menlo Park, USGS Pasadena and the California Governor's Office of Emergency Services.
- *The Disaster Information Management Research Center*, the U.S. Department of Health & Human Services (https://disasterinfo.nlm.nih.gov/), develops and offers access to the information resources and to the information technology in the health field for emergency situations in case of disasters and public health.

It also aims to connect the people to relevant information in terms of disasters, to create knowledge regarding disasters, and to increase the resilience of the neighbours in case of disasters. It offers bibliographic information in data bases of the National Medical Library: Disaster Lit® and PubMed.

- *The Cornerstone On Demand Foundation* (https://www.disasterready.org/) offers free of charge resources for professional development.
- *Global Disaster Alerting and Coordination System* (http://www.gdacs.org/alerts/) manages and monitors worldwide disasters to improve the alerts, information exchange and the coordination after major catastrophes with sudden occurrence. It contributes to efficient humanitarian feedback based of the coordination and humanitarian information services for disaster situations. Many governments and organizations use the tools and services provided by GDACS within their national feedback plans in case of catastrophes. The information provided by GDACS is available through the interfaces on the GDACS platform and can be directly integrated in other web portals or in other sites through RSS feeds or other standard formats.
- *Google Person Finder* (http://google.org/personfinder/) helps people to reconnect after the occurrence of natural disasters (first used after the earthquake that took place in Haiti). The service offers the possibility to locate friends and families from the affected areas during the disaster. The platform was successfully used in 2010 when the earthquakes in Chile and Yashu occurred.
- *Global Seismographic Network* (https://earthquake.usgs.gov/monitoring/gsn/) is partnership between the USGS, the National Science Foundation, USA, and the Incorporated Research Institutions for Seismology (IRIS). GSN has a permanent digital network of seismic geophysical sensors connected in a telecommunication network. GSN has over 150 seismic modern stations distributed globally.
- *The Inter Agency Standing Committee* supports to the humanitarian operations. (http://interagencystandingcommittee.org/). It is a unique inter-institutional forum used for coordination, political development and decision-making processes that involve the main UN and non-UN organisms.
- National Earthquake Information Center is a part of the Department of the Interior, U.S. Geological Survey (https://earthquake.usgs.gov/contactus/golden/neic.php), used to locate and to determine the magnitude of significant earthquakes in the USA and globally.
- NetQuakes (https://earthquake.usgs.gov/monitoring/netquakes) is a new type of digital seismograph that connects itself to a local Wi-Fi network to send data related to an earthquake to USGS. These instruments can be installed anywhere there is available a broadband Internet connection.
- The United Nations Office for the Coordination of Humanitarian Affairs is (https://www.unocha.org/) a part of the UN secretariat which is responsible with joining the decision-making factors that provide a coherent feedback to emergencies. It includes several digital services and mobile applications (Android and iOS). The complete list can be found at https://www.unocha.org/ocha-digital-services.

- Plan Ahead for Disasters, Department of Homeland Security (https://www.ready.gov), is a company of public services. It educates and prepares people in the US to face emergency situations, including natural and human-made disasters.
- The Platform on Disaster Displacement (https://disasterdisplacement.org/the-platform/our-response) has the main objective to consolidate the protection of people located abroad when disasters happen, and to prevent or reduce the displacement risks in case of catastrophes.
- U.S. Geological Survey: it is the only scientific agency of the Interior Department (https://www.usgs.gov/connect/locations?location_types=5046). It has 297 research centres. Develops new methods and instruments to provide updated, relevant and useful information about the Earth processes. It works together with many partners; it deploys activities on a large range of disasters.

A.2 Mobile applications (Android and iOS) to warn and manage the disaster and crisis situations

- Disaster Alert (https://www.pdc.org/) – It offers early warning for automatic alerts before the disaster cases and information on up to 18 different types of active dangers while taking place around the globe.
- Earthquake Alert – It displays news and summaries related to earthquakes and shows details from U.S. Geological Service. It also warns the users regarding new possible earthquakes, according to their reporting manner.
- Earthquake Network – Real-time Alerts – It's the most generous application used for the earthquakes in most countries. It is the only system that quickly alerts the occurrence of earthquakes because it allows real-time detection of earthquakes by using the smartphones networks and emits alerts prior the earthquake.
- Federal Emergency Management Agency (FEMA) App – It contains a guidance set and suggestions regarding the personal security in case of disasters. It also includes a list of interactive emergency kits, information about proper locations for use as a meeting place in case of emergency, as well as a map with the nearest shelters used for salvage operations in case of disasters.
- Global Emergency Overview – The application offers the possibility of quick web surfing to obtain useful information from different countries.
- Guardly – It offers the possibility to receive emergency and operational alerts from the aimed company or from the security team in case of emergency at the place of employment.
- H.Kiosk – It offers a series of humanitarian last minute information regarding emergency situations all around the world.
- Humanitarian ID – H.I.D. offers the 'self-administrated' approach of the useful contact lists in case of humanitarian crisis situations.
- iGDACS European Commission – It offers almost real-time information about natural disasters, as well as the possibility of sending feedback.
- INSARAG (International Search and Rescue Advisory Group) Guidelines – It includes a methodology that may guide the actions carried on in sudden

affected countries. INSARAG represents a useful guide in the preparation process, in the cooperation and coordination process of the national and international participants in case of disasters.

- KoBo Collect – It is used for collecting primary data in case of humanitarian emergencies, including difficult areas and circumstances in terms of access.
- Life360 – It allows establishing of a private network. One's family can be informed about the location of the endangered person and if the person is safe or not – by a click.
- Red Panic Button – By activating the red panic button, the GPS coordinates and a link towards Google Maps are sent by SMS (short message service) or by e-mail to a list of contacts previously created.
- Relief Central – A free of charge mobile resource that contains the most recent versions of The World Factbook from the CIA, CDC Health Information for International Travel (The Yellow Book), Ebola Guidelines, the Field Operations Guide from USAID, MEDLINE Journals, and Relief News from the Red Cross, UN, CDC, FEMA, etc.
- SirenGPS – It facilitates the access to the emergency services just by pressing a button by instantly finding the exact location and determining the personal data.
- The American National Red Cross – It provides the possibility of monitoring critical and emergency meteorological alerts.
- The Weather Channel – It displays information about real-time weather reports and forecasts for the iPhone, Android, Blackberry and Windows Phone users.
- The United Nations Disaster Assessment and Coordination (UNDAC) allows the user to download and automatically synchronize on the mobile device the chapters referring to the documents that one is interested in.

References

[1] Cohen O., Goldberg A., Lahad M., and Aharonson-Daniel L. 'Building resilience: The relationship between information provided by municipal authorities during emergency situations and community resilience'. *Technological Forecasting and Social Change*. 2016; vol. 121, pp. 119–125.

[2] Parsons M., Glavac S., and Hastings P. *et al.* 'Top-down assessment of disaster resilience: A conceptual framework using coping and adaptive capacities'. *Int. J. Disaster Risk Reduct*. 2016; vol. 19, pp. 1–11.

[3] Ahmed, W., Bath, P.A., Sbaffi, L., and Demartini, G. 'Novel insights into views towards H1N1 during the 2009 Pandemic: A thematic analysis of Twitter data'. *Health Information and Libraries Journal*. 2019; vol. 36, pp. 60–72.

[4] Asadzadeh A., Kötter T., Salehi P., and Birkmann J. 'Operationalizing a concept: The systematic review of composite indicator building for measuring community disaster resilience'. *Int J Disaster Risk Reduction*. 2017; vol. 25, pp. 147–162.

[5] Teodorescu H.N.L. 'Revisiting models of vulnerabilities of the networks'. *Studies in Informatics and Control*. 2016; vol. 25, no. 4, pp. 469–478.

[6] Teodorescu H.N.L., and Pirnau M. 'Analysis of requirements for SN monitoring applications in disasters – a case study'. *Proceedings of the 8th International Conference on Electronics, Computers and Artificial Intelligence (ECAI);* Ploiesti, Romania, 30 Jun–2 Jul, 2016, pp. 1–6.

[7] Teodorescu H.N.L., and Saharia N. 'A semantic analyzer for detecting attitudes on SNs'. *Proceedings of the International Conference on Communications (COMM)*, Bucharest, Romania, 9–11 Jun 2016, pp. 47– 50.

[8] LODGD (Linked Open Data for Global Disaster Risk Research) Available from http://www.codata.org/task-groups/linked-open-data-for-global-disaster-risk-research/ [Accessed 10 Jun 2019].

[9] Laylavi F., Rajabifard A., and Kalantari M. 'Event relatedness assessment of Twitter messages for emergency response'. *Information Processing and Management*. 2017; vol. 53, no. 1, pp. 266–280.

[10] JPL, NASA Demonstrates Tsunami Prediction System – 14 June 2010. Available from: https://www.jpl.nasa.gov/news/news.php?release=2010-198/ [Accessed 1 Jun 2019].

[11] Okada Y. 'Preliminary report of the 2011 off the Pacific coast of Tohoku Earthquake'. *National Research Institute for Earth Science and Disaster Prevention*. 2011. Available from: http://www.bosai.go.jp/e/pdf/Preliminary_report110328.pdf/ [Accessed 13 Jun 2019].

[12] Alexander D. 'Disaster and emergency planning for preparedness, response, and recovery'. *Oxford Research Encyclopedia of Natural Hazard Science*. Oxford University Press, Oxford, 2015, pp. 1–20.

[13] Han R., Zhao X., Yu Y., *et al.* 'Emergency communication and quick seismic damage investigation based on smartphone'. *Advances in Materials Science and Engineering*. 2016; vol. 2016, no. 1, pp. 1–15.

[14] Hu Q.-S., Zhang S., Wu L.-X., and Ding E. 'Localization techniques of mobile objects in coal mines: Challenges, solutions and trends'. *Journal of China Coal Society*. 2016; vol. 41, no. 5, pp. 1059–1068.

[15] Rahman M., Rahman S., Mansoor S., Deep V., and Aashkaar M. 'Implementation of ICT and wireless sensor networks for earthquake alert and disaster management in earthquake prone areas'. *Procedia Computer Science*. 2016; vol. 85, pp. 92–99.

[16] Cheng J.W., Mitomo H., Otsuka T., and Jeon S.Y. 'The effects of ICT and mass media in post-disaster recovery – A two model case study of the Great East Japan Earthquake'. *Telecommunications Policy*. 2015; vol. 39, pp. 515–532.

[17] Linders D. 'From e-government to we-government: Defining a typology for citizen coproduction in the age of social media'. *Government Information Quarterly*. 2012; vol. 29, pp. 446–454.

[18] Kumar S., and Shah N. 'False information on Web and social media: A survey'. *Social Media Analytics: Advances and Applications*. Boca Raton, FL:

CRC Press, 2018. Available from: https://arxiv.org/pdf/1804.08559.pdf [Accessed 10 Jun 2019].

[19] Giroux J., Roth F., and Herzog M. 'Using ICS & social media in disasters: Opportunities & risks for government'. *Center for Security Studies/ETH Zurich*. 2013.

[20] Leelawat N., Suppasri A., Latcharote P., and Imamura F. 'The evacuation of Thai citizens during Japan's 2016 Kumamoto earthquakes: An ICT perspective'. *Journal of Disaster Research*. 2016; vol. 12, pp. 669–677.

[21] Cutter S.L., Burton C.G., and Emrich C.T. 'Disaster resilience indicators for benchmarking baseline conditions'. *Journal of Homeland Security and Emergency Management*. 2010; vol. 7, no. 1, pp. 1–24.

[22] Teodorescu H.N.L. 'On models of "having friends" and SN friends distribution'. *Proc. 3rd International Conference on Control, Decision and Information Technologies (CoDIT)*, St Pauls Bay, Malta, 6–8 Apr 2016, pp. 659–664.

[23] Shan S., Zhao F.R., Wei Y., and Liu M. 'Disaster management 2.0: A real-time disaster damage assessment model based on mobile social media data – A case study of Weibo (Chinese Twitter)'. *Safety Science*. 2019; vol. 115, pp. 393–413.

[24] Teodorescu H.N.L. 'Using analytics and social media for monitoring and mitigation of social disasters'. *Procedia Engineering*. 2015; vol. 107C, pp. 325–334.

[25] Bolea S.C. 'Vocabulary, synonyms and sentiments of hazard-related posts on social networks'. *Proceedings of the 8th Conference Speech Technology and Human-Computer Dialogue (SpeD);* Bucharest, Romania, 14–17 Oct, 2015, pp. 1–6.

[26] Bolea S.C. 'Language processes and related statistics in the posts associated to disasters on Social Networks'. *Int. Journal of Computers Communications and Control*. 2016, vol. 11, no. 5, pp. 602–612.

[27] Teodorescu H.N.L., and Bolea S.C. 'On the algorithmic role of synonyms and keywords in analytics for catastrophic events'. *Proceedings of the 8th International Conference on Electronics, Computers and Artificial Intelligence. ECAI;* Ploiesti, Romania, 30 Jun–2 Jul 2016, pp. 1–6.

[28] Teodorescu H.N.L. 'Emergency-related, social network time series: description and analysis', Rojas I., and Pomares H. (Eds.) *Time Series Analysis and Forecasting. Contributions to Statistics*. Springer, Cham, 2016, pp. 205–215.

[29] Teodorescu H.N.L. 'Survey of IC&T in disaster mitigation and disaster situation management', Chapter 1 in: H.-N. Teodorescu, A. Kirschenbaum, S. Cojocaru, and C. Bruderlein (Eds.), *Improving Disaster Resilience and Mitigation – IT Means and Tools. NATO Science for Peace and Security Series – C*. Springer, Dordrecht, 2014, pp. 3–22.

[30] Pirnau M. 'Tool for monitoring Web sites for emergency-related posts and post analysis'. *Proceedings of the 8th Speech Technology and Human-Computer Dialogue (SpeD)*; Bucharest, Romania, 14–17 Oct, 2015, pp. 1–6.

[31] Teodorescu H.N.L. 'Social signals and the ENR index – Noise of searches on SN with keyword-based logic conditions'. *Proceedings of the International Symposium on Signals, Circuits and Systems (ISSCS 2015)*, Iasi, Romania, 2015.

[32] Takahashi B., Tandoc E.C., and Carmichael C. 'Communicating on Twitter during a disaster: An analysis of tweets during typhoon Haiyan in the Philippines'. *Computers in Human Behavior*. 2015; vol. 50, pp. 392–398.

[33] Houston J.B., Hawthorne J., and Perreault M.F. *et al.* 'Social media and disasters: A functional framework for social media use in disaster planning, response, and research'. *Disasters*. 2014; vol. 39, pp. 1–22.

[34] Eriksson M., and Olsson E.K. 'Facebook and Twitter in crisis communication: A comparative study of crisis communication professionals and citizens', *J. Contingencies and Crisis Management*. 2016; vol. 24, no. 4, pp. 198–208.

[35] Pirnau M. 'Analysis of the use of social networks in acquiring information on disastrous events'. *Proceedings of the 15th RoEduNet Conference: Networking in Education and Research*. Bucharest, Romania, 2016, pp. 236–240.

[36] Apopei V. 'Detection dangerous events in environmental sounds – A preliminary evaluation'. *Proceedings of the 8th Speech Technology and Human-Computer Dialogue (SpeD)*. Bucharest, Romania, 14–17 Oct, 2015, pp. 1–5.

[37] Teodorescu H.N.L., and Dascalescu L. 'Statistical characterization of images – Anisotropy'. *Proceedings of the 21st International Conference on Applied Electronics (AE)*, Pilsen, Czech R., 6–7 Sep 2016, pp. 269–272.

[38] Kaltsa V., Briassouli A., Kompatsiaris I., Hadjileontiadis L.J., and Strintzis M.G. 'Swarm Intelligence for detecting interesting events in crowded environments'. *IEEE Transactions on Image Processing*, 2015, vol. 24, no. 7, pp. 2153–2166, doi: 10.1109/TIP.2015.2409559.

[39] Kaltsa V., Briassouli A., Kompatsiaris I., and Strintzis G. 'Swarm-based motion features for anomaly detection in crowds', *Proceedings of the IEEE International Conference on Image Processing (ICIP)*, Paris, 2014, pp. 2353–2357.

[40] Pirnau M. 'Analysis of data volumes circulating in SNs after the occurrence of an earthquake', *Romanian Journal of Information Science and Technology*, 2017; vol. 20, no. 3, pp. 286–298.

[41] Teodorescu H.N.L. 'On the responses of social networks to external events'. *Proceedings of the 7th International Conference on Electronics, Computers and Artificial Intelligence (ECAI 2015)*. Bucharest, Romania, 2015, pp. 13–18.

[42] Vocabularies for EmotionML Available from: https://www.w3.org/TR/emotion-voc/ [Accessed 9 Jun 2019].

[43] Teodorescu H.N.L., and Saharia, N. 'An Internet slang annotated dictionary and its use in assessing message attitude and sentiments'. *Proceedings of the 8th Speech Technology and Human-Computer Dialogue (SpeD)*. Bucharest, Romania, 2015, pp. 1–8.

[44] Stanford CoreNLP – Natural language software. Available from: https://stanfordnlp.github.io/CoreNLP/ [Accessed 06 Jun 2019].
[45] Pirnau M. 'Sentiment analysis for the tweets that contain the word earthquake'. *Proceedings of the 10th International Conference on Electronics, Computers and Artificial Intelligence (ECAI);* Iasi, Romania, 2018, pp. 1–6.
[46] Cao L, and Zhou G. 'Research on emergency response mechanisms for meteorological disasters'. *Proceedings of the Computing, Communication, Control, and Management, CCCM, ISECS International Colloquium on;* Guangzhou, China, 2008, pp. 649–653.
[47] Teodorescu H.N.L. 'SN voice and text analysis as a tool for disaster effects estimation – a preliminary exploration'. *Proceedings of the 7th Conference Speech Technology and Human-Computer Dialogue (SpeD);* Cluj Napoca, Romania, 16–19 Oct 2013, pp. 1–8.
[48] Liben-Nowell D., and Kleinberg J. 'The link prediction problem for social networks'. *Proceedings of the Twelfth International Conference on Information and Knowledge Management*, ACM, New York, NY, USA, 2003, pp. 556–559.
[49] Kleinberg, J. M. 'Navigation in a small world'. *Nature*. 2000; vol. 406, p. 845.
[50] Ugander J., Backstrom L., Marlow C., and Kleinberg J. 'Structural diversity in social contagion'. *PNAS*. 2012; vol. 109, no. 16, pp. 5962–5966.
[51] Liben-Nowell D., Novak J., Kumar R., Raghavan P., and Tomkins A. 'Geographic routing in social networks'. *PNAS*. 2005; vol. 102, no. 33, pp. 11623–11628.
[52] Rutherford A., Cebrian M., Dsouza S., Moro E., Pentland A., and Rahwan I. 'Limits of social mobilization'. *PNAS*. 2013; vol. 110, no. 16, 6281–6286.
[53] Lü L, and Zhou T. 'Link prediction in complex networks: A survey'. *Physica A: Statistical Mechanics and its Applications*. 2011; vol. 390, no. 6, pp. 1150–1170.
[54] Abran A., Khelifi A., Suryn W., and Seffah A. 'Usability meanings and interpretations in ISO standards'. *Software Quality Journal*, 2003; vol. 11, pp. 325–338.
[55] Islam M.N., and Bouwman H. 'An assessment of a semiotic framework for evaluating user-intuitive Web interface signs'. *Universal Access in the Information Society*. 2015, vol. 14, no. 4, pp. 563–582.
[56] Seffah A., Donyaee M., Kline R.B., and Padda H.K. 'Usability measurement and metrics: A consolidated model'. *Software Quality Journal*. 2006; vol. 14, no. 2, pp. 159–178.
[57] Islam M.N. 'Towards designing users' intuitive web interface'. *Proceedings of the 6th International Conference on Complex, Intelligent, and Software Intensive Systems,* Sicilia, IEEE CS, 2012, pp. 513–518.
[58] Banujan K., Banage Kumara T. G. S., and Paik I. 'Twitter and online news analytics for enhancing post-natural disaster management activities'. *Proceedings of the 9th International Conference on Awareness Science and Technology (iCAST), Fukuoka*, 2018, pp. 302–307.

[59] Bevan N. 'International standards for HCI and usability', *Int. J. Hum. Comput. Stud.* 2001, vol. 55, pp. 533–552.

[60] Gardner-Stephen P. 'Sustaining telecommunications capability and capacity during acute phase of disasters and disaster responses'. *Prehospital and Disaster Medicine*. 2011; vol. 26, no. S1, Supplement S1, pp. s94–s95.

[61] Miller R., Xu W., Kamat P., and Trappe W. 'Service discovery and device identification in cognitive radio networks'. *Proceedings of the 2nd IEEE Workshop on Networking Technologies for Software Define Radio Networks*, San Diego, CA, USA, 2007, pp. 40–47.

[62] Raschellà A., Umbert A., Pérez-Romero J., and Sallent O. 'A testbed platform to demonstrate spectrum selection in opportunistic networks'. *Proceedings of the 6th Joint IFIP Wireless and Mobile Networking Conference (WMNC)*, Dubai, 2013, pp. 1–8.

[63] Lin Y., Wang T., and Wang S. 'UAV-assisted emergency communications: An extended multi-armed bandit perspective'. *IEEE Communications Letters*, 2019, vol. 23, no. 5, pp. 938–941.

[64] UNDP, A guide to radio communications standards for emergency responders. Prepared under United Nations Development Programme (UNDP) and the European Commission Humanitarian Office (ECHO) through the Disaster Preparedness Programme (DIPECHO), Regional Initiative in Disaster Risk Reduction, March, 2010 Maputo – Mozambique.

[65] Tan M.L., Prasanna R., Stock K., Hudson-Doyle E., Leonard G., and Johnston D. 'Mobile applications in crisis informatics literature: A systematic review'. *Int. J. Disaster Risk Reduction*. 2017; vol. 24, pp. 297–311.

[66] Nakamura H., Umeki H., and Kato T. 'Importance of communication and knowledge of disasters in community-based disaster-prevention meetings'. *Safety Science*. 2017; vol. 99, Part B, pp. 235–243.

[67] Rappaport T.S. 'Engineers need to preserve amateur radio growth and transparency'. *WIRELESS*, 15 April 2019, Available from: https://www.edn.com/electronics-blogs/anablog/4461799/Engineers-need-to-preserve-amateur-radio-growth-and-transparency/ [Accessed 6 Jun 2019].

Chapter 7

Use cases of blockchain technology for humanitarian engineering

Arvind W. Kiwelekar[1], Sanil S. Gandhi[1], Laxman D. Netak[1] and Shankar B. Deosarkar[1]

Humanitarian engineers need innovative methods to make technological interventions for solving societal problems. The emerging blockchain technology has the enormous potential to provide effective interventions in various developmental sectors, including agriculture, education, health and transportation. In these sectors, mediators have been considered as one of the impediments for developmental work. Blockchain technology facilitates peer-to-peer (P2P) business transactions, thus eliminating the role of mediators. Hence, the blockchain technology is emerging as an alternative to conventional mediator-centred solutions adopting client/server-based Internet technologies.

A combination of blockchain technology with other technologies can be used to address domain-specific challenges. For example, the combination of blockchain technology and Internet of Thing (IoT) has the potential to monitor the usage of scarce resources such as the level of ground water and amount of energy consumption.

The aims of this chapter are twofold. First, it describes the primary building blocks of blockchain technology. Second, it illustrates various use case scenarios of blockchain technology in the fields of agriculture, energy health and others.

7.1 Humanitarian engineering: an example

The goal of Humanitarian Engineering is to help underprivileged and marginalized people. It does so by designing and implementing technology-based solutions to address the challenges faced by such people. The *Jaipur foot* [1] is one of the best examples of humanitarian engineering products. The *Jaipur foot* is an artificial leg designed to help people with below-knee leg amputation. It uses a new material of that time called *polyurethane* to develop near-natural but artificial leg. Since the

[1]Department of Computer Engineering, Dr. Babasaheb Ambedkar Technological University, Lonere, Raigad, Maharashtra, India

last 50 years, the *Jaipur foot* has been assisting thousands of physically challenged people to compete with ordinary people, bringing a smile on their faces.

Durability, flexibility and convenience of use are some of the critical factors behind the successful and widespread use of *Jaipur foot* which was the result of innovative use of material technology.

The example of *Jaipur foot* illustrates two significant aspects of humanitarian engineering. The first one is the use of technology, i.e. material technology in case of *Jaipur foot*, and the second one is relieving pain or grief of underprivileged people.

The focus of this chapter is on blockchain technology, one of the emerging information and communication technologies (ICTs). The objective of this chapter is to illustrate how blockchain technology has promising applications to address the diverse needs of various developmental sectors as envisaged by the United Nations (UN) sustainable development programme.

7.2 A framework for understanding humanitarian engineering

Before delving into details of technological alternatives for realizing humanitarian engineering, let us develop a framework to understand the dynamics of humanitarian engineering as a discipline. The three main pillars of humanitarian engineering are (1) stakeholders, (2) goals and (3) sectors for humanitarian engineering.

7.2.1 Stakeholders for humanitarian engineering

The main stakeholders for humanitarian engineering projects are the people and organizations that play a critical role in the development of underprivileged communities. These are also known as *development agencies*. These agencies operate at different geographical levels from global to local level.

The first kinds of development agencies are *international development agencies*. Their primary function is to address the problems faced by humankind. These agencies set the agenda for development programmes, identify the problems, monitor and evaluate the performance of development programmes, collect data and evidence to study the impact of development programmes, and fund the developmental activities in various countries.

Table 7.1 describes the role of three leading international development agencies. Besides this, the World Bank and Asian Development Bank (ADB) are some of the other international development agencies which sponsor developmental projects.

Many countries in the world have development agencies at the national level to formulate country-specific development policies, align it with policies formulated by UNDP and implement developmental projects. For example, in India, the National Initiative for Transforming India (NITI) Aayog is one such organization.

Also, there are some regional development agencies whose primary role is to implement the developmental projects in coordination with national and

Table 7.1 International development agencies and their roles

Sr. no.	Development agencies	Activities
1	UNDP	United Nations Development Programme (UNDP) is one of the subunits of the United Nations headquartered in New York City. The primary role of the UNDP is to formulate policies, set priorities and sponsor development programmes to address the challenges faced by humanity. It does so by proactively working with all the member nations of United Nations Organization (UNO) [2].
2	UNWFP	United Nations World Food Programme (UNWFP) is a branch of UNO raising the war against hunger and malnutrition. It provides food assistance to member countries so that the nutritional needs of the people leaving in poverty can be met. It aims to reduce child mortality, improve maternal health and fight against diseases [3].
3	WHO	World Health Organization (WHO) is a subsidiary of UNO dealing with matters of public health at the international level. It articulates policies for ensuring the highest possible health for individuals by identifying social and economic factors affecting health. It closely works with national government agencies and monitors the effective implementation of health policies [4].

international agencies. In a few countries, universities and academic institutes are slowly emerging as a regional knowledge centre to provide information necessary to get insights about the local developmental needs [5].

7.2.2 Sustainable development goals

The sustainable development goals (SDGs) shown in Figure 7.1 capture the expectations and aspirations of people to live in a prosperous society without compromising on the needs of future generations. These SDGs are a set of ambitious goals put forward by the UN in the year 2015 for the entire development of humanity [6,7]. This set of goals is linked to specific targets to be achieved by the end of the year 2030. Table 7.2 lists out all 17 goals categorized into 5 labels, namely People, Prosperity, Peace, Partnership and Planet [8].

Achieving these goals is an enormous task. Development agencies need to devise social, legal, financial and technological interventions. A development engineer responsible for technological interventions should know these SDGs so that project objectives and outcomes can be aligned with the SDGs. Knowledge of these SDGs is also essential for any engineer to become conscious of societal needs and environmental responsibilities.

7.2.3 Sectors for humanitarian engineering

The set of SDGs described in the last section aims at the overall development of humanity. The three main sectors which directly capture the needs of the human

Figure 7.1 Sustainable development goals (Source: news.un.org/en/story/2015/12/519172-sustainable-development-goalskick-start-new-year)

being responsible for achieving SDGs are economic systems, social systems and environmental systems. However, some additional institutes and sectors are responsible for laying the operational and legal framework necessary for business transactions. A legal system which enacts the laws protecting the environment is one such example. This section reviews such business sectors responsible for achieving the target indicators set for each SDG.

1. **Agriculture.** The agriculture sector is the most crucial sector for sustainable development. The SDGs like *End Poverty* and *Zero Hunger* are directly related to agricultural production. Feeding the evergrowing population with sufficient food and nutrient is a huge task. As most of the rural population depends on income from agricultural produce, the goal of ending poverty is also correlated with agriculture. The sustainable agricultural production is constrained by factors like growing population, water shortages, declining soil fertility and climate change [8]. Recently scientists have been exploring technologies such as wireless sensor technologies [9] and ICT [10] for sustainable agricultural production.
2. **Banking and finance.** Banking and financial markets drive economic development. It provides capital to start new businesses. It provides various avenues for income growth and wealth accumulation [11]. The number of banks and non-banking financial institutes present in the community is one of the indicators of economic development. Trust in financial institutes, efficiency of processing business transactions (e.g. remittances and payment made to farmers on selling upon crops), diversity of financial instruments (e.g. crop insurance, loans and micro-credits) are some of the factors necessary for decent

Table 7.2 Sustainable development goals

Category	SDGs	Descriptions
People	(SDG 1) No Poverty (SDG 2) Zero Hunger (SDG 3) Good Health and Well-being (SDG 4) Quality Education (SDG 5) Gender Equality	End poverty and hunger, in all their forms and dimensions, ensure all human beings can fulfil their potential in dignity, equality and healthy environment
Prosperity	(SDG 8) Decent Work and Economic Growth	Enjoy prosperous and fulfilling lives, economic, social and technological harmonic progress
Peace	(SDG 10) Reducing Inequality (SDG 16) Peace, Justice and Strong Institutions	Foster peaceful, just and inclusive societies, free from fear and violence
Partnership	(SDG 17) Partnerships for the Goals	Revitalized global partnership, the participation of all countries, stakeholders and people
Planet	(SDG 6) Clean Water and Sanitation	Protect degradation through sustainable consumption, production, natural resource management
	(SDG 7) Affordable and Clean Energy (SDG 9) Industry, Innovation and Infrastructure (SDG 11) Sustainable Cities and Communities (SDG 12) Responsible Consumption and Production (SDG 13) Climate Action (SDG 14) Life below Water (SDG 15) Life on Land	Stakeholders and people

work and economic growth. Emerging technologies such as cryptocurrencies have to play a significant role in making business transactions simpler.

3. **Education.** Education brings changes in behaviour. Education makes people more knowledgeable and skilful to get decent jobs. Education brings awareness about the environment and one's social responsibility. It also has an indirect impact on the health of an individual and reduction in the rate of population growth. The task providing education to all is becoming more challengeable because of a shortage of trained teachers, difficulties in providing better learning experiences to students on account of a shortage of playground in urban areas and oversized classrooms. Educators are gradually adopting technologies to overcome some of these challenges.
4. **Energy.** The goal of *affordable and clean energy* (SDG 7) is directly related to the energy sector. This sector also drives industrial development and economic growth. The demand for energy is increasing exponentially because of increasing population and widespread use of electro-mechanical devices to carry out routine works. Conventional energy generation methods that use natural resources such as coal and fossil fuels fail to meet this increasing

energy demand. Hence use of non-conventional and renewable energy sources need to be increased on a large scale for sustainable development. In this context, it is required to provide technological solutions that would reduce energy demand, techniques for efficient energy production and replacement of conventional energy sources [12].

5. **E-governance.** E-governance is the use of ICTs into day-to-day government processes. It is aimed to bring effectiveness in administration, transparency in the services provided by the government and increasing participation of citizens in implementing government policies. The sector of E-governance is directly linked to Goals 16 and 17. This sector plays a critical role in building accountable institutes necessary for enforcing peace and justice in societies (Goal 16). It also lays the technological infrastructure necessary to strengthen partnership at the global level required to implement various SDGs (Goal 17) [13,14].
6. **Environment science and engineering.** The policymakers formulating the developmental goals have realized the drawbacks of uncontrolled development that took place in the last century. The economic growth that pollutes air and water that reduces the portion of forest and agricultural land never satisfy the needs of the future generation. Hence the SDGs such as the provision of clean water and sanitation (SDG 6), actions for regulating climate change (SDG 13) and protecting life below water and on land (SDG 14 and SDG 15) are directly linked to environmental science and engineering. Newer clean and green technologies need to be developed and adopted, especially in the areas of civil engineering, construction, water management, urban development and for maintaining biodiversity [15].
7. **Health.** To provide good health and well-being is one of the sustainable goals directly linked to the health sector. Besides the lack of primary healthcare mechanisms, there exist many factors which affect the health and wellness of individuals. Some of these are the by-products of uncontrolled development. For example, industrialization and modern urban centred life are causing stress and respiratory system-related diseases. Technologies such as information and communication and genetic engineering have been found useful in preventing, monitoring and diagnosis of diseases [16].

7.3 Technological perspective of humanitarian engineering

The SDGs listed in the previous section are ambitious in terms of number, scope and indicators used to measure the attainment of these goals. One of the effective ways to achieve the set targets is through innovative use of existing and emerging technologies. Table 7.3 lists some of the emerging technologies that will have a positive impact on sustainable development.

These technologies can help to combat climate change and degradation of natural resources, to achieve good health and well-being, to speed up economic growth and to promote all-inclusive and equitable social development.

Table 7.3 Emerging technologies for sustainable development

Sr. no.	Emerging technologies	Benefits of technology
1	Bioplastic, biodegradable plastic and sustainable plastic	To protect from climate change and prevent degradation of natural resources (SDG 13)
2	Renewable energy sources (e.g. wind, solar and biofuels)	Prevents degradation of natural resources, clean and green energy (SDG 7)
3	Electric vehicle	Effective fuel consumption and pollution-free automobiles (SDG 7 and SDG 13)
4	Artificial intelligence and machine learning	Personalized medicine, outcomes based public health and diagnosis of cancer, pneumonia and other diseases (SDG 3)
5	Communication technologies (e.g. mobile communication, wireless communication, smartphone and Internet)	Dissemination of knowledge between doctors, patients and caretakers (SDG 3)
6	3D Printing	Lowers cost of personalized medicine and organ transplant (SDG 3)
7	Genetic technology	Facilitates precision medicine (SDG 3)
8	Digital financial technologies (e.g. cryptocurrencies and blockchain)	Supports micro-payments and creates a trustworthy environment for business transactions (SDG 8 and SDG 17)

Data are one of the common threads across all the technologies listed in Table 7.3. It has been envisaged that in the coming decade, data will drive sustainable development. At the same time, data-driven development is continuously raising many issues, such as security of information, privacy and ethical concerns.

In the following sections, we discuss and evaluate one the emerging technologies called blockchain technology in this context.

7.4 A blockchain primer

The necessity of blockchain technology can be understood through the knowledge of potential and pitfalls of the Internet as a platform for business.

The Internet has introduced an information-centric model of business, and it has revolutionized the way people transact online. For example, the emergence of e-commerce sites (e.g. Amazon) has been attributed to the growth and widespread presence of the Internet.

The Internet has bridged the information gap that exists between a service provider and service consumer by creating a third party for information exchange called intermediaries or agents or service providers. These agents, which are e-commerce sites, hold the information about who sells what, i.e. seller's information and who wants what, i.e. buyers profile and their needs thus bringing together consumers of services or goods with that of producers.

The advantages of doing online business are that it simplifies the process of business transactions, reduces the time required for businesses, and as a result, it has brought prosperity in the society.

In the context of SDGs, the Internet as a platform for business has created various opportunities for decent work and economic growth. The Internet as a platform for communicating information has reduced the impact of natural disasters such as cyclones and spread of epidemics by timely disseminating useful information. The Internet as a learning platform has increased the accessibility of education fostering the goal of education to all.

Despite the various benefits of the Internet, it has always remained an unreliable platform to share valuable personal information because of its mediator-centric or client/server model for information exchange. As shown in Figure 7.2(a), personal information is exchanged through mediators or servers. A server or mediator may be a payment gateway or an e-commerce site. The information shared with such sites is always susceptible to breach of security and privacy attacks.

The emerging blockchain technology removes these pitfalls by laying a trust layer on top of the existing Internet technology. It replaces the mediator-centric model of information exchange with the P2P model or decentralized model, as shown in Figure 7.2(b) of information exchange. It transforms the Internet into a trustworthy platform for doing business when transacting parties do not trust each other. It eliminates the role of mediator responsible for authenticating the identities of transacting parties. Table 7.4 differentiates the client/server and decentralized model of interactions.

Initially emerged as a platform to exchange digital currency over the Internet, now the blockchain technology is gradually emerging as a general-purpose platform for doing business over the Internet. Due to potential applications of blockchain technology in various fields, the UN has included it as one of the frontier technologies to realize SDGs [17].

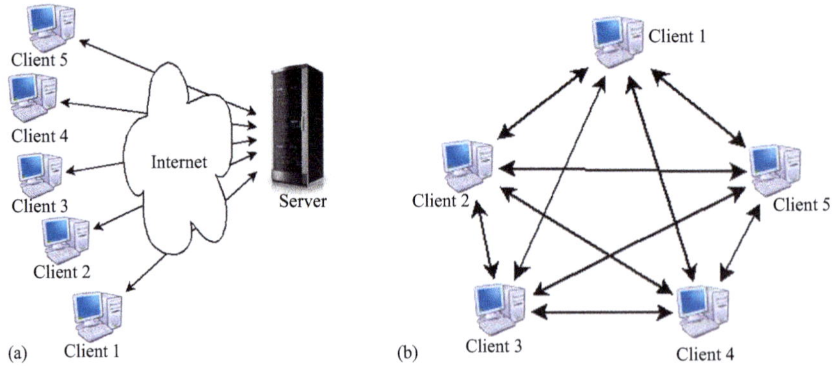

Figure 7.2 Client/server versus decentralized systems. (a) Client/server architecture. (b) Decentralized architecture

Table 7.4 Differences between client/server and decentralized architecture

	Client/server architecture	**Decentralized architecture**
Mode of communication	Client and server communicate with each other via inter-networking protocols	P2P communication via inter-networking protocols
Components	Nodes, server and communication links	Nodes and communication links
Architecture style	Layered	Object-based
Availability	Need to achieve high availability	No need to achieve high availability
Servers	Only single server is present to serve clients	Servers are distributed in a system
Privacy	It is easy to determine who is handling the content	Difficult to determine who is handling the content
Failure	Bottleneck as the sever fails down	No bottleneck as services can be provided through other nodes in case of failure
Cost	Implementation cost is high	Implementation cost is negligible

This section provides an overview of essential elements of blockchain and how it achieves the various quality attributes that make it as one of the promising technologies. The blockchain technology can be understood at the conceptual level and specific instance level. The Bitcoin, Ethereum and Hyperledger are a few common examples of specific blockchain. This section reviews the blockchain technology at the conceptual level. The four fundamental concepts common across the blockchain implementation are [18] (i) Distributed Ledger, (ii) Cryptography, (iii) Consensus Protocols and (iv) Smart Contracts.

7.4.1 Distributed ledger

In a conventional sense, ledgers are the registers or logbooks employed for account keeping or bookkeeping operations. Similarly, in the context of a blockchain-based information system, ledgers are the databases storing up-to-date information about business transactions. These are distributed among all the nodes participating in the network. In a blockchain environment, ledgers are not stored at a central place. They are distributed among all the nodes. So, multiple copies of a ledger exist in a business network. Hence, these are referred to as distributed ledgers. When a node in a network updates its local copy, all other nodes synchronize their copy with the updated one. Hence, each copy is consistent with each other.

These ledgers are used to store information about valuable assets. In the Bitcoin implementation, the first blockchain-based system, ledgers are used to store digital currencies. It may be used to store information about other valuable assets such as land records, diamonds, student's academic credentials and others.

In a blockchain-based information system, records in a distributed ledger are arranged in a chain format, as shown in Figure 7.3, for storage purpose. Here,

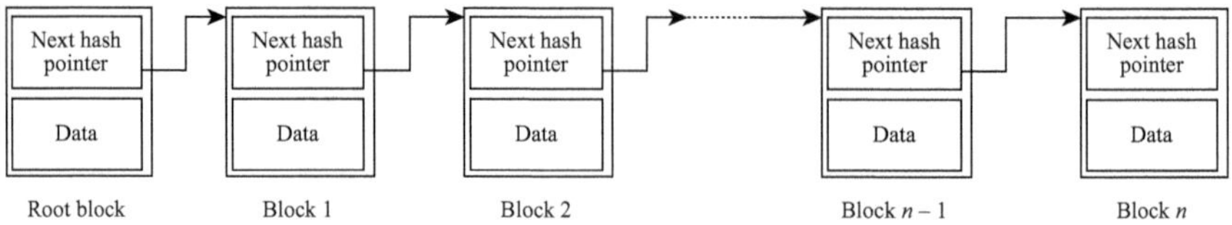

Figure 7.3 Blockchain

multiple transactions related to an asset are grouped in a block. The $(n + 1)$th block in the chain links to the nth block and the nth block links to the $(n - 1)$th block and so on. The first block in a chain is called a genesis block or the root block. Due to this peculiar storage arrangement, the distributed ledgers are also known as blockchain. Here it is worth to note that all blockchain-based systems contain distributed ledgers but not all distributed ledgers employ blockchain-based storage mechanism. In the blockchain data structure, new records are permitted to append and insertion or deletion operations are prohibited.

The most critical design feature of blockchain-based information system is the use of hash pointers instead of physical memory-based pointers to link blocks in a chain. A hash pointer is a message digest calculated from the information content of a block. Whenever a node attempts to tamper the information content, a small change in the information leads to a ripple effect of changes in hash pointers, making it impossible to change the information once it has been recorded in the blockchain.

Facilitating mediator-less business transactions and supporting immutability of stored information are the two significant quality attributes associated with blockchain-based information systems. These quality attributes are derived from replicating ledgers on all the nodes in a network and linking blocks in a chain through hash pointers.

Typically, blockchains are of two kinds based on how blockchains are accessed, i.e. private and public blockchains. In a public blockchain, any node can join and leave the network and validate the business transactions. While in the case of private blockchain, the network is small and requires permission to join and leave the network. Hence private blockchains are also known as permissioned blockchain, and public blockchain is known as permission-less blockchain. For example, Bitcoin is a public blockchain, and Hyperledger is an example of private blockchain.

7.4.2 Cryptography

Blockchain technology makes heavy use of cryptographic functions to assure trust among the users transacting over a blockchain-based business network. A typical business network includes many un-trustworthy elements. In a conventional banking domain, an agent issuing the cheque without having sufficient balance in the account, or an agent forging a signature are typical examples. In a digitized economy, these challenges are aggravated because of information transfer over an unreliable communication medium. Hence, cryptographic functions, a set of mathematical functions, are used to encode messages to assure the security of information in a network containing malicious or untrustworthy agents. These cryptographic functions address various purposes. Some of them are the following:

1. *Authenticating the identity of agents involved in a business transaction:* Blockchain-based systems use a kind of asymmetric key cryptography. These protocols use two different keys called public and private keys. The public keys are open and used as addresses for performing business transactions while private keys are secret and used for validating the transactions. SHA (Secure Hash Algorithms)-256 (e.g. Bitcoin) and ECDSA (Elliptic Curve Digital Signature Algorithm) (e.g. Hyperledger) are some of the cryptographic protocols used for this purpose. The private blockchain uses another service called membership service, which authenticates the identity for business transactions.
2. *Ensuring privacy:* Maintaining the privacy of transactions is a challenge, especially in public blockchains (e.g. Bitcoin). In such systems, transactions are possible to trace to real-life identities. Advanced cryptography-based techniques such as cryptographic mixers (e.g. Zerocoin) and Zero-knowledge proof (e.g. Zerocash) have been found useful to address this challenge. Cryptographic functions such as digital signature are also used to authenticate a particular transaction.

7.4.3 Consensus protocol

In decentralized systems, agreeing upon the global state of the transaction is a challenge. In a centralized system, this is not an issue because only one copy of transaction history is present at the central authority (e.g. bank's main server machine). Blockchain being a decentralized system holds multiple replicas of transactions at several nodes. Agreeing upon the unique state of the transaction is an issue which is solved by executing a consensus process involving all the nodes in the system.

This process is typically carried out in three stages. In the first phase, a node is elected/selected as a leader node to decide upon a unique state. In the second stage, transactions are validated. In the third stage, transactions are committed. A variety of consensus algorithms exist in blockchain-based system. These are often compared based upon how scalable the algorithm is and several malicious nodes it tolerates.

The Proof-of-Work (PoW) algorithm used in Bitcoin is one example of the consensus protocol. It selects the leader node responsible for deciding upon a global state by solving a cryptographic puzzle. It takes about 10 minutes for solving the puzzle requiring extensive computational work and much electric energy. It can work in the presence of 50% of malicious nodes in the network. The Proof of Stake (PoS) is another consensus protocol in which a leader is selected with the highest stakes in the network. It has been found as scalable as compared to PoW, and it also works in the presence of 50% of malicious nodes in the network.

The Practical Byzantine Fault Tolerant (PBFT) is the third example of consensus protocol which has been found scalable and works in the presence of 66% (2/3) malicious nodes in the network.

7.4.4 Smart contracts

Smart contracts are the most significant element in the blockchain-based system because it provides configuring the behaviour of such systems. Blockchain programmers can customize the working of blockchain systems by writing programs called *smart contract*. The smart contracts are scripts which are executed when a specific event occurs in a system. For example, in the context of Bitcoin, a coin may be released when more than one signatures are validated, or when miners solve a cryptographic puzzle.

These scripts can be written in a native language provided by blockchain systems or general-purpose programmable language. For example, Bitcoin provides a simple and less expressive native language to write a smart contract while Ethereum provides a Turing complete native language called Solidity to write smart contracts. In Hyperledger, blockchain programmers can write a smart contract in a general-purpose language such as Java/Go.

7.5 Sectoral applications of blockchain technology

7.5.1 Blockchain technology in agriculture

Numerous business factors motivate the adoption of blockchain technology in agriculture sector.

The first and the topmost key factor behind the adoption of blockchain technology in agriculture is to eliminate bad actors and poor processes involved in the food supply chain. For example, *AgriDigital (agridigital.io)* is a cloud-based platform that is designed using blockchain technology to provide an efficient interface for the agriculture supply chain.

The second key factor is to create trust among consumers and retailers about the originality and authenticity of food products. For example, *ripe.io* is an organization which provides a blockchain-based platform with this sole purpose.

Blockchain technology has the potential to open up new market areas for food producers, specifically farmers from the developing nations. For example, *AgriLedger (agriledger.io)* adopts the distributed ledger technology to open up new market areas for the farmers.

We broadly classify the typical use cases of blockchain technology in the agriculture sector as follows:

1. **Crowdfunding for the development of agricultural products.** Entrepreneurs and start-up organizations have been increasingly using blockchain as cryptocurrencies or digitized tokens to raise capital for their entrepreneurial ventures. This mode of application is generally known as Initial Coin Offerings (ICOs) [19].

   The ICOs are a type of digitized financial instruments which can be monetized through cryptocurrencies or fiat currencies. The *NagriCoin (nagricoin.io)* is one such application, which has raised capital to develop smart fertilizers that stimulate plant growth.

2. **Development of agricultural supply chain management.** Various business factors drive the use of blockchain technology for the agricultural supply chain management. First, paying the farmers their due share in food production is the key motivating factor behind the adoption of blockchain technology for the food supply chain. The second most important factor is to address the concerns about food safety in the minds of consumers. For example, the *AgriDigital* is a cloud-based supply chain management platform primarily supporting the supply chain for grains.

   The objective of such platforms is to effectively connect the various operators in the supply chain such as farmers, distributors, storage operators, retailers and consumers.

   Creation of digital assets representing physical products, recording and tracking their flow within the supply chain are the main functionalities offered by such platforms. They implement it by providing a set of smart contracts executing over a blockchain implementation such as Ethereum.
3. **Tracking and traceability.** Assuring about the quality of food products and agricultural commodities is one of the main challenges faced by agriculture and food processing industries [20]. Consumers expect authenticity about food products in terms of composition, origin and purity. These expectations are valid when the food products are costly, consumed at a distant place far from its origin, and they are directly linked to the health and safety of consumers.

   Providing accurate and verifiable information about the composition and origin of the food product is one of the solutions to address this challenge. The blockchain technology plays this role precisely while tracking and tracing the origin of food production.

   The blockchain-based solutions to address this problem are typically built around the immutability feature of the blockchain along with other technologies such as DNA marking, Radio-frequency identification (RFID) tagging, Quick Response (QR) coding and isotope testing.
4. **Blockchain-based financial products for farmers.** Conventional financial institutions such as banks, insurance companies and cooperative credit societies face the challenge of including small farmers in the mainstream financial sectors. This situation leads to undue exploitation of farmers by illegal money lenders and traders.

   Blockchain, as a technology initially emerged as a payment system through digital currencies (e.g. Bitcoin), can be used to provide financial products specially designed for farmers.

   The *AgriWallet* is one such example of saving scheme designed and implemented by *COIN22 (coin22.com)* for smallholder farmers to manage the risk associated with droughts, floods and low yield of crops [21].

   Also, the use of blockchain technology is currently being explored by many companies and government agencies to provide agriculture insurance, micro-credit and subsidy transfer [22].

Mapping of the physical commodity to the digital artefact is one of the challenges faced while designing such platforms. This challenge can be addressed by using the IoT and various sensors along with blockchain technology. The sustainable goal of ensuring zero hunger (SDG 2) and no poverty (SDG 1) is indirectly supported by extensive adoption blockchain technology in the agriculture sector.

7.5.2 Blockchain for financial services

The factors such as to eliminate the role of mediators between two transacting parties, to reduce transaction time and to simplify the transaction process motivate the use of blockchain in providing financial services.

In such scenarios, blockchain technology mainly plays the role of validating transactions. It helps to reduce the transaction charges by offering an efficient and secure way to transfer the funds either within a country or between two different countries.

The use cases of blockchain technology in finical sectors belong mainly to two categories, first, in the area of payment and transfer, and other record-keeping and verification.

For example, the *Mojaloop (mojaloop.io)* is one of the examples of the blockchain-based payment transfer platform developed for *Gates Foundation* (*gatesfoundation.org*) to connect financial service providers and customers from all over the world. In addition to offering efficient payment mechanism, the blockchain technology reduces the possibilities of frauds such as corruption during the disbursement of funds.

Second, the features of blockchain technology, such as distributed ledger and smart contracts, have been in use for effective aid distribution. Each step involved during the aid distribution from donation to receipt of the amount can be tracked and monitored using distributed ledgers. Donations are released upon meeting specific criteria and occurrence of pre-specified events encoded in the smart contracts. For example, *Amply* [23] is a blockchain-based protocol developed by IXO Foundation which links the release of government subsidies to the attendance of children in a school.

Applications of blockchain technology in the banking and finance sectors enable the attainment of SDGs such as SDG 1, SDG 9 and SDG 8. For example, the World Food Programme (WFP) used the blockchain technology to provide food assistance to refugees from Syria.

7.5.3 Blockchain in education

The ICTs have revolutionized the mode of delivering education. It has realized the goal of open education to all and created multiple ways to equip oneself and upgrade the skill set. The blockchain technology is now providing value-added services in the education sector, which includes credit transfer, issuing a transcript, digital certificates and payment for online courses.

We use the following categories to classify the uses of blockchain technology in the education sector:

1. **Record verification and validation.** In this mode, developers have been using blockchain as distributed ledgers which store and verify academic credentials

of students. For example, *BlockCert* (*https://www.blockcerts.org*) is a set of tools that provides the functionality to create, store and validate blockchain-based digital certificates.
2. **Information sharing.** Universities need to share and transfer the academic credentials of a student when a student migrates from one university to another. In this mode of application, blockchain technology enables effective transfer of academic information between two universities. For example, [24] describes the design of one such platform called *EduCTX* that enables credit transfer among institutes.
3. **Identity management.** Online learners use multiple applications and services such as Coursera, Google Classroom and note-taking apps (e.g. Keep). Securely sharing and authenticating identity credentials across the applications remain a big challenge. Blockchain technology addresses this challenge by providing a unique identifier accessible and verifiable across the applications, for example, *BlockStack* (*https://www.blockstack.org*), one such cloud-based identity management service.

These applications of blockchain technology have realized the goal of providing equal learning opportunities to everyone, i.e. (SDG 4).

7.5.4 Blockchain in energy sector

The energy sector worldwide is undergoing structural as well as functional changes. The decentralized network of renewable energy generators has been gradually expanding, which is currently dominated by the conventional centralized grid structure. The functionalities of the conventional power grid are not only limited to the functions of power generation and distribution but are providing additional services such as analysing the data generated through the deployment of smart devices and offering intelligent services to consumers.

The blockchain technology is the driving force behind all these changes. The features of blockchain, namely distributed ledger and smart contracts, are being increasingly used to provide additional services. Andoni *et al.* present a comprehensive survey of such services in [25]. Some of these services are the following:

1. **Wholesale energy trading and supply.** The technologies such as distributed ledgers and smart contract are used to simplify the business processes involved in energy trading and supply. In this context, developers use blockchain technology primarily for record-keeping, autonomous agents executing business logic and for payment and transfers. Elimination of intermediaries and reduced transaction times are the benefits gained by adopting the blockchain-based approach.
2. **Imbalance settlement.** It is a process executed by grid operators. It aims to remove discrepancies between energy units sold by energy suppliers and the energy sold by distributors to the consumers. Blockchain technology provides

an efficient solution to this problem which normally takes months for settlement. Here also, distributed ledgers are used to keep track of energy generation and consumption.

3. **P2P trading.** P2P trading is a form of trading in which energy providers or generators directly sell the energy units to consumers and thus eliminate the role of energy distributors. Here, blockchain technology is used as a platform that facilitates energy trading in exchange of digital currencies.

Providing affordable access to reliable and modern energy (SDG 7) is one of the sustainable goals that blockchain technology realizes through the use cases, as mentioned above. Further blockchain technology is providing an effective platform for energy trading in micro-grids and community-owned energy systems.

7.5.5 Blockchain for e-governance

Government agencies face the challenge of providing transparent and citizen-centric services. These agencies are looking at blockchain technology as a medium to improve the quality of their service. These agencies have been adopting blockchain for multiple application areas. Some of which are listed as follows:

1. **A platform for offering integrated services.** The government of Estonia has been using blockchain as a platform to provide more than a thousand services to its citizen. Here, the distributed ledgers are used for logging and auditing each transaction of citizens with the government. The government of Estonia has issued digital ID and signatures to all citizens for this purpose. The use of blockchain has brought transparency and trust in the services offered by the Estonian government.
2. **E-voting.** Democratic societies face the challenge of conducting free and fair elections. Conventional ballot-paper-based elections are prone to be rigged and manipulated. Blockchain technology is currently being explored as an alternative to address this challenge. The *Votem* is a cloud service that uses blockchain to securely cast votes by mobile or remote users. The system uses distributed ledger technology to record and verify the votes.
3. **Land records.** Recording the information of landholding and ownership is one of the obvious usages of blockchain technology. The recording of such information in the immutable registry and for archival purpose is essential to avoid land-related frauds. Some countries, such as India and Sweden, have initiated the use of blockchain technology for land records. In cities such as Chicago, civic administrators have been using blockchain technology to automate the complete workflow involved in land registry.

The blockchain technology, in some cases, have been directly realizing the goal of Peace, Justice and Strong Institutions (SDG 16). While in other cases, it has been indirectly supporting the goal of equitable development (SDG 5).

7.5.6 Applications of blockchain in environment and climate change

Developers have started thinking about innovative applications of blockchain technology in the areas of environmental protection and climate change. Some of these novel applications include the following:

1. **Incentive mechanism for recycling project.** The blockchain technology is useful in rewarding digitized tokens or cryptocurrencies for joining in the recycling project. A digital token can be exchanged for the deposit of plastic wastage.
2. **Enforcing environmental treaties.** Government agencies face the challenge of enforcing environmental obligations while sanctioning developmental project. Developers have started applying the feature of the smart contract embedded in various blockchain platforms for this purpose.
3. **To implement carbon tax.** The carbon footprint is a measure used for assessing carbon dioxide emission caused by an individual, event or an organization. The blockchain technology is useful to track and record the carbon footprint. Further, it can levy a tax based on carbon footprint or build a reputation system to provide incentives for low carbon footprint products and services.

These applications illustrate the potential of blockchain technology in protecting the environment and ensure responsible consumption and production (SDG 12 and SDG 13).

7.5.7 Blockchain in health sector

The applications of blockchain technology in health sector belong to the following categories:

1. **Record keeping and ensuring confidentiality of information.** The use of blockchain technology in the health sector is driven by the motive to overcome the negligence of health-related documents and privacy of these documents (e.g. medical prescriptions and health reports). Preserving this information is useful not only to an individual but in carrying out research related to genetic diseases. All such documents are valuable to medical researchers, healthcare providers, public health authorities and health insurance providers.

   Blockchain technology provides secure access to store existing medical reports. These reports are made available on-demand through various blockchain platforms. One of the typical applications is *MedRec* [26], which stores *Electronic Medical Records* on blockchain. This information is accessible in a secure way to various stakeholders, such as medical researchers and public health authorities.
2. **Drug supply chain and traceability.** In the supply chain of drugs and medicinal products, it needs to track the initial details of medical products while it

is in transit. The *Modum.io*, a European company, has implemented a hardware sensor-based technology to monitor the temperature of drugs in pharmaceutical supply chain. In such applications, drug-related information exchanged between drug manufacturers, wholesalers, pharmacists and the patients are recorded and tracked to authenticate the originality of the drug. The features like a distributed ledger and smart contracts are used to implement these functionalities.

The combination of technologies such as artificial intelligence, blockchain and IoT are providing various solutions and products to realize the goal of smart technologies for a healthy life (SDG 3).

7.6 A cost–benefit analysis of adopting technology

The blockchain technology is an emerging technology. It is yet to reach the level of maturity of many Internet-based technologies such as web applications and mobile applications. In this scenario, it becomes essential to know the costs and benefits associated with the deployment of blockchain-based solutions. From the humanitarian engineering point of view, it becomes more vital because it offers numerous opportunities to realize SDGs and will have an impact on all aspect of development.

The strengths of blockchain technologies include the following:

1. **It provides a platform to build trustworthy services.** Many government and development agencies face the challenge of creating trust offered by their services which may include aid transfer. Blockchain technology addresses this challenge and enables the strengthening of institutes and services.
2. **It provides a platform for open market.** As observed in Section 7.5.1, blockchain technology is creating an open platform to sell and buy products without any mediators. Thus, it has the potential to bring small producers in the mainstream global market leading to multiple avenues for prosperity and growth.
3. **It simplifies business processes and transactions.** The feature of *smart contract* in blockchain technology allows to code the business logic and execute it upon the occurrence of a specific event. For example, the release of payment upon successful delivery of goods. Smart contracts are an effective mechanism to implement business logic when business processes are well defined, and it involves multiple business operators crossing the organizational boundaries.

The following risks associated with the blockchain-based services need to be considered while implementing services on top of it:

1. **No central authority means greater responsibility on users.** A blockchain system, by its design, is a decentralized system. It means there is no central authority to contact in case of loss of public keys or loss of passwords. Hence,

restoring users' credentials becomes impossible in such events. Also, the data entered in a system are immutable one and needs to be entered carefully. The implications of all these architectural elements put a greater responsibility on users to be careful.

2. **High energy requirement of some consensus protocols.** Blockchain-based systems use a consensus protocol to validate a business transaction. Some consensus protocols such as PoW used in Bitcoin have been found ineffective in terms of energy consumption. It needs a high amount of electric power to validate a business transaction. Hence, this cost needs to be considered when one selects a public blockchain for implementation.
3. **Maturity of code to replace business logic.** Blockchain-based system uses smart contracts to encode business logic and well-defined business rules. These contracts are automatically executed, and it replaces the need for human interventions. However, the smart contracts thus implemented need to be tested rigorously and formally verified. Otherwise, it will lead to a situation called *hard fork* of blockchain, i.e. replacement of older code with newer code for the smart contract, which is a costly affair.
4. **Legislation related to data privacy.** In most of the countries, the use of public blockchain such as Bitcoin is legally prohibited for payments systems because of lack of provision to trace the identity of transacting parties. Also, blockchain-based systems use immutable distributed ledgers, which means they are *write-once-never-forget* systems. Such systems violate the privacy principle of right to forget. Such legal and ethical concerns need to be considered before adopting a blockchain-based approach.

7.7 Conclusion

The chapter reviews the existing and potential applications of blockchain technol ogy for humanitarian engineering. The scope of humanitarian engineering is vast. It includes every aspect of humanity, from agriculture to finance. ICTs have played a vital role to address the challenges in these sectors by sharing and exchanging information. However, many of the challenges need to be addressed, such as information privacy and security, which require more effective solutions. Humanitarian engineers are looking towards blockchain technology as a potential solution to address many of these challenges.

The chapter defines the scope of humanitarian engineering actions in terms of achieving the SDGs formulated by UNO. It identifies seven sectors crucial for the attainment of SDGs. The chapter further describes the challenges in these sectors, which can be addressed by the use of blockchain technology.

In general, across all these sectors, blockchain technology provides an effective mechanism to track and record information exchanged in a chain of business transacting entities. It automates the business workflow by reducing human intervention. It opens up new global market opportunities by creating a sound digital payment system crossing the boundaries of nations.

Blockchain being a technology yet to achieve the maturity level of web and mobile technologies, the deployment of blockchain-based solutions come with risks and adverse consequences. The chapter briefly illustrates such implementation risks by doing cost–benefit analysis from a humanitarian engineering point of view.

Two main contributions of the chapter include the following: (1) identifying the scope of human humanitarian engineering in terms of SDGs and sectors responsible for the achievement of these goals; (2) a functional evaluation of blockchain technology by describing potential and existing applications of blockchain for achieving SDGs.

The functional evaluation presented in the chapter highlights the promising role for blockchain technology to build a global society free from hunger and diseases, with equal opportunities for growth and decent work. However, for more theoretical and foundational role of technologies and engineers in a society, one can refer to [27].

References

[1] Arya A, and Klenerman L. The Jaipur Foot. The Journal of Bone and Joint Surgery (British volume). 2008; 90(11):1414–1421.

[2] McCall B. UNDP to re-engineer funding strategy. The Lancet. 2013;381 (9867): 613.

[3] Fao I. WFP, 2015, The state of food insecurity in the world. Food and Agriculture Organization of the United Nations; 2015, pp. 1–62.

[4] Organization WH, Zhang Xiaorui, World Health Organisation, *et al.* WHO traditional medicine strategy 2002–2005. Geneva: World Health Organization; 2002.

[5] Goddard J, and Chatterton P. Regional development agencies and the knowledge economy: harnessing the potential of universities. Environment and Planning C: Government and Policy. 1999; 17(6):685–699.

[6] Desa, UN. Transforming our world: the 2030 agenda for sustainable development. United Nations General Assembly; 2015.

[7] Nilsson M, Griggs D, and Visbeck M. Policy: map the interactions between sustainable development goals. Nature News. 2016; 534(7607):320.

[8] Wu J, Guo S, Huang H, *et al.* Information and communications technologies for sustainable development goals: state-of-the-art, needs and perspectives. IEEE Communications Surveys & Tutorials. 2018; 20(3):2389–2406.

[9] Culibrina FB, and Dadios EP. Smart farm using wireless sensor network for data acquisition and power control distribution. In: 2015 International Conference on Humanoid, Nanotechnology, Information Technology, Communication and Control, Environment and Management (HNICEM). IEEE 2015, pp. 1–6.

[10] Singh D, Pande A, Kulkarni S, *et al.* Innovation for crop quality certification using ICT. In: 2015 7th International Conference on Communication Systems and Networks (COMSNETS), IEEE 2015, pp. 1–6.

[11] Rioja F, and Valev N. Finance and the sources of growth at various stages of economic development. Economic Inquiry, 2004; 42(1):127–140.

[12] Lund H. Renewable energy strategies for sustainable development. Energy. 2007; 32(6):912–919.

[13] Ndou VE. Government for developing countries: opportunities and challenges. The Electronic Journal of Information Systems in Developing Countries. 2004; 18(1):1–24.

[14] Lim J-h. E-government for sustainable development in SIDS, United Nations Project Office on Governance (UNPOG), 2016.

[15] Clayton T, and Radcliffe N. Sustainability: a systems approach. New York: Routledge; 2018.

[16] Deen MJ. Information and communications technologies for elderly ubiquitous healthcare in a smart home. Personal and Ubiquitous Computing. 2015; 19(3-4):573–599.

[17] Department of Economic and Social Affairs, World Economic and Social Survey 2018 Frontier technologies for sustainable development, the United Nations New York, 2018.

[18] Dinh TTA, Liu R, Zhang M, *et al.* Untangling blockchain: a data processing view of blockchain systems. IEEE Transactions on Knowledge and Data Engineering. 2018; 30(7):1366–1385.

[19] Catalini C, and Gans JS. Initial coin offerings and the value of crypto tokens. National Bureau of Economic Research; 2018.

[20] Kamilaris A, Fonts A, and Prenafeta-Boldú FX. The rise of blockchain technology in agriculture and food supply chains. Trends in Food Science & Technology. 2019.

[21] Sylvester G. E-agriculture in action: blockchain for agriculture: opportunities and challenges, The food and agriculture organization (FAO), United Nations, 2019.

[22] Bermeo-Almeida O, Cardenas-Rodriguez M, Samaniego-Cobo T, *et al.* Blockchain in agriculture: a systematic literature review. In: International Conference on Technologies and Innovation. Guayaquil, Ecuador: Springer; 2018, pp. 44–56.

[23] Schweiz Z. The blockchain for impact: technical white paper. https://ixo-foundation/ 2018.

[24] Turkanović M, Hölbl M, Kŏsič K, *et al.* EduCTX: A blockchain-based higher education credit platform. IEEE access. 2018; 6:5112–5127.

[25] Andoni M, Robu V, Flynn D, *et al.* Blockchain technology in the energy sector: A systematic review of challenges and opportunities. Renewable and Sustainable Energy Reviews. 2019; 100:143–174.

[26] Ekblaw A, Azaria A, Halamka JD, *et al.* A Case Study for Blockchain in Healthcare: MedRec prototype for electronic health records and medical research data. In: Proceedings of IEEE open & big data conference. vol. 13; 2016. p. 13.

[27] Riley D. Engineering and Social Justice. Synthesis Lectures on Engineers, Technology, and Society. 2008; 3(1):1–152. Available from: https://doi.org/10.2200/S00117ED1V01Y200805ETS007

Chapter 8

Big data-driven disaster resilience

Md Nazirul Islam Sarker[1], Md Saiful Islam[2], Md Enamul Huq[3], GM Monirul Alam[4] and Md Lamiur Raihan[5]

Disaster management is now a global concern and priority sector of almost all countries in the world. Resilience is a key tool for disaster management and development which can use big data for enhancing the individual's as well as system's capacity through developing a link between information, response to disaster and aid. The use of big data technologies for disaster management is a debating topic among the researchers, academicians and policymakers. This chapter intends to contribute to this debate by highlighting the potential of big data for disaster management through increasing resilience against the socio-ecological vulnerability. A qualitative approach focusing desk review and secondary data has been used to substantiate the arguments. This chapter argues that disaster resilience is a function of the adaptive, absorptive and transformative capacity of an individual or society to withstand and cope with the adverse effects of the disaster. Big data provides an opportunity to provide huge information to develop these capacities so that a social system can prepare themselves against disasters. This chapter also emphasizes the major principles of big data for effective use for disaster management like open source tools, strong infrastructure, developing local skills, context-specific data sources, data sharing with ethics, awareness about the right of data and learning from experience. This chapter also argues that big data is a potential tool for policymakers, administrators and related stakeholders to take necessary actions during and after disasters like early warning system, weather forecasting, emergency evacuation, immediate responses, relief distribution, training need assessment and increasing trained individuals.

[1]School of Political Science and Public Administration, Neijiang Normal University, Neijiang, China
[2]Department of Soil Science, Patuakhali Science and Technology University, Patuakhali, Bangladesh
[3]State Key Laboratory of Information Engineering in Surveying, Mapping and Remote Sensing, Wuhan University, Wuhan, China
[4]Faculty of Agricultural Economics and Rural Development, Bangabandhu Sheikh Mujibur Rahman Agricultural University, Gazipur, Bangladesh
[5]Graduate School of Global Environmental Studies, Kyoto University, Japan

8.1 Introduction

Disasters are increasing gradually all over the world. Though developed countries can handle it but developing countries suffers a lot. The poorer society of the developing countries is the major sufferer of natural disasters due to their less resilient capacity [1]. People can see only direct damages but it is very difficult to realize the indirect damages. Various factors are responsible for natural disasters. Climate change is one of the major factors that cause frequent natural disasters. Climate change is now is a growing concern all over the world. Though people of different communities have adopted their context-specific adaptation strategies but it fails to save them from devastation effect. Similar cases are happened in case of natural disasters. Adaptation strategy of vulnerable communities cannot help properly from huge direct and indirect damages of natural disasters [2]. It is now well established that technology can help to handle the situation of natural disasters. Technology can help to develop strategy for disaster management from early warning about disaster to post-disaster management. In every step of disaster management, policymakers, leaders, researchers and administrators can use technology for disaster management.

Disaster resilience is a popularly used concept that indicates an ability of an individual, group or community to bounce back to the previous normal condition after facing natural disasters [3]. It is an integration of adaptive, absorptive and transformative capacity. Resilience actually focuses on the multi-dimensional capacity of a unit of community to face disasters and successfully manage it through reducing vulnerability and enhancing adaptive, absorptive and transformative capacity [4]. It can address all the root causes of disasters through finding out and proper adaptation strategy. Resilience concept actually focuses on the way of capacity building of people so that people can reduce vulnerability, potential threat, stress, challenges and risk related to natural disasters [5]. It also acts as a motivation to develop internal capacity of people and prepare them to tackle any undesirable situation. Big data technology can help to enhance resilience in terms of policymaking, decision-making, leadership, research and administration [6].

Big data is a comparatively new paradigm of technology and approach to disaster resilience. It helps academicians, researchers and administrators to make an efficient analysis and effective decision by using a huge amount of available data [7]. It is an arrangement of scientific tools and approaches for the maximum use of huge data. It also focuses data accessibility, openness, analysis and meaningful presentation. Big data provides a big opportunity of communication which can help susceptible community people about upcoming threat, challenges, risks and disasters [8]. Communication provides a way to communicate each other before, during and after disaster to inform the condition to one another and make a preparation and also acts as a source of big data [9]. Big data encourages researchers and policymakers to conduct in-depth analysis on communication data from mobile phone, social media and other communication devices which has a big rationality for disaster resilience [6].

Some studies have already been done on disaster management [10], disaster resilience [5,11] and big data application in environmental management field [12–14] but the potential of big data for disaster resilience is still lacking. Since natural disasters are the effect of natural hazards, vulnerability and scarcity of natural resources, big data can be an effective approach to enhance disaster resilience. Natural disasters are also result of frequently changing dynamic environmental conditions. Big data-driven effective strategy can be helpful to increase resilience. Considering the importance of the issue, this chapter intends to explore the potential of big data for enhancing disaster resilience.

8.2 Methodology

8.2.1 Research design

A desk systematic literature review has been done in this study covering the last 10 years. Recent data has been collected for participating ongoing debate on the potential of big data for disaster resilience. This study mainly considers to the big data approaches which have potential contribution for enhancing disaster resilience.

8.2.2 Search strategy

Desk literature review is considered as an indispensable part for developing a new paradigm of potential field. Therefore, recent related data has been searched extensively in the popular databases like web of science and Scopus. A number of keywords such as disaster, resilience, vulnerability, adaptive, absorptive and transformative capacity, big data, and disaster management have been used. The data collection has been done from January to April 2019.

8.2.3 Inclusion and exclusion criteria

The desk review has been guided by certain criteria such as (a) Does this study focus on big data for disaster resilience? (b) Is this study articulating disaster management using big data approaches? And (c) is full text of this study available? Certain exclusion criteria have also been followed such as articles other than English language, duplication and articles having similar concept.

8.3 Results of the study

Systematic Review and Meta-Analysis (PRISMA) guidelines have been followed in this study for systematic literature review [15]. At first, 397 documents have been obtained with 8 related documents from bibliography. Then 167 documents have been detected after abstract screening; 134 documents have been removed from 155 documents due to lack of full text, duplicate of concept and unavailability of big data approach for disaster management. Finally, most relevant 21 documents comprising journal articles, books, book chapters and working papers (Appendix A) have been considered for in-depth analysis (Figure 8.1). Strengthening the Reporting of Observational Studies in Epidemiology (STROBE) checklist has been followed for qualitative selection of documents [16].

8.4 Discussion

This section mainly provides a detail explanation of the potential of big data for ensuring disaster resilience obtaining from existing literature.

8.4.1 Big data sources

8.4.1.1 Satellite imagery

Satellite imagery provides quantitative and qualitative data for disaster management which can help to conduct management operation as well as risk reduction. It can be frequently used for assessing the condition of post-disaster [17]. The major contribution of remote sensing such as high-resolution, multi-technical and multi-dimensional imagery provides a support for planning pre- and post-disaster assessment. Satellite imagery provides information about changing land use system, waterbodies, direction and damaged items of affected area [18]. This information can help to take proper decision about rescue method. It is providing not only general images but also three-dimensional (3D) images with attitude that can easily help to detect the affected areas and level of damages [12]. It also helps to identify damaged building and volumes of disaster-affected areas. Satellite

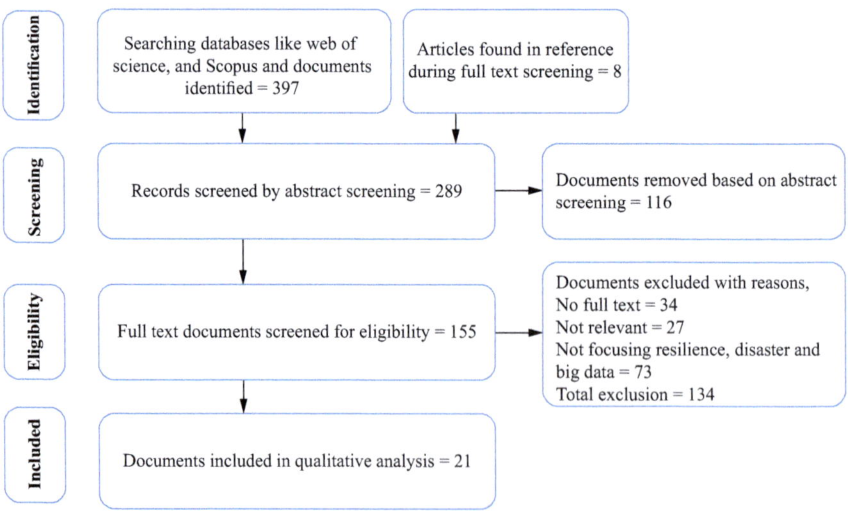

Figure 8.1 PRISMA flow diagram of document selection

imagery is considered one of the major methods for disaster management due to its usage on reduction of risk related to flood, landslide and human settlement [19].

8.4.1.2 Aerial imagery and videos

Unnamed aerial vehicles (UAVs) have been used for capturing aerial image which can play a vital role for creating situational awareness [12]. It is considered a better method than satellite imagery due to speed and spatial resolution of image. It can be used as an advanced level tool for detecting fine cracks, damaged structure and the extent of damages. UAVs comprise different kinds of sensors such as camera, infrared, ultraviolet, radiation and weather sensors along with spectrum analysers [9]. It is a tool which can supply useful information to transportation planning related to real-time and situational information. UAVs are considered as an authentic data source that can help to identify real damage caused by disasters.

8.4.1.3 WSW network

Technology related to wireless sensor web (WSW) can be used for easy warning system which helps to take preparation for saving asset from natural disasters [20]. Situational awareness can be done by using WSW [21]. Integrated use of WSW network can enhance response time, reducing the latency as well as increasing success delivery. These technologies also ensure connection between affected population and rescue team. WSW-based Internet of things technology provides a better communication in the disaster-affected areas where the communication structure is damaged by natural disasters [22].

8.4.1.4 Light detection and ranging

Exact ground condition of disaster-affected areas can be easily detected by light detection and ranging (LiDAR) by using advanced elevation model [12]. Though it is a little bit time-consuming and expensive, but it provides authentic and reliable information. It explores the real condition by providing high resolution [23]. The ability of LiDAR is very helpful for geological, features and mapping. It is well recognized that LiDAR can provide accurate data for water and flood assessment as well as prediction of future flooding [13]. It also provides reliable information about structural damage as well as elevation changes by natural disasters.

8.4.1.5 Simulation data

Simulation is a key approach for prediction. Numerical simulation can be a good approach for predicting future natural disaster by analysing meteorological and land surface phenomena as well as different kinds of pollutions [24]. It also provides 3D modelling that can help to predict probable damage of natural disasters. Generally huge data is generated at the time of disasters. Disaster management requires proper production, verification, validation and improvement of data for exploring real complexity caused by natural disasters [25]. Simulation data is also helpful for assessing environmental changes through agent-based modelling. An ecological model can provide realistic information about landslide by using simulation data [13].

8.4.1.6 Spatial data

Spatial data is helpful for disaster management especially for vulnerability assessment and prediction of natural hazards. Tomaszewski *et al.* [18] conducted a study on geographic information system (GIS) and mentioned that GIS data such as Federal Emergency Management Agency (FEMA) data feeds, World Bank data and national as well as open street map is helpful for disaster management. Spatial data is frequently used for disaster resilience in the developing countries.

8.4.1.7 Crowdsourcing

Crowdsourcing provides an opportunity to work a large number of people in an online platform for achieving common goal. In disaster-related crowdsourcing, a number of affected people can share their idea, experience and practices for disaster management [26]. In that platform, disaster victim can share real-time information. Though it is a good source for big data but it has still some challenges especially from credibility of the data to decision-making [17]. The collection, processing and analysis of crowdsourced data requires advanced tools because of its nature and volume. Crowdsourced data is also helpful for finding out the location of the disaster-affected area [27]. It is convinced to collect crowdsourced data by using mobile and online platform.

8.4.1.8 Social media

Nowadays social media is playing a vital role in almost all aspects of life. It is one of the main big data sources. Social media is considered one of the top communication tools for disaster management information [26]. It provides multi-dimensional information regarding disaster events. Though social media has shortcomings due to its different kinds of data but it is still effective for disaster management [28]. Communication for disaster management was dealt with participating organization, victims and affected population, vulnerable community and areas in traditional model [29]. But big data-driven technologies can easily handle the issues very accurately and timely. Since various disaster genres such as early warning, caution instruction and immediate interaction are connected with disaster type, phases and causes, technology-based communication can speed up the process of disaster management [30]. Various social media such as Facebook, Twitter, WhatsApp, IMO (In My Opinion), WeChat and QQ have a great impact on almost all phases of disaster management. Social media can be used in various ways for disaster management. Social media data should be collected carefully and then processed and analysed for decision-making necessary for disaster management [21]. Similarly, decision about disaster can be easily spread out using social media. Scholars recognized the importance of the social media for disaster management phases and ensuring resilience.

8.4.1.9 Mobile-based GPS and record of call data

Mobile phone acts as an important instrument in disaster situation to contact family, relatives and friends as well as to know the location for moving to safer place. Integrated sensors of mobile phone help to identify the most affected people as well as urgent needs of resources [17]. But sometimes natural disaster disrupts the electricity connection which causes an interruption of getting the mobile phone services in disaster-affected places. Mobile-based Global Positioning System (GPS) is a vital way to collect mobile-based sensing data for detection of people's behavior and movement at the time of natural disasters [18]. It helps to get real-time data of disasters regarding human reaction to the effect of natural disasters as well as warning and evacuation process [12]. GPS helps to identify the location, altitude, magnitude and related issues of natural disasters. GPS is generally working based on three basic criteria such as proper location, comparative movement and real time.

Call detail records (CDRs) of mobile companies can record all the call during disasters, which provide a huge number of data that are real-time and need-based. Disaster management personnel can easily use this big data to take quick decision for ensuring immediate services to disaster-affected people [11]. As the size of CDR data is large big data approaches can take opportunity to handle it. CDR tools can also collect the data related to human movement as well as behaviour in societal network regarding natural disasters. This approach can also collect the identity of the sender and receiver as well as the data of calls and SMS. From CDR data, concerned personnel can know the population density and the size of total population in the disaster-affected region [9]. Since all the subscribers are under the cellular network, it is useful to get accurate data by using this approach.

8.4.2 Big data approach for disaster management

Vulnerability is an emerging concept across disciplines, useful in understanding and assessing the status of people's condition in the face of natural hazards. The major characteristics of vulnerability are dynamic and influence people's social and biophysical processes and systems. Significant mobilization is necessary from the government, non-governmental organizations (NGOs), researchers and farmers to develop successful adaptation strategies [31]. The people of developing countries are the vulnerable communities due to excessive dependency on agriculture and having low income. However, these burdens may fuel the exploration of potential adaptive capacities of resource-poor communities [32]. The extent of people's susceptibility is increased due to the increasing vulnerability to natural hazards of almost all spheres of life, like the social, physical, human, financial and natural dimensions [33]. Though the effect of natural hazards may be occasional, seasonal or year-round, the extent of exposure is not the same for all communities.

The word 'resilience' originates from the Latin word *resilio* that means 'to jump back'. Walker and Salt [34] have claimed that 'resilience' originated from ecological research where Holling [35] sought to differentiate between an ecological system that persists in a condition of equilibrium or stability, and response of dynamic systems when they are stressed and move from this equilibrium. A resiliency perspective is an understanding of a system's adaptive capacity [36]. For livelihood systems, the four pillars of this perspective relate to activities and processes allowing for effectively (a) anticipating livelihood challenges and potential for surprises, (b) minimizing the impact of present vulnerabilities, (c) recovering from the impact of past as well as present vulnerabilities and (d) thriving from a complex livelihood situation [37]. Livelihood resilience is a top policy concept in development context research which emerging from various disciplines. It actually focuses the ability of a community to bounce back to normal condition of livelihood, i.e. previous conditions after suffering shocks. Resilience is a process which tackles a wide range of shocks, vulnerability and stress-like vulnerable livelihood, food insecurity, social protection, disasters and social conflicts.

Disaster management is an integrated process of preparedness, response, recovery and mitigation. All the four processes are tools of enhancing adaptive, absorptive and transformative capacity of an individual, group or community and ensure resilience (Figure 8.2). Preparedness indicates a short-term preparation for disaster management while mitigation indicates a long-term preparation. Response and recovery are the phases during and after disaster.

8.4.2.1 Preparedness

Detection and disaster monitoring

Proper monitoring of disaster is the first step of disaster management. An effective detection and monitoring can reduce the damage of disaster. Remote sensing data is

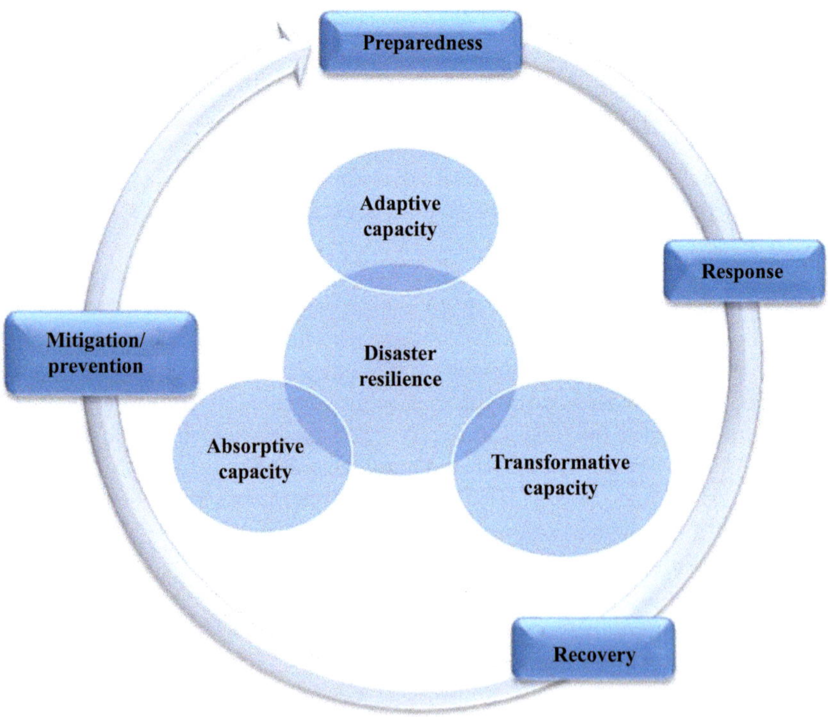

Figure 8.2 Conceptual framework of disaster resilience

usually a big source of big data which helps to detect any abnormalities of weather and disaster probability. Similarly, disaster detection can be done by satellite remote sensing that provides a real-time data with more accuracy. Some natural disasters like flooding and fire can be monitored by remote sensing imagery that helps to take proper measure for mitigation. Early detection helps to provide basic information to people so that people can prepare themselves to minimize the damages. Social media acts as a big source of big data that is easily generated by people's communication. In this case, people of social media share disaster-related data to each other and act as a sensor. Social media supplied real-time data which can help to identify hotspot of disaster. It also facilitates to provide information related to probability of damages, location, duration, distance and the extent of natural disasters. Musaev *et al.* [38] conducted a study on landslide and developed a detection system which integrates manifold physical sensors as well as social media (Instagram, YouTube and Twitter) to explore the real origins and positions of natural hazards. Real-time satellite and social media data have a vital role to detect the upcoming disasters like flood. The activity of disaster can be also monitored by a crowdsourcing-based detection system. The main challenge of disaster management and resilience is early detection and prediction about upcoming disaster. Only accurate data and authentic information are helpful for prediction about natural disaster. The prediction message should be conveyed to the public in a way that they can understand the possible extent of natural disasters.

Early warning

Data from sensor can be used for developing early warning process. Early warning is useful for flood management. It is usually seen from prevailing experience that several devices have been developed based on sensor data and integrated satellite such as Tropical Rainfall Measuring Mission (TRMM) rainfall, Radarsat Synthetic Aperture Radar (SAR) and Namibia Flood SensorWeb. Flood model have been also developed based on the big data gained from sensors and satellite. These tools are able to provide various services related to observation, planning of sensor, web notification and processing. Social media acts as a convenient medium of early warning to inform people about disasters. Carley *et al.* [13] conducted a study on role of social media particularly Twitter on early warning system of disasters, pattern on spatiotemporal network, coverage area and people's awareness about early warning. They mentioned that Twitter was a powerful social media which had a vital role for early warning system of disasters in Indonesia.

8.4.2.2 Mitigation/prevention

Risk assessment

Some risks of natural disasters are related to geographical position which can be easily detected by satellite images. Many scholars developed various tools and system such as satellite-based flood mapping, and Moderate Resolution Imaging Spectroradiometer (MODIS) for effective monitoring disaster events [12]. Vulnerability and time series analysis are also helped to assess the possible risk of natural disasters. Risk assessment can be done by using user-generated data and predict the probable events of natural disasters. It can save a huge damage such as people's livelihood assets, infrastructure, health and other basic public service systems. Crowdsourcing system is also used for risk assessment particularly to the oil industry through assessing potential exposures like smell and smoke. Mobile metadata and call times can be a great source of big data and helpful for decision-makers to avoid any unexpected situation related to natural disasters. It is also used for risk assessment and decision-making. Flood-related risk can be minimized by using spatial data and decision support system.

Forecasting and prediction

Forecasting is a vital component of disaster resilience which increases adaptive capacity of the people, group or community. Though it is very difficult to forecast about natural disaster due to its complex nature but it has a great impact to save the people and asset from damages. The prevailing forecasting is dependent on manual process of parameters as well as physical models. But accurate and rapid prediction requires modern approaches like big data analytics. Advanced model on resolution prediction requires huge meteorological data for solving upcoming natural disasters such as cyclones, heavy rain, storms, flood and hurricanes. Goldenberg *et al.* [39] introduced H212 model (Hurricane Weather research and Forecasting) and Muwakami *et al.* [40] developed FLOR (Forecast-Oriented Low Ocean Resolution) model for prediction of hurricane and cyclones, respectively. These two models provide accurate and timely information regarding natural disasters like hurricane, wind storms and cyclones. Both models are developed based on realistic simulation of big data. NASA also launched CYGNSS (Cyclone Global Navigation Satellite System) in 2016 that can predict cyclones and convection. This CYGNSS helps to predict the intensity, way/track and approximate time of tropical cyclones and storm surges. Similarly, airborne radar resolution can also help to forecast cyclones and hurricanes at the advent of 4 or 5 days.

8.4.2.3 Response

Assessment of damage

Remote sensing imagery is the key tool to assess the damages caused by natural disasters. The requirement of imagery depends on the technique and process. Rapid initial assessment required large-scale but low-resolution data of remote sensing which can help to recommend for proper initiative for prioritized area. 3D resolution imagery is necessary for assessing damages of roads and buildings. Damages of people's shelter and transportation network require high-resolution remote sensing imagery. It is also used for detecting open spaces of disaster-affected areas. Manned aircraft is also used for getting a good imagery based on UAV-based aerial system that can be easily used for detecting disaster-affected areas. Crowdsource is a vital tool for damage assessment. It actually works based on the data provided by the people of the disaster-affected areas to the platform of crowdsource. Some of the crowdsource platforms have been used to detect damages like GEN-CAN (Global Earth Observation Catastrophe Assessment Network) for Haiti earthquake in 2010, Open Streetmap for Typhoon detection in China, Twitter in Hurricane detection in the US. All these systems and models are using the big potential of big data for assessment of damage caused by natural disaster and enhance resilience.

Coordination for post-disaster management
Disaster affected areas are facing various problems such as lack of communication, coordination of rescue team and lack of stakeholder's awareness. These problems create barrier in post-disaster phase and reduce resilience. Disaster usually causes damages of the local resources that also create barrier to post-disaster management. These barriers can be easily solved by using big data analytics through assessment of condition of affected areas and maximum utilization of limited resources. Mobile networks are hampered in some affected areas, which also causes barrier for post-disaster resilience but this problem can be solved by aerial ad-hoc network that ensures the connectivity among users of affected areas. Felice *et al.* [41] checked small UAVs' potentiality to establish connectivity among mesh networks and found that it was a quality tool to make links among huge number of devices. Lu *et al.* [25] also introduced 'team Phone' device to provide message and self-rescue system through establishing a good smartphone communication system during disaster recovery. Team Phone is able to make better communication among rescue team members through cellular, ad-hoc and opportunistic seamless networking system. The self-rescue system can make a connection among phones of victimized survivors as well as send them rescue message in order to make efficient operation. Big data analytics can play a vital role in disaster management through identifying hotspot for urgent response, coordination of people to move out from risks and identifying proper response method.

8.4.2.4 Recovery

Recovery is also an important stage of disaster management. This phase starts during and after disaster. In this stage, the recovery teams try to collect information and data related to the damage caused by natural disasters. The recovery teams help to victim and rebuild damaged infrastructure to bounce back to previous normal condition. The aim of this phase is to recover the people and their assets and bringing them to normal condition which enhances resilience. Various kinds of infrastructure are necessary for making proper plan for quick recovery such as information related to relief distribution, confirmation of safety, coordination of volunteer activities and logistic supply. The damaged communication networks are required to recover and improve rapidly. Big data can help to improve regular as well as adaptive optimization for increasing infrastructure network as an essential part of disaster recovery. Communication and adaptation mechanism can be done properly by utilizing limited resources under the approaches of big data. Satellite imagery may be a vital source of big data in the post-disaster stage. Disaster-affected areas can be easily detected by remote sensing imagery. Recovery evaluation is easily done by using remote sensing data. Contras *et al.* [42] conducted a study on recovery stage assessment on earthquake in Italy by using QuickBird imagery and find out the performance of recovery phase from affected areas. Social media data is also useful in recovery phase as a user-generated data and source of big data. But handling the huge data during disaster is a challenge. Social network analysis is also helpful to find out influential opinion leader to fast the recovery process.

Big data can enhance disaster resilience in all stages of disaster management by using multi-dimensional analytics such as descriptive, predictive, prescriptive and discursive analytics. Descriptive analytics focuses on the description of the condition and critical issues of disasters. It generally uses social media and remote sensing imagery. Predictive analytics deals with inferences related to unobservable issues which may happen in future such as early warning and forecasting. Prescriptive analytics deals with disaster management policy-related issues for disaster resilience. Discursive analytics focuses on community resilience through raising awareness, timely response and feedback by using big data.

8.4.3 Challenges for the implementation of big data approach

The expected outcomes of big data approaches are challenged due to some barriers, gaps and threats. This study found some implementation challenges for big data.

8.4.3.1 Data collection, privacy and accuracy

Disaster resilience needs many types of heterogeneous data from various sources. Heterogeneous data can fill the gap among various components of disaster management. Social networking sites, weather department, NGOs and remote sensing imagery usually act as a large source of big data. Collection and analysis of the huge volume of data is very difficult to handle but it is necessary for taking realistic decision for disaster resilience. Using various sources help to enhance the quality of data but validation is another problem. After ensuring validation, data integration may bring expected accurate result. There are always some invalid data associated with social networking and crowdsourcing sites. It requires necessary filtering for proper validation. Privacy and accuracy of data are also the major challenges of big data implementation. Privacy should be maintained according to the policy of the government. Accuracy can be maintained by removing various misinformation as well as noise.

8.4.3.2 Analytics of big data

Adaptive capacity for disaster resilience can be enhanced through predictive modelling. The results of predictive modelling are solely dependent on the data and information provided by disaster-affected people. Though data from disaster-victimized people is very helpful for predictive model and taking adaptive measures such as forecasting and early warning but it should be ensured to collect accurate and valid data. Crowdsourced data is effective for raising disaster awareness but analytical method should be selected carefully for context-specific result. Accurate results can help to make proper decision to response, rescue and recovery phases of disasters.

8.4.3.3 Infrastructure for big data

A good infrastructure is necessary for collection, analysis and interpretation of big data. Since big data collection is dependent on all the phases of disaster management, a well-designed and developed infrastructure is essential for its maximum utilization. A well-designed cyber infrastructure can act as a platform for various stakeholders of disaster resilience to response, share and exchange their ideas, plan and experiences. This kind of sharing can help to make proper decision for disaster resilience.

8.4.3.4 Gaps between the capacity of human and technology

There is always a gap between human and technological capacity. This gap creates a problem in handling technology. It requires a technically sound team for running the technological approaches properly. Big data-based disaster resilience requires a technical expert team to collect, analyse and interpret the data which can help to make accurate decision. These challenges can be resolved by hiring technically expert people and smart technology. This process requires a big investment. So, it is one of the big challenges for big data approach implementation for disaster resilience.

8.4.3.5 Lack of coordination among related organizations

Various organizations are working in different stages of disaster management. However, it is usually seen that there is a lack of coordination among the organizations working for disaster management in the same area. This lack of coordination creates barrier and reduces the efficiency of the work and effectiveness of the project implemented for disaster resilience. Similar cases may happen in case of big data-based disaster resilience. Since big data-based approaches require various kinds of support from different organizations, it needs coordination among all related stakeholders.

8.4.3.6 Challenges related to ethical and political aspect

In near future, there will be a big challenge related to the ethical and political aspects of the use of big data. These challenges are related to collection, storage, analysis, interpretation and privacy of data. So, it is necessary to ensure the ethical issues in all stages of the use of big data-driven disaster resilience.

8.5 Recommendations for enhancing resilience

In this section, some recommendations have been made based on the previous studies for reducing disaster vulnerability and enhancing resilience. These recommendations should be implemented by various stakeholders such as governments, NGOs, public–private partnerships and related voluntary organizations. These are as follows:

1. Technological intervention is necessary in every step of disaster management through enhancing systemization and standardization. The low-cost and

time-saving technology should be selected. The standardization can be improved by using big data. The sharing protocol should be transparent and understandable to others with advanced software installation. Specific and standard signs or hashtags should be used in social networking sites so any disaster-related individual can easily detect the information.

2. Disaster management techniques are always dependent on digital technologies which can help to make a better communication among various stakeholders. Most of the people use social media, so it is a great source of big data. The attitude of people can be easily explored by social media analysis. Disaster management team can get an idea about the impact and effects of disaster from social media and plan to distribute reliefs to the victimized community. It should be worth mentioning that it will not be wise to depend on only one media for getting accurate information.
3. Awareness regarding the information repository and how can we use effectively should be raised in the concerned organizations. Various technologies are using by various experts for same purpose. So, all related technologies should be assessed and find out suitable one for effective disaster management. It is known that much information regarding technologies is available in the newspaper and research reports but only a few of them get concentration from research community and policymakers. So, suitable context-specific technology should be selected for specific disaster and areas.
4. Public–private partnership should be developed for understanding the consequence of disaster to each other and also need to share experience to manage disaster. It will be helpful for assessing the feasibility of project related to disaster management and enhance resilience. Steps should be taken by the government for enhancing disaster resilience through inviting NGOs, academic professionals, experts, voluntary organizations, etc. Private organizations may be a great source of big data because of their nature of work with various local communities.
5. Research organizations should come forward to conduct research on the big data for enhancing disaster resilience. Various related research organizations can make a collaboration for field-based research and exchange their idea to each other. This sharing experience can help to develop a number of strategies effective for specific local community. Simultaneously, research organizations can use each other's data for effective use and time- and cost-saving purposes.
6. Context-specific customization is necessary for specific approach and technology for getting full benefits from approach and technology for disaster resilience. Since the extent and nature of natural disasters are not same in all the regions of the world, it requires context-specific strategy and technology to control and manage for reducing vulnerability and enhancing resilience. Appropriate approach should be found out for effective management and made compatible to various events in a specific area. Local adaptation strategy should be considered for designing and selecting any approach for enhancing resilience. In decision-making process, the representative of local people should be present for making a participatory decision process.

7. An awareness-building programme should be taken by the government for creating awareness among local people so that they can prepare themselves to save from the destructive event of natural disasters. Training can be an effective approach to train the local people on awareness among them about the consequences about natural disasters. Not only local people but also disaster management-related personnel, volunteers and local leaders can get an idea about disaster management through training. Training manual should be provided to the participants so that they can use it when necessary.
8. A strong legal support is necessary to handle the sensitive issues like big data use and disaster resilience. Privacy- and data protection-related legal issues should be informed to the concerned personnel so that they can always become cautious about privacy of individual and government data. Sometimes privacy law creates burden for disaster management personnel, so it should be amended in such a way that they can get access easily to big data sources of the relevant departments. Personal data should be kept carefully. It should be clearly explained in the information communication technology law that any violation of the privacy should be punishable. A strong protocol should be developed for data protection, privacy and effective use.

Disaster management agencies should need to select the appropriate technologies for effective management of all phases of disasters. They take help from big data technology experts through recruitment or outsourcing. The appointment of experts, timing, budget and other related issues should be clearly defined in the disaster management plan so that it can be implemented easily. For smooth operation, coordination is necessary by involving experts from related fields, which can help to develop adaptive, absorptive and transformative capacity of the community people as well as organizations.

8.6 Conclusion

Considering the importance of the potentiality of leveraging big data for enhancing resilience for disaster management, this study focused on the recent technology explored by various scholars on big data-based disaster resilience which can help people in all stages of disaster management. This study reveals that disaster resilience is a combined function of the adaptive, absorptive and transformative capacity of an individual or society to withstand and cope with the adverse effects of the disaster. This study explores the potential of big data in various stages of disaster management which can ensure the resilience capacity of a social unit. This chapter also highlights the major principles of big data for effective use for disaster management like open source tools, strong infrastructure, developing local skills, context-specific data sources, data sharing with ethics, awareness about the right of data and learning from experience. It also argues that big data is a potential tool for policymakers, administrators and related stakeholders to take necessary actions during and after disasters like early warning system, weather forecasting, emergency evacuation, immediate responses, relief distribution, training need assessment and increasing trained individuals. For getting maximum benefit from big data approach for disaster resilience, this study suggests to solve the related problems such as challenges in data collection, analytics, infrastructure, gaps between human and technological capacity, ethical and political anomaly, poor coordination, privacy and accuracy. This study recommends implementing proper infrastructure, technologies, tools and expertise for ensuring proper utilization of big data for disaster resilience. This study also focuses on further research on big data approaches for enhancing disaster resilience in context-specific cases by collecting primary data which can help to extend the use of it over the world.

References

[1] Sarker MNI, Wu M, Alam GM, and Shouse RC. Livelihood vulnerability of Riverine-Island dwellers in the face of natural disasters in Bangladesh. Sustainability [Internet]. 2019;11(6):1623. Available from: https://www.mdpi.com/2071-1050/11/6/1623

[2] Kumar SAP, Bao S, Singh V, and Hallstrom J. Flooding disaster resilience information framework for smart and connected communities. J Reliab Intell Environ [Internet]. 2019;5(1):3–15. Available from: https://doi.org/10.1007/s40860-019-00073-2

[3] Folke C. Resilience: The emergence of a perspective for social–ecological systems analyses. Glob Environ Chang. 2006;16(3):253–67. Available from: https://linkinghub.elsevier.com/retrieve/pii/S0959378006000379

[4] Adger WN. Social-ecological resilience to coastal disasters. Science (80-) [Internet]. 2005;309(5737):1036–39. Available from: http://www.ncbi.nlm.nih.gov/pubmed/16099974

[5] Agrawal N. Disaster resilience. In: Agrawal N, editor. Natural Disasters and Risk Management in Canada. Springer Publishers; 2018. pp. 147–91. Available from: http://link.springer.com/10.1007/978-94-024-1283-3_4

[6] Yang C, Su G, and Chen J. Using big data to enhance crisis response and disaster resilience for a smart city. In: 2017 IEEE 2nd International Conference on Big Data Analysis, ICBDA 2017. Institute of Electrical and Electronics Engineers Inc.; 2017. pp. 504–7. Available from: https://www.engineeringvillage.com/share/document.url?mid=cpx_bb6fc77160ff7e10bdM70cb10178163176&database=cpx

[7] Oussous A, Benjelloun F-ZZ, Ait Lahcen A, and Belfkih S. Big data technologies: A survey. J King Saud Univ – Comput Inf Sci. 2018;30(4):431–48. Available from: https://doi.org/10.1016/j.jksuci.2017.06.001

[8] Ragini JR, Anand PMR, and Bhaskar V. Big data analytics for disaster response and recovery through sentiment analysis. Int J Inf Manage. 2018;42(May):13–24. Available from: https://doi.org/10.1016/j.ijinfomgt.2018.05.004

[9] Anbarasan M, Muthu B, Sivaparthipan CB, *et al.* Detection of flood disaster system based on IoT, big data and convolutional deep neural network.

Comput Commun. 2020;150(November 2019):150–57. Available from: https://doi.org/10.1016/j.comcom.2019.11.022

[10] Alam GMM, Alam K, Mushtaq S, and Filho WL. How do climate change and associated hazards impact on the resilience of riparian rural communities in Bangladesh? Policy implications for livelihood development. Environ Sci Policy [Internet]. 2018;84(January 2017):7–18. Available from: https://doi.org/10.1016/j.envsci.2018.02.012

[11] Gupta A, Deokar A, Iyer L, Sharda R, and Schrader D. Big data & analytics for societal impact: Recent research and trends. Inf Syst Front. 2018;20(2):185–94.

[12] Yu M, Yang C, and Li Y. Big data in natural disaster management: A review. Geosciences. 2018;8(5):1–26.

[13] Carley KM, Malik M, Landwehr PM, Pfeffer J, and Kowalchuck M. Crowd sourcing disaster management: The complex nature of Twitter usage in Padang Indonesia. Saf Sci [Internet]. 2016;90(5):48–61. Available from: http://dx.doi.org/10.1016/j.marpol.2017.05.032

[14] Lv Z, Li X, and Choo KKR. E-government multimedia big data platform for disaster management. Multimed Tools Appl. 2018;77(8):10077–89.

[15] Moher D, Liberati A, Tetzlaff J, *et al.* Preferred reporting items for systematic reviews and meta-analyses: The PRISMA statement. PLoS Med. 2009;6(7):1–10.

[16] Vandenbroucke JP, Von Elm E, Altman DG, *et al.* Strengthening the Reporting of Observational Studies in Epidemiology (STROBE): Explanation and elaboration. PLoS Med. 2007;4(10):1628–54.

[17] Qadir J, Ali A, Rasool R, Zwitter A, Sathiaseelan A, and Crowcroft J. Crisis analytics: Big data-driven crisis response. J Int Humanit Action. 2016; 1(12):1–21. Available from: http://dx.doi.org/10.1186/s41018-016-0013-9

[18] Tomaszewski B, Judex M, Szarzynski J, Radestock C, and Wirkus L. Geographic information systems for disaster response: A review. J Homel Secur Emerg Manag. 2015;12(3):571–602.

[19] Park CH, and Johnston EW. An event-driven lens for bridging formal organizations and informal online participation: How policy informatics enables just-in-time responses to crises. In: Policy Analytics, Modelling, and Informatics. 2018. pp. 343–61. Available from: http://link.springer.com/10.1007/978-3-319-61762-6

[20] Adeel A, Gogate M, Farooq S, *et al.* A survey on the role of wireless sensor networks and IoT in disaster management. In: Durrani TS, Wang W, Forbes SM, editors. Geological Disaster Monitoring Based on Sensor Networks. Singapore: Springer Singapore; 2019. pp. 57–66. (Springer Natural Hazards). Available from: http://link.springer.com/10.1007/978-981-13-0992-2

[21] Ogie RI, Clarke RJ, Forehead H, and Perez P. Computers, environment and urban systems crowdsourced social media data for disaster management: Lessons from the PetaJakarta.org project. Comput Environ Urban Syst. 2019;73(March 2018):108–17. Available from: https://doi.org/10.1016/j.compenvurbsys.2018.09.002

[22] Ha K. Integrating the resources of Korean disaster management research via the Johari window. Eval Program Plann. 2019;77(June):101724. Available from: https://doi.org/10.1016/j.evalprogplan.2019.101724
[23] Shan S, Zhao F, Wei Y, and Liu M. Disaster management 2.0: A real-time disaster damage assessment model based on mobile social media data – A case study of Weibo (Chinese Twitter). Saf Sci. 2019;115(8):393–413. Available from: https://doi.org/10.1016/j.ssci.2019.02.029
[24] Hyslop M. Comments on standards in information security, disaster recovery, business continuity and business resilience. In: Critical Information Infrastructures. Boston, MA: Springer US; 2007. pp. 94–144. Available from: http://link.springer.com/10.1007/978-0-387-71862-0_7
[25] Lu Z, Cao G, and La Porta T. TeamPhone: Networking SmartPhones for disaster recovery. IEEE Trans Mob Comput. 2017;16(12):3554–67. Available from: http://creativecommons.org/licenses/by/3.0/igo/
[26] Ogie RII, Clarke RJJ, Forehead H, and Perez P. Crowdsourced social media data for disaster management: Lessons from the PetaJakarta.org project. Comput Environ Urban Syst [Internet]. 2019;73(9):108–17. Available from: https://doi.org/10.1016/j.compenvurbsys.2018.09.002
[27] Clark N, and Guiffault F. Seeing through the clouds: Processes and challenges for sharing geospatial data for disaster management in Haiti. Int J Disaster Risk Reduct [Internet]. 2018;28(2):258–70. Available from: https://linkinghub.elsevier.com/retrieve/pii/S2212420918301997
[28] Schempp T, Zhang H, Schmidt A, Hong M, and Akerkar R. A framework to integrate social media and authoritative data for disaster relief detection and distribution optimization. Int J Disaster Risk Reduct. 2019;(4):101143. Available from: https://doi.org/10.1016/j.ijdrr.2019.101143
[29] Enenkel M, Saenz SM, Dookie DS, Braman L, Obradovich N, and Kryvasheyeu Y. Social media data analysis and feedback for advanced disaster risk management. In: Social Web in Emergency and Disaster Management 2018. 2018. pp. 1–5. Available from: http://arxiv.org/abs/1802.02631
[30] Resnyansky L. Social media data in the disaster context. Prometh (United Kingdom). 2015;33(2):187–212.
[31] Cinnamon J, Jones SK, and Adger WN. Geoforum evidence and future potential of mobile phone data for disease disaster management. Geoforum. 2016;75: 253–64. Available from: http://dx.doi.org/10.1016/j.geoforum.2016.07.019
[32] Ragini JR, Anand PMR, and Bhaskar V. Big data analytics for disaster response and recovery through sentiment analysis. Int J Inf Manage. 2018; 42(9):13–24. Available from: https://doi.org/10.1016/j.ijinfomgt.2018.05.004
[33] Monirul Alam GM, Alam K, Mushtaq S, and Clarke ML. Vulnerability to climatic change in riparian char and river-bank households in Bangladesh: Implication for policy, livelihoods and social development. Ecol Indic. 2017;72:23–32.
[34] Walker B, and Salt D. Practicing resilience in different ways. In: Walker B, Salt D, editors. Resilience Practice. Washington, DC: Island Press/Center for

Resource Economics; 2012. pp. 145–67. Available from: http://www.springerlink.com/index/10.5822/978-1-61091-231-0_9

[35] Holling CS. Resilience and stability of ecological systems. Annu Rev Ecol Syst. 1973;4(1):1–23. Available from: http://www.annualreviews.org/doi/10.1146/annurev.es.04.110173.000245

[36] FAO. Analysing resilience for better targeting and action. Food and Agriculture Organization of the United Nations, Rome, Italy; 2016. pp. 1–80. Available from: http://www.fao.org/3/a-i5665e.pdf%0Ahttp://www.preventionweb.net/publications/view/48984

[37] Gwimbi P. Linking rural community livelihoods to resilience building in flood risk reduction in Zimbabwe. Jàmbá J Disaster Risk Stud. 2009; 2(1):71–89. Available from: http://www.ndmc.gov.za/LinkClick.aspx?fileticket=n_5rPYFPsgc%3D&tabid=267

[38] Musaev A, Wang D, and Pu C. LITMUS: A multi-service composition system for landslide detection. IEEE Trans Serv Comput. 2015;8(5):715–26. Available from: http://ieeexplore.ieee.org/document/6971222/

[39] Goldenberg SB, Gopalakrishnan SG, Tallapragada V, *et al.* The 2012 triply nested, high-resolution operational version of the Hurricane Weather Research and Forecasting Model (HWRF): Track and intensity forecast verifications. Weather Forecast. 2015;30(3):710–29. Available from: https://www.iisd.org/library/big-data-resilience-storybook

[40] Masood T, So E, and McFarlane D. Disaster Management Operations – Big Data Analytics to Resilient Supply Networks. Vol. 28. 2015. Available from: http://euroma2017.eiasm.org/userfiles/HKJGLML_GDFJMK_AW7YG3VT.pdf

[41] Di Felice M, Trotta A, Bedogni L, Chowdhury KR, and Bononi L. Self-organizing aerial mesh networks for emergency communication. In: 2014 IEEE 25th Annual International Symposium on Personal, Indoor, and Mobile Radio Communication (PIMRC) [Internet]. IEEE; 2014. pp. 1631–36. Available from: http://flash.org/intersectionofresilience.pdf

[42] Contreras D, Forino G, and Blaschke T. Measuring the progress of a recovery process after an earthquake: The case of L'aquila, Italy. Int J Disaster Risk Reduct. 2018;28:450–64. Available from: https://linkinghub.elsevier.com/retrieve/pii/S2212420917302868

Appendix A

Sl no.	Author/researcher	Title of research	Sources/journal
1	Holling, C. S. (1973)	Resilience and stability of ecological systems	Annual Review of Ecology and Systematics
2	Adger, W. N. (2005)	Social-ecological resilience to coastal disasters	Science
3	Folke, C. (2006)	Resilience: The emergence of a perspective for social–ecological systems analyses	Global Environmental Change,
4	Agrawal, N. (2018)	Disaster resilience	Natural Disasters and Risk Management in Canada
5	Alam *et al.* (2017)	Vulnerability to climatic change in riparian char and river-bank households in Bangladesh: Implication for policy, livelihoods and social development	Ecological Indicators
6	Alam *et al.* (2018)	How do climate change and associated hazards impact on the resilience of riparian rural communities in Bangladesh? Policy implications for livelihood development	Environmental Science & Policy
7	Carley *et al.* (2016)	Crowd sourcing disaster management: The complex nature of Twitter usage in Padang Indonesia	Safety Science
8	Contreras *et al.* (2018)	Measuring the progress of a recovery process after an earthquake: The case of L'Aquila, Italy	International Journal of Disaster Risk Reduction
9	FAO (2016)	*Analysing Resilience for Better Targeting and Action*	Food and Agriculture Organization
10	Gupta *et al.* (2018)	Big data & analytics for societal impact: Recent research and trends	Information Systems Frontiers
11	Gwimbi (2009)	Linking rural community livelihoods to resilience building in flood risk reduction in Zimbabwe	Jàmbá: Journal of Disaster Risk Studies
12	Kumar *et al.* (2019)	Flooding disaster resilience information framework for smart and connected communities	Journal of Reliable Intelligent Environments

(Continues)

(*Continued*)

Sl no.	Author/researcher	Title of research	Sources/journal
13	Lu *et al.* (2017)	TeamPhone: Networking SmartPhones for disaster recovery	IEEE Transactions on Mobile Computing
14	Lv *et al.* (2018)	E-government multimedia big data platform for disaster management.	Multimedia Tools and Applications
15	Murakami *et al.* (2015)	Simulation and prediction of Category 4 and 5 hurricanes in the high-resolution GFDL HiFLOR coupled climate model	Journal of Climate
16	Musaev *et al.* (2015)	LITMUS: A multi-service composition system for landslide detection	IEEE Transactions on Services Computing
17	Ragini *et al.* (2018)	Big data analytics for disaster response and recovery through sentiment analysis	International Journal of Information Management
18	Walker *et al.* (2012)	Practicing resilience in different ways	Resilience Practice
19	Yang *et al.* (2017)	Using big data to enhance crisis response and disaster resilience for a smart city	IEEE 2nd International Conference on Big Data Analysis, ICBDA 2017
20	Yu *et al.* (2018)	Big data in natural disaster management: A review	Geosciences
21	Fallis (2015)	Big data for climate change and disaster resilience: Realising the benefits for developing countries	Journal of Chemical Information and Modeling

Chapter 9

Design and development of a rapid damage assessment for Albay elementary schools in the Philippines

Maria Charmy A. Arispe[1]

Albay has zero casualty goals on every disaster that comes its way. Along with this, it is important that elementary schools, especially those schools that served as evacuation centers, receive and act upon the timely delivery of a warning system and make their school buildings as safe as possible. The main objective of the study was to design and develop the rapid damage assessment system for public elementary school buildings of Albay, Philippines. Specifically, it achieved the following: (1) identify the current procedures in conducting rapid damage assessment in public elementary school buildings of Albay; (2) determine the design features of the proposed rapid damage assessment system; and (3) generate appropriate report formats to represent the rapid damage assessment information. Rational Unified Process methodology phases and disciplines were used to design and develop the proposed system. Data were gathered through key informant interview and document analysis that covers Albay Province with four school divisions and 602 public elementary schools. The study is an efficient rapid damage assessment tool capable of processing and extracting information from the database and locates the public elementary school buildings of Albay through geographical information system support. It gathers information on school building damages, saves and views the data in real time, generates a report from the data collected, and provides quick and accurate information to the Department of Education officials to enable them to determine the type and extent of damages to the school building. Findings show that this new technique of assessing school damages is fast and efficient. The main advantage is the quickness and automation of the procedure. An evaluation study was also conducted to assess the effectiveness and efficiency of the proposed damage assessment system and found that the system was effective, reliable, useful, and efficient to the end users.

[1]Bicol University Polangui Campus, Computer Studies Department, Polangui, Albay, Philippines

9.1 Background of the study

A natural disaster occurs when an extreme meteorological, hydrological, or geological event exceeds the ability of a community to withstand the event [1]. Disasters are emergencies that cannot be handled by those affected without outside assistance. They are caused by natural or man-made events wherein communities experience severe danger and incur loss of lives and properties disrupting their social structure and all or some of the affected communities' essential functions. Disasters are inevitable. They are caused by unsustainable development that has not taken into account of possible hazard impacts in that location. They can be less damaging if the population has a better understanding of locally experienced hazards and implements preventive or mitigating measures against them [2].

9.1.1 Hyogo Framework for Action for 2005–15

The Hyogo Framework for Action for 2005–15 [3] emphasized that the disasters can be substantially reduced if people are well informed and motivated toward a culture of disaster prevention and resilience, which in turn requires the collection, compilation, and dissemination of relevant knowledge and information on hazards, vulnerabilities, and capacities.

In their approach to disaster risk reduction, states, regional, and international organizations and other actors concerned should consider the key activities listed under each of these five priorities and should implement them, as appropriate, to their circumstances and capacities.

Ensure that disaster risk reduction is a national and a local priority with a strong institutional basis for implementation. Countries that develop policy, legislative, and institutional frameworks for disaster risk reduction and that can develop and track progress through specific and measurable indicators have a greater capacity to manage risks and to achieve widespread consensus for engagement in and compliance with disaster risk reduction measures across all sectors of society.

Identify, assess, and monitor disaster risks and enhance early warning. The starting point for reducing disaster risk and for promoting a culture of disaster resilience lies in the knowledge of the hazards and the physical, social, economic, and environmental vulnerabilities to disasters that most societies face and of how hazards and vulnerabilities are changing in the short and long term, followed by action taken on the basis of that knowledge.

Use knowledge, innovation, and education to build a culture of safety and resilience at all levels. Disasters can be substantially reduced if people are well informed and motivated toward a culture of disaster prevention and resilience, which in turn requires the collection, compilation, and dissemination of relevant knowledge and information on hazards, vulnerabilities, and capacities.

Reduce the underlying risk factors. Disaster risks related to changing social, economic, environmental conditions, and land use, and the impact of hazards associated with geological events, weather, water, climate variability, and climate change, are addressed in sector development planning and programs as well as in post-disaster situations.

Strengthen disaster preparedness for effective response at all levels. At times of disaster, impacts and losses can be substantially reduced if authorities, individuals, and communities in hazard-prone areas are well prepared and ready to act and are equipped with the knowledge and capacities for effective disaster management.

All states are encouraged to implement this Framework for Action for the period 2005–15 and should endeavor to undertake the following tasks at the national and local levels, with a strong sense of ownership and in collaboration with civil society and other stakeholders, within the bounds of their financial, human, and material capacities, and taking into account their domestic legal requirements and existing international instruments related to disaster risk reduction. States should also contribute actively in the context of regional and international cooperation.

9.1.2 Damage assessment

There are pre- and post-disaster actions designed to keep the losses at the minimum in human, structural, and economic aspects. Pre-disaster loss management is activities focusing on reducing the community vulnerability to hazards. Actions include improving the resistance of physical structures such as school buildings, developing improved safety plans for the occupants, and increasing/diversifying the network of social support mechanisms available to communities in threatened areas. Post-disaster loss management focuses on improving the emergency response and broadening the range of support given to victims that include the facilitation of relief delivery and stimulating a rapid recovery [2].

In the assessment of activities, it provides to decision-makers the information they need to set the objectives and policies for emergency assistance. The assessments provide information on the progress of recovery; highlighting areas that require further analysis and intervention; additional general information about the disaster; the damage; urgent needs and priorities; and actual response measures being taken [4]. Disaster management is not necessarily a full-time activity but it is intended to be useful even for those who expect to be active only during some aspects of disaster-related operations [5].

The immediate effects of the disaster are typically classified as damages. Damages refer to the total or partial destruction of physical assets. It typically occurs during or immediately after a disaster and is measured in physical units and valued at replacement cost [6]. Conducting a rapid damage assessment is one of the most important pieces of information that can be gathered during or immediately after a disaster has occurred. Emergency response officials can utilize personnel and equipment more effectively, and provide a faster response to the areas that have the greatest need if a preexisting rapid damage assessment procedure is provided and performed [7]. A rapid damage assessment is an immediate estimate of the damage that has occurred in a defined area or a specific site. It is used to quickly comprehend the magnitude of the disaster situation [8].

Assessing an affected area immediately after a severe natural hazard event and saving the resulting data are vitally important in any effort to reduce future economic losses from natural hazards. These data are used as a record of buildings performance and as a major component for statistical analysis and damage modeling studies. To make building damage assessment more effective and more accurate, a systematic process to standardize assessment data is needed. Additionally, to ensure that data are correctly assessed and collected, a standard protocol implemented in damage assessment activities is vitally needed [9].

9.1.3 Geographic information system

Recent developments in information technology have been proved critical in meeting emergency management requirements. Globally used technologies involved geographic information system (GIS), remote sensing, decision support system, warning systems, and hazards modeling systems. Rising of the technologies has its significance in emergency management, and studies showed that GIS is the second most effective technology after the internet in emergency management functions [10].

Relatively, GIS web services provide applications such as basic geoprocessing tasks, address matching, map image display, and routing, without maintaining GIS tools or the associated geographical data [11]. The GIS provides integration to data mining by adding the spatial dimension. The spatial data is available in latitude and longitude form. It is in large volumes and various formats. The interoperability of data is a big issue; the web service provides a convenient way to analyze and integrate data, so there is a need for spatial data service to integrate the geospatial data rising from diverse locations. Web services technology has emerged as a cornerstone for interoperability solutions by providing the standard for communication between hybrid systems and thus combining web service technology with spatial data infrastructure and GIS technologies [12].

In the recovery phase after the initial relief has been provided and the goal is returning life to normal or improved circumstances, the GIS can serve as a spatial inventory system for coordinating recovery activities. Some of the challenges during recovery include assessing the damage, educating the public, rebuilding, and preventing reoccurrence [13].

Sheehan [14] in his study improved disaster assessment with the use of ArcGIS Online. It copes with increasing demand and decreasing resources for collecting damage assessments from a paper-based to a tablet-based system. ArcGIS Online allows for a more holistic approach to GIS, marrying web, mobile, and desktop technologies, which have allowed the company to extend its GIS client solutions. If Wi-Fi or cellular access has been destroyed by the disaster, offline capabilities, data visualization, and collection are still available in disconnected mode. Once back in the Wi-Fi range, collected data can be uploaded to a map layer hosted centrally. For attachments, the tablet's camera was integrated into the app. Users can, after taking a picture of a parcel, attach the image to the feature or take a new photo.

Additionally, the OAK Ridge National Laboratory [15] developed the RunMobile Damage Assessment Reporting Solution (DARS) as an emergency management application that enables remote field agents to collect, organize, and submit damage assessment reports gathered from natural and man-made disasters. The information is submitted to back-end servers where Command Center users can perform analytics on the data and use it to kick off workflows that can initiate targeted response efforts and issue escalation. RunMobile DARS expedites the accurate collection and aggregation of damage assessments, which facilitates reporting and escalation to state and federal agencies.

9.1.4 The Philippine Disaster Management System [2]

The Philippine Islands are prone to all kinds of natural hazards because of their geographical location and physical environment. The country is strategically located in the path of turbulent and destructive cyclones in the Pacific, and the "Ring of Fire." This situation has adverse effects, not only on the lives and properties of the Filipino people but also on the economy of the nation, as hazard impacts may result in widespread environmental and property damages. Natural hazards may cause danger to people, structures, or economic assets and may lead to a disaster if they are not mitigated against and prepared for.

To be able to cope with the worsening effects of hazards impacting the country, the government developed a Natural Disaster and Calamities Plan in 1969. The guiding principle for the development of the Natural Disaster and Calamities Plan was to use all available government resources and encourage all concerned agencies to work together in addressing the issue of disasters and calamities. The plan assigned specific tasks or emergency functions to member agencies including the Department of Education, which is one of the members, in addition to their primary day-to-day tasks. From 1946 to 1970, the Philippine Disaster Management System was reactive in nature wherein disaster responses were limited to emergencies or situations after the disasters have already affected part of the country. Management efforts were highly centralized with minimum participation from the local government officials. The organization of Civil Defense Units at the local level was mostly on paper and people were content to wait for assistance from the national government. The Office of Civil Defense set up field stations in the 12 Administrative Regional Centers outside of Metro Manila. Field personnel started to convert the local civil defense units into local disaster coordinating councils and retained the leaders and members of these councils.

From 1969 to 1973, government agencies with training funds started conducting disaster preparedness programs to prepare the populace to respond to any emergency, including the Department of Education with primary concern for the welfare of schoolchildren and school personnel. The Office of Civil Defense started organizing and training the chairpersons and members of the councils at different levels. The Department of Social Welfare and Development (DSWD) trained community leaders on relief distribution, the Philippine National Red Cross (PNRC) trained community leaders on relief distribution and their volunteers on

First Aid, while the Department of Health (DOH) focused their training on barangay health workers. In 1976, the Ministry of Education and Culture (MEC) issued a Memorandum No. 303, s. 1976 on Survival through Disaster Preparedness.

In 1978, the National Disaster Coordinating Council (NDCC) was formally created to further strengthen the system and formalize the ad hoc organizations at the national, regional, and local levels and to allocate emergency tasks to the different governmental units pursuant to Presidential Decree 1566 of 1978. The Decree was issued given the need to revitalize the system to enhance the survival capability and economic stability of the country, supported by an LOI 453 regarding all types of disasters. Finally, a cogent system for pre-disaster planning, community disaster preparedness and positive, precise disaster control action for rescue evacuation, relief, and rehabilitation to ensure the survival of every Filipino was developed.

In May 1994, a World Conference on Natural Disaster Reduction was held by the United Nations to reduce the loss of life, property damage, social and economic disruptions caused by natural disasters through a concerted effort in the international and local fronts. It produced the Yokohama Strategy and Plan of Action for a Safer World, called Guidelines for Natural Disaster Prevention, Preparedness and Mitigation. In the Philippines, the NDCC and its member agencies were identified as those responsible for addressing the concerns brought out at the conference. In support of the Plan objectives, it created four committees: on structural measures, nonstructural measures, disaster research, and disaster legislation.

Even with the emergence of many ideas, both local and international, relative to disaster risk management, the Philippine institutional arrangements and disaster management systems tend to rely on a response or reactive approach, in contrast to a more effective proactive approach. There is a widespread emphasis on post-disaster relief and short-term preparedness rather than mitigation or post-disaster support for sustainable economic recovery. This much shorter-term focus does not adequately emphasize natural hazards as a potential obstacle to long-term sustainable development. Thus, to evolve to a more proactive role, the NDCC in cooperation with all member agencies prepared and implemented a Comprehensive Disaster Management Framework for comprehensive disaster risk management. The framework provides for political leadership and policy support at the highest levels while facilitating the active engagement and implementation of all relevant stakeholders at the national, local, and household levels. The basic components of the framework are mitigation and preparedness for pre-event and response and rehabilitation for the post-event. The framework includes the essential steps of integrated risk management, which include risk identification, risk reduction, and risk sharing/financing in the total risk reduction management system. Current approaches adapted the risk management as a disaster planning framework that encompasses disaster mitigation, preparedness, response, and rehabilitation.

9.1.5 Disaster Risk Reduction Management Office [2]

The Disaster Risk Reduction Management Office (DRRMO) shall be headed by the Secretary of the Department of Education (DepEd) as Chairman and the

Undersecretary for Regional Operations as the Vice-Chairman. The DRRMO, being empowered with policy-making, coordination, integration, supervision, and evaluation functions, shall have the following responsibilities:

(a) Develop a DepEd Disaster Risk Management Framework (DRMF), which shall provide for comprehensive, all-hazards, multi-sectoral, interagency, and school-based approach to disaster risk management. The framework shall serve as the principal guide to disaster risk management efforts in the DepEd System and shall be reviewed on a 5-year interval, or as may be deemed necessary to make it relevant and flexible to current changes.
(b) Develop, formulate, and lead the implementation of the DepEd Disaster Risk Management Plan (DRMP) and ensure that this shall be considered and integrated into the Medium-Term Philippine Development Plan (MTPDP), in the DepEd Budget, and in the Regional, Division, and School Disaster Risk Management Plans nationwide.
(c) Ensure that all regions, divisions, and schools take adequate and appropriate measures on disaster management, including the preparation and upgrading of the School Mapping Exercise through the GIS-Based School Profiling System to facilitate disaster preparedness and mitigation.
(d) Ensure that disaster preparedness, mitigation, response, and rehabilitation activities such as training, education and public information, repair, and rehabilitation are given priority.
(e) Advice the President on the declaration of suspension of classes in areas extensively damaged by disasters and submits remedial proposals.
(f) Develop a Disaster Risk Management Information System and GIS-Based School Profiling System for storage, easy retrieval of relevant data for policy formulation, resource mobilization, planning, and decision-making tools.
(g) Establish a DepEd Disaster Operations Center that shall be managed by DepEd regular employees on a twenty-four hour/seven days (24/7) basis as the need arises.
(h) Coordinate and meet with concerned government agencies and other stakeholders in education as often as necessary to effectively manage national/regional/division and school efforts on disaster risk management.
(i) Conduct monitoring and evaluation to ensure the system's efficiency and provide interventions and corrective measures for the effective implementation of DRRM programs and projects.

There shall be a DRRMO at the regional level that shall establish an operating facility or a Disaster Management Center in their areas of jurisdictions. The DRRMO at the region, division, district, and school levels shall also be organized, which shall have the following functions:

(a) Identify, assess, and manage the hazards and risks that may occur in their locality;
(b) Communicate about those hazards and risks, their nature, effects, early warning signs, and countermeasures;

(c) Identify and implement cost-effective risk reduction measures or strategies;
(d) Take all necessary steps on an ongoing basis to maintain, provide or arrange the provision of a trained and competent personnel for effective and efficient disaster risk management in their areas of jurisdiction;
(e) Respond to and manage the adverse effects of emergencies in their area of jurisdictions;
(f) Carry out recovery activities;
(g) Promote and raise public awareness of and compliance with policies and directives issued by the Chairman of DRRMO relative to disaster risk management;
(h) Develop, approve, implement, and monitor School DRMP and regularly review and test the plan consistent with other national and local planning programs;
(i) Establish linkage and network with other local government units for disaster risk reduction and emergency response purposes;
(j) Formulate, prepare, and issue orders, memoranda, and issuances consistent with the requirements of the NDCC;
(k) Integrate risk reduction into school development plans, programs, and budgets as a strategy in schools' sustainable development and improvement plans;
(l) Establish an operating facility to be known as the Region/Division/District and School Disaster Operations Center;
(m) Prepare and submit to the Regional DRRMO, National DRRMO, and the NDCC damage and needs assessment reports; and
(n) Include as part of the School Improvement Plan disaster risk reduction measures to ensure safety, and security of all teaching, nonteaching, and schoolchildren.

For DepEd Damage Analysis Needs Analysis Team, it has the following functions:

(a) Evaluates crises and determines courses of actions to be followed and formulates guidelines in assessing the situation;
(b) Assesses information and advises the Chairman of DRRMO on possible measures to be undertaken to lessen the impact of the crisis;
(c) Submits recommendation for allocation of needed resource;
(d) Coordinates the plans and actions of the group with the proper authorities whenever a crisis occurs;
(e) Monitors the probable consequences of potential, ongoing and past disasters or emergency around the country in close coordination with internal and external stakeholders in education;
(f) Coordinates predefined pre- and post-disaster operational activities being undertaken by relevant agencies and ensures that all key actors are taken on board;
(g) Initiates and leads the conduct of multiagency damage and needs assessment mission as the post-disaster situation warrants;

(h) Facilitates the conduct of a multiagency debriefing of a past disaster situation to look into areas of strength and areas for improvement;
(i) Allocates working stations for pre-identified cluster members who will be working at the Disaster Operations Center during emergencies;
(j) Promotes synergistic multiagency approach in managing the potential consequences of disasters in the country and does related work.

9.1.6 School Evacuation Center

Evacuation is the process of moving or transferring people from hazard threatened or stricken areas to a safe place or places. There are two types of evacuation: (1) precautionary, that is, done before impact; and (2) after impact, which involves the movement of victims from hazard-stricken areas into safer surrounding conditions. It is a temporary shelter where survivors in emergencies can take refuge. It is a venue where evacuees can be helped in coordination with concerned government and nongovernment agencies in their basic needs. DepEd officials may allow the use of the school as an evacuation center when there is no other safer place where the students/pupils can be accommodated. People from the community may be allowed to use the school as a temporary holding center. In case the school is used as an evacuation center, DepEd officials and personnel shall be responsible for the preservation of the school facilities and properties for the duration the school is being occupied by evacuees, so that after the emergency period the school can resume operation as soon as possible [2].

For some of the tasks faced by the education sector during and after a disaster, United Nations Children's Fund (UNICEF) and DepEd stated that teachers and school administrative staff play an important role in managing the influx of evacuees in schools during the emergency phase. They provide unwavering assistance and assume various important functions during the relief phase. After the lapse of the critical period and things are getting back to normal, they still have to contend with the rehabilitation of damaged facilities and account for students affected by the disaster [16].

According to Former DepEd Assistant Secretary Laguda, the rapid damage assessment is a quick accounting from all the school divisions affected by the typhoon on how many classrooms have been completely or partially damaged and how many can be repaired at the school level. The use of the schools as evacuation centers shall also be factored into the assessment [17].

9.1.7 Elementary schools of Albay Province

It is important to point out that, in the Philippines, it is very common for schools to serve as evacuation centers [18]. School safety is the responsibility of the entire school community [19]. School administrators, staff, teachers, and students should be prepared in case of emergencies and disasters to protect themselves from personal injury and loss of life and protect the school property from damage [20].

In July 2014, typhoon Glenda hit the Philippines. The schools having sustained major damages were from Calabarzon, Mimaropa, and the Bicol Region. The cost of damage to school buildings amounted to P28.08 million, with 115 of the 162 school divisions. Of the 25,856 schools in affected regions, 187 were used as evacuation centers [21].

The Bicol Region, particularly the Albay Province, is one of the most disaster-prone areas in the Philippines. The region commonly experiences typhoons, floods, and volcanic eruptions. It was hit by super typhoons Milenyo and Reming in 2006, which killed thousands of lives and tremendously damaged properties. In Albay alone, 702 out of 704 schools were damaged when roofs were ripped off and instructional materials were washed away affecting 357,400 enrolled children. Similarly, about 21,500 preschoolers were also affected in 595 learning centers. The damage to education infrastructure was estimated to have reached $66 million [22].

In Albay Province, rapid damage assessments for public schools were conducted through the use of DepEd Rapid Assessment of Damages Report (RADaR) form. For each school incurred any damages and used as evacuation centers, assigned focal persons for Disaster Risk Reduction Management (DRRM) were required to conduct the rapid assessment and submit the accomplished RADaR form. The DepEd also requires the concerned school to send via SMS/text message to the cellphone number of DepEd Regional DRRM focal person the details of the conducted rapid damage assessment.

Indeed, the design and development of rapid damage assessment system for elementary school of Albay, Philippines is proposed to enhance the current procedures in conducting rapid damage assessment to design and develop a standardized rapid damage assessment system that shall generate appropriate report formats to represent the rapid damage assessment and establish strong links with better integration between various schools and Albay's Disaster Risk and Reduction Management Council (DRRMC). Albay has zero casualty goal on every disaster that comes its way, but it is important that individual schools, especially those schools that serve as evacuation centers, receive and act upon the timely delivery of a warning system, and make their school buildings as safe as possible.

In the Provincial Development and Physical Framework Plan for 2011–16 [23], the Province of Albay is described to be located in the eastern seaboard of the country and subjected to the pressures and consequent effects of the Pacific Jinx. It is referred to as such because of its geographic location, that of being situated along the Pacific Ring of Fire making it vulnerable to earthquake, tsunami, and volcanic hazards and along the Western Pacific Basin that is a generator of climatic conditions such as typhoons, monsoon rains, and thunderstorms, among others. These cause the province to experience a more pronounced distribution of precipitation and no pronounced dry season all-year-round. Because of its geographic location, volcanism, physiographic and hydrogeologic nature, the province becomes vulnerable to disasters and the effects of climate change as well.

Natural hazards in the province are generally classified into two categories: (1) the geologic hazards, and (2) the hydrometeorological hazards. Geologic hazards include the earthquake-induced hazards and the volcanic hazards while hydrometeorological include the typhoons/tropical cycles, storm surge, floods, and flash floods; tsunami and landslides. The province lies at the southern tip of Luzon Island and about 550 km from Manila. It is approximately 13–13.5 degrees north latitude and 123.25–124.25 degrees east longitude. It is bounded by Lagonoy Gulf and the province of Camarines Sur in the northwest, the Pacific Ocean in the east, the Province of Sorsogon in the south, and the Burias Pass in the southeast. It has a total land area of 2,552.6 square kilometers politically subdivided into 3 cities and 15 municipalities [24].

Moreover, to provide a strong institutional basis for DRRM, the DepEd as a member of DRRM councils, adopted the Philippine Disaster Risk Reduction and Management Act of 2010 or RA 10121 to transform the Philippine disaster management system from disaster relief and response toward risk reduction. DRRM has an end-goal: "Safer, adaptive and disaster-resilient Filipino communities towards sustainable development" and institutionalize with four phases: (1) prevention and mitigation, (2) preparedness, (3) response, and (4) recovery and rehabilitation. Rapid damage assessment for public schools was initiated by the DepEd through the creation of DRRMO. It is properly implemented by public schools through their DRRM focal persons.

One of their tasks is to assess school damages and identify affected areas. DepEd also institutionalizes the four DRRM phases: (1) prevention and mitigation, (2) preparedness, (3) response, and (4) recovery and rehabilitation. They adopted a Comprehensive School Safety (CSS) Framework to standardize their DRRM efforts across levels: national, region, division, and schools. The CSS Framework provides specific action points for education policymakers, managers, and administrators in building school resilience to disasters and it has three core areas: (pillar 1) safe school facilities; (pillar 2) school disaster management; and (pillar 3) risk reduction and resilience education. The Damage Assessments of Schools are under the recovery and rehabilitation phase of DRRM and pillar 2—school disaster management of CSS Framework. Immediately after a disaster occurs, public elementary school buildings may have damages. Thus, the assessment of the safety of the buildings needs to be addressed. Safety assessment involves damage assessment to ensure the safety of the public, noting that, once a given public elementary school building is deemed safe, the evacuees shall utilize the building. Figure 9.1 shows the picture of school damages.

9.1.8 Importance of the study

The proposed design and development of rapid damage assessment system for elementary school of Albay shall establish strong links between various schools and Albay's DRRMC and provide quick and accurate information on the extent and cost of damages due to disaster. Specifically, the proposed project study shall be beneficial to the following:

Public Elementary School Administrators—This study shall help increase their awareness of rapid assessment of damages, improve their ability to manage operations during the emergency response phase of a disaster, and provide an effective aid on damage assessment.

Figure 9.1 School damages

Albay Public Safety and Emergency Management Office (APSEMO)—This study shall support the office's disaster preparedness program and strengthens the communication link between the schools and Albay DRRMC for immediate disaster mitigation.

Local Government Unit (LGU)—This study shall help LGU officials to determine when it is safe for students and staff to return to the school buildings after a disaster. The results of this study may be used to recover costs from the responsible party or as a basis for requesting financial assistance for certain allowable response and recovery costs from the LGU.

People of Albay—This study shall support the activities on public awareness to enhance a culture of prevention and community engagement. People would know better the risk of public-school building and which areas are highly vulnerable.

Future Researchers—The study shall serve as an additional reference in the design and development of similar studies.

9.1.9 Objectives of the study

The main objective of the study was to design and develop a rapid damage assessment system for the elementary school of Albay. Specifically, it achieved the following:

1. Identify the current procedures in conducting rapid damage assessment in public elementary school buildings of Albay;
2. Determine the design features of the proposed rapid damage assessment system; and
3. Generate appropriate report formats to represent the rapid damage assessment information.

9.2 Methodology

The researcher used the Rational Unified Process (RUP) Methodology as a guide in the design and development of the proposed rapid damage assessment system for the

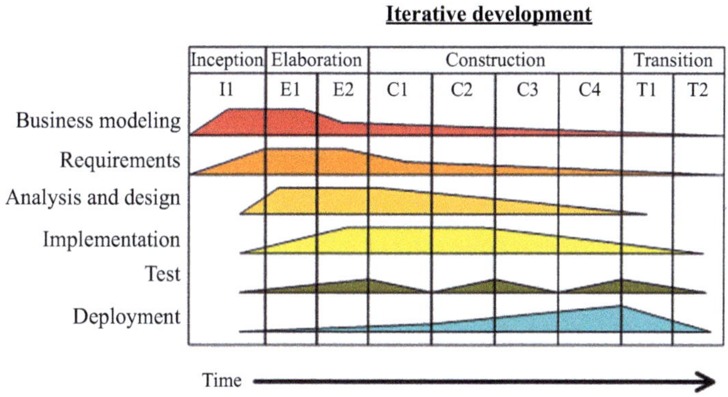

Figure 9.2 RUP methodology phases and disciplines over iterations

elementary school of Albay. Among the methodologies, RUP methodology is deemed appropriate on the proposed system because it includes gathering, documenting, maintaining of requirements, and incorporating changes systematically. RUP is a system development process for developing systems with objects and/or component-based technologies, see Figure 9.2. It consists of four phases and nine disciplines. The four phases are executed sequentially while the disciplines are performed iteratively throughout the four phases (inception, elaboration, construction, and transition). RUP is based on a set of building blocks and content elements, describing what is to be produced, the necessary skills required and the step-by-step explanation describing how specific development goals are to be achieved [25].

Inception (project objectives milestone). This is the explanatory phase of the project, iterate through business modeling and requirements disciplines. The project objective and description are described at this stage. The purpose of this phase is to collect and understand business requirements, detail the project plan, and agree upon a high-level statement of work. This phase identifies the project's primary objectives, assumptions, constraints, deliverables, and acceptance criteria.

In this phase, the researcher researched the current procedures in conducting rapid damage assessment in elementary school of Albay. The researcher gathered the needed rapid damage assessment rules, requirements, needs, and understanding and identified the roles and responsibilities of the DepEd and APSEMO in implementing DRRM. The researcher also identified and evaluated potential strategies for reengineering the existing process, including the ability of the researcher to design and develop a new system. The scope of the project was defined. The specification documents were prepared to identify the functional and nonfunctional requirements of the project.

Elaboration (product architectural milestone). The architecture of the system is designed, iterate through analysis and design and implementation disciplines. The goal is to translate requirements and specifications into a technical solution to produce technical design. The primary objective is to mitigate the key risk items identified by analysis up to the end of this phase.

In this phase, the researcher gathered necessary documents based on the scope of the study, determined the scope of the system, and identified the needs of the stakeholders in terms of rapid damage assessment. These requirements were analyzed, findings of the study were discussed, and the architecture design of the system was made, as well as the designing of the database, components, program source code, and modules of the system.

Construction (operational capability milestone). The complete development of the system based upon the baseline architecture takes place. Emphasis is laid on optimizing costs, schedules, and quality. Iterate through implementation and test disciplines.

For this phase, the researcher analyzed the requirements for the system and designed a solution to be implemented taking into consideration the requirements, constraints, and all applicable standards and guidelines. The program source codes were developed, and unit testing was done to ensure that the system works based on the design of the system and meets all the requirements specified in the specification documents.

Transition (product release milestone). This phase includes testing the product in preparation for release and fine-tuning the product and taking care of issues like configuring, installing, and usability issues. The focus of this phase is to ensure that software is available for its end users and iterate through deploy discipline. The primary objective is to "transit" the system from development into production, making it available to and understood by the end user. The activities of this phase include training the end users and maintainers and beta testing the system to validate it against the end users' expectations.

This phase for the deployment of the system was not covered by the researcher, but on the development process, the researcher developed a prototype system and still transformed the design of the proposed study into executable code with supporting documentation and performs a basic level of testing. Additionally, an evaluation study was also conducted to assess the effectiveness and efficiency of the proposed damage assessment system and found that the system was effective, reliable, useful, and efficient to the end users.

9.2.1 Scope and limitation

The proposed study covered Albay Province that has four school divisions with a total of 602 public elementary schools. Albay school division has 465 public elementary schools with 32 designated evacuation centers; Legazpi City school division has 43 public elementary schools with 12 designated evacuation centers; Tabaco City school division has 39 public elementary schools with 17 designated evacuation centers; and Ligao City school division has 55 public elementary schools with 6 designated evacuation centers.

The basis and instrument used in the assessment of damages for school building came from the RADaR of DepEd. School Mapping data and GIS were also employed to be able to have a visual representation of school location and damages. However, damages on equipment, communication, power supply, watery, and sanitation facilities; number of learners and personnel affected; and the like were not covered by the system. The focus of the proposed system was for the building safety evaluation process on immediate public safety, not the provision of an engineering assessment service to building owners.

9.2.2 Sources of data

The primary sources of data are the information gathered from DepEd-DRRMC and APSEMO. The main respondents are the Education Program Supervisor II Focal Person of DepEd Region V, DRRMC, and Action Officer of APSEMO. They guide the researcher where to gather the tangible data needed for the proposed study and provide necessary inputs that help the researcher to fully understand the existing procedures in conducting rapid damage assessment in the public elementary school building of Albay.

9.3 Results and discussion

9.3.1 Rapid damage assessment current procedures

The DepEd Executive Committee (EXECOM) as a member of the National Disaster Risk Reduction and Management Council (NDRRMC) and lead of the Education Cluster provides strategic directions for the DRRM in Education system. Through the DRRMO and DRRM focal persons, the strategic directions are being translated into operational directions. They have an organizational structure that provides an overview of the coordination and reporting mechanism, not only within DepEd but also with other partner agencies.

Monitoring of hazards and emergencies. DepEd DRRM monitors hazards and emergencies experienced by schools and offices, depending on the office's jurisdiction as a preparedness measure. The regional focal person monitors the effects of hazards through division focal person that monitors the emergency needs of the division office and schools. The school focal persons identify the effects of a hazard, identify the emergency needs of the school, and prepare monitoring reports that shall be evaluated by the division focal person and submitted to the regional office. The regional focal person receives the monitoring reports, monitors the emergency needs of the regional office, division offices, and schools and then makes a report to DRRMO the status of the regional office, division offices, and schools.

Identification of the affected public elementary schools. The starting point for any assessment is the identification of the users of the information and their particular information needs. DepEd identifies the number of public elementary schools affected by the disaster through proper coordination and collaboration with local DRRM councils and other education partners. The focal person provides real-time data during disasters to appropriate offices to facilitate a faster response.

For the flow of activities in identifying the affected public elementary schools, the school focal person shall provide a real-time update to the division office. The division focal person shall identify the affected public elementary schools and provide a real-time update to the regional office. The regional focal person shall identify the affected public elementary schools per division, provide a real-time update to DRRMO, and submit a report to DRRMO the total number of affected public elementary schools.

Determination of the level of damages. The school DRRM focal person identifies the level of damages used by DepEd for the school building. Level I refers to minor damage or no damage with 0%–10% damages to the school building and usable for evacuation center; Level II refers to major damage or heavy damage with 11%–49% damages to the school building and unusable for evacuation center; and Level III refers to total damage or entirely destroyed with 50% and above damages to the school building and unusable for evacuation center. The estimated amount of damages is based on the standard regional costing of the school building.

Conduct rapid damage assessment. The rapid damage assessments for public elementary schools should be conducted by the school DRRM focal persons within 12–36 h after the event through the use of DepEd RADaR form. The RADaR form requires the school focal person to identify the division name, school id, incurred damages, number of totally damaged classrooms, number of classrooms with major damage, number of classrooms with minor damage, number of makeshift classrooms needed, number of classrooms used as an evacuation center, and if the school has evacuees. It also requires the signature, full name, position, date of report, and contact number of the school focal person.

For the flow of activities in conducting a rapid damage assessment. The regional focal person disseminates memorandum order to accomplished RADaR form, whereas the division focal person ensures that the communication reaches all schools in the division. If the school focal person receives the memorandum order, the RADaR form needs to be accomplished and submitted to the division office; then, the division focal person collates and submits it to the regional office. Then, the regional focal person request for, consolidate, and validate RADaR reports and submit it to DRRMO and APSEMO.

For immediate consolidation of data, the DepEd also requires the concerned school to use social media sites to post the school damages photos or send it through e-mail or via text message to the cellphone number of the DepEd Regional DRRM focal person.

9.3.2 System design

The following diagrams show the factors and events that were considered in developing the system requirements and constraints.

Figure 9.3 shows the use case diagram of the proposed system. The system admin manages the user's account. The school DRRM focal person views the disaster report and creates a damage assessment report. The division DRRM focal person searches and views the school location, views the damage assessment

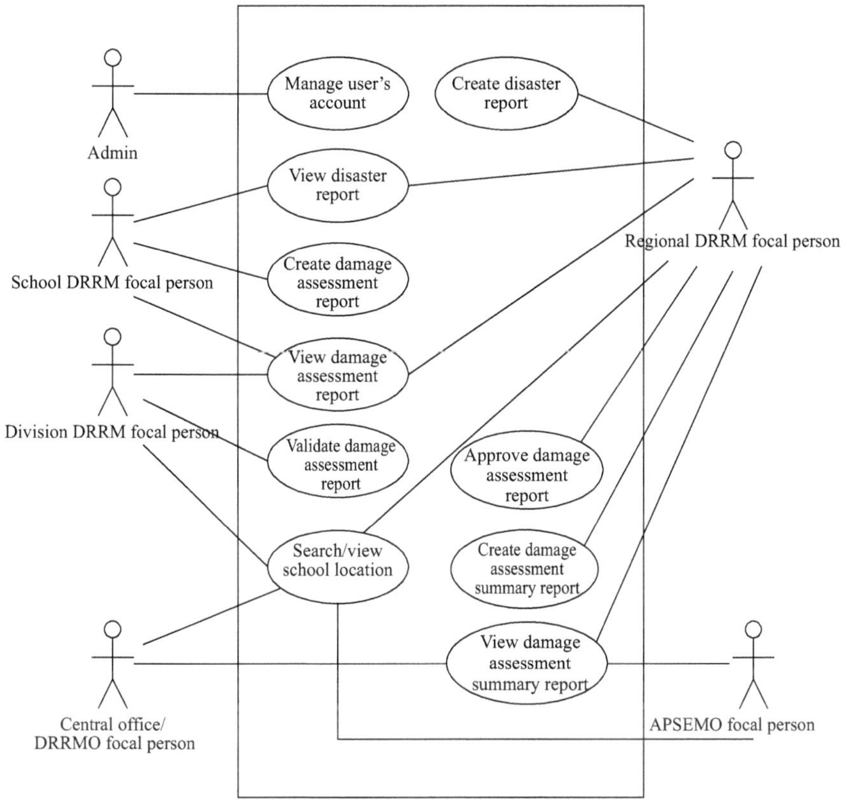

Figure 9.3 Use case diagram

report, and validates the damage assessment report. The regional DRRM focal person searches and views the school location, creates the disaster report, views the damage assessment report, approves the damage assessment report, and creates the damage assessment summary report. The DRRMO and APSEMO focal persons search and view the school location and view the damage assessment summary report. The damage assessment reports have document tracking and coordination services to provide proper coordination to DepEd officials and allow users to streamline how they complete and update the reports. It provides more flexibility to users' work schedules and provides accountability in tracking their reports.

The Unified Modeling Language (UML) class diagram of the proposed system in Figure 9.4 shows the type of static structure diagram that describes the structure of a system by showing the system's classes, their attributes, operations (or methods), and the relationships among objects.

9.3.3 Physical design

This portion of the project study presents the system requirements and system architecture to be used in the implementation of the proposed system.

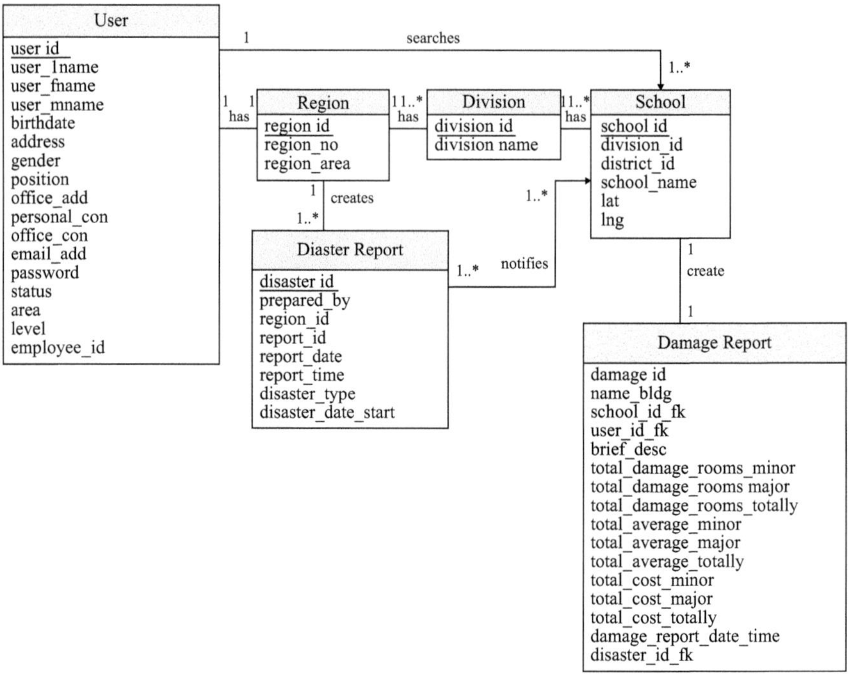

Figure 9.4 UML class diagram

Table 9.1 System requirements

Minimum hardware requirements	**Minimum software requirements**
Processor—Intel core 2 duo	Apache
Memory—2GB RAM	PhP
Hard disk—at least 500GB	MySQLWindows 8

System requirements. Table 9.1 shows the minimum hardware and software requirements of the proposed system. It presents the required programming languages and tools, and characteristics of hardware as size and operating conditions.

System architecture. Figure 9.5 shows the network connectivity design, where it enables the rapid damage assessment to minimize the timeline and provides reliability to the users.

9.3.4 Rapid damage assessment system features

The following are the features of the proposed project study:

User registration. The proposed project study has signup and log-in system for focal persons. Users can easily register by filling id number, last name, first name, middle

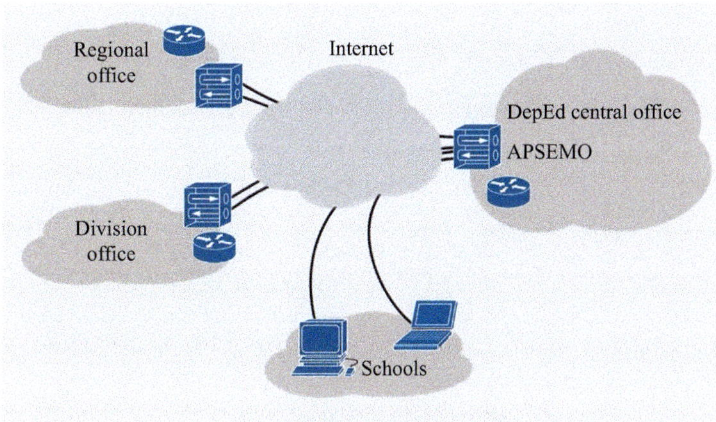

Figure 9.5 Network connectivity design

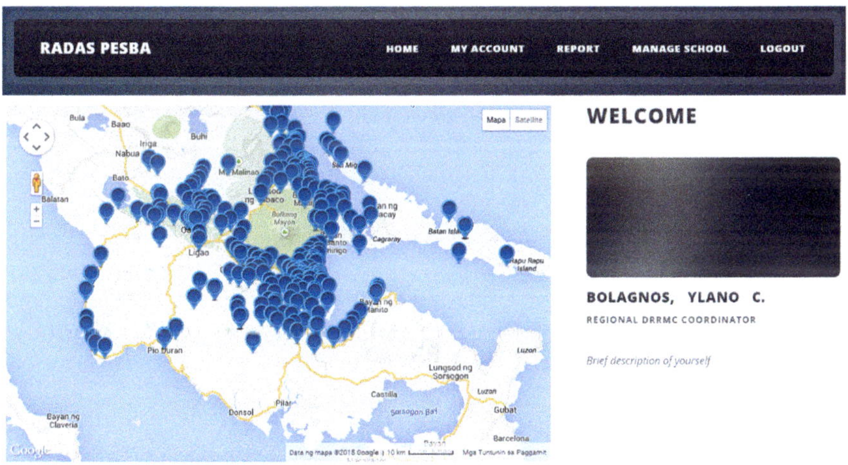

Figure 9.6 GIS school location

name, gender, birthday, address, contact number, user type, office address, office phone number, email address, and password. After submitting the form, data are stored in the database table and users are redirected to the school information page.

Geocoding facilitation of school location. Geocoding is the process of transforming a description of location through a pair of coordinates, address, or the name of a place. The proposed project study integrates with Google Maps and enables users to view the public elementary school location accurately. The user locates the public elementary school of Albay through GIS technology support or

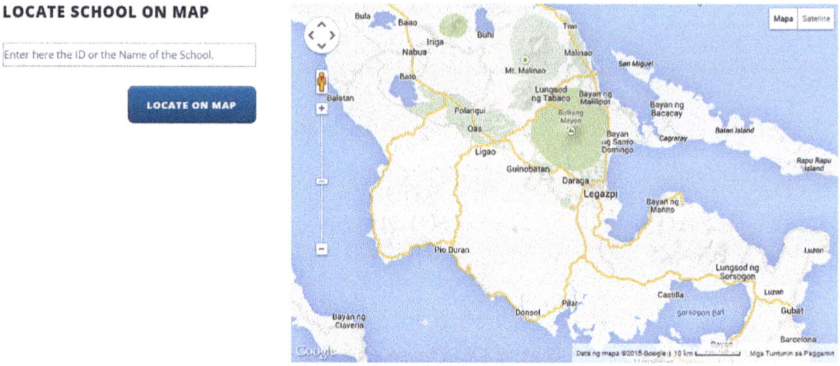

Figure 9.7 Locate school through school name or school id

through locating the school by school name and school id. Figures 9.6 and 9.7 show the sample screenshot in locating public elementary schools in Albay.

Report status. The proposed system provides a link to the users about the status of disaster and the needed reports to formulate and validate.

Standardized rapid damage assessment form. The proposed project study can move from paper-based damage assessments to a standardized rapid damage assessment form; it improves the speed and consistency of field data collection and enables the availability of information to decision-makers. It stores data to the local server, has customized access for individual users, enables users to upload school damage pictures, improves data analysis, and eliminates the duplicate record, which, in turn, reduces the amount of repetitive data entry. Figure 9.8 shows the proposed RApid DAmage System of Public Elementary School Building of Albay (RADAS-PESBA) sample form for damage assessment. It provides school building names, school id, school name, brief description, upload photos, building types, classrooms cost, damage from each level, and damage percentage.

9.3.5 *Rapid damage assessment system report formats*

The following are the report formats of the proposed project study:

Informative school information. The proposed project study is an efficient rapid damage assessment tool capable of processing and extracting information from the database and locates the elementary school building with GIS. It has a database for receiving and recording information on the list of public elementary schools of Albay. Figure 9.9 shows the sample screenshot of school information.

Accurate RADAS-PESBA report. The proposed project study provides technological improvements in conducting a rapid damage assessment. It gathers information on school building damages, saves and views the data in real time, generates a report from the data collected, and provides quick and accurate information to DepEd officials to enable them to determine the type and extent of damages to the

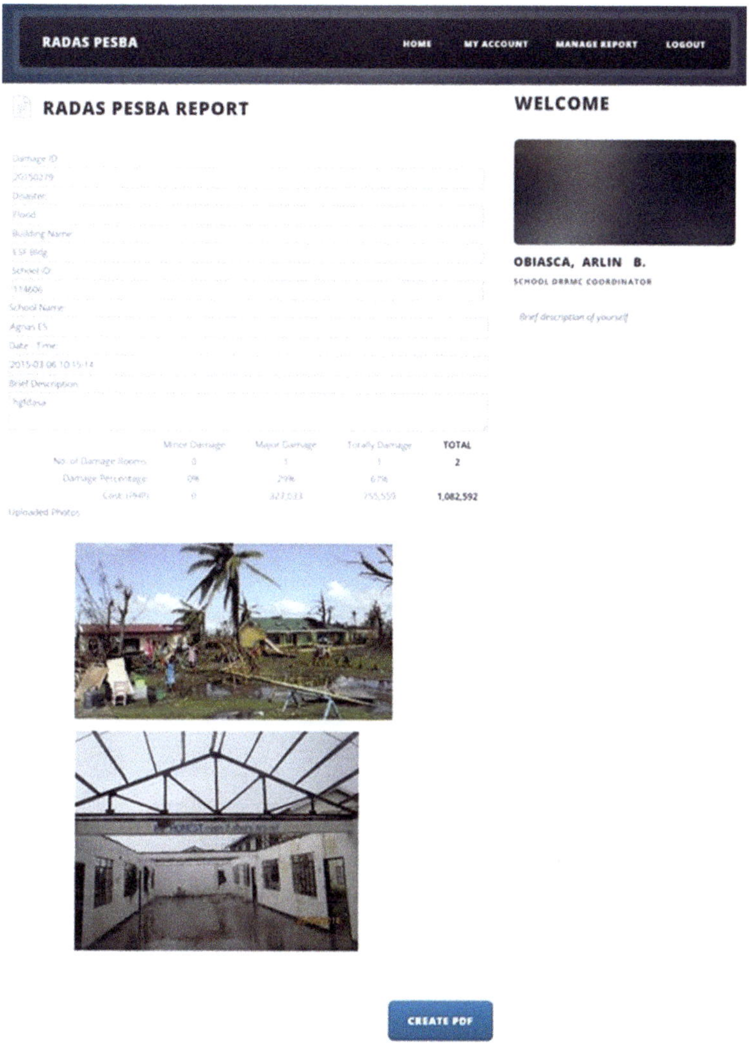

Figure 9.8 RADAS-PESBA sample form

school building. Depending on the access of the users to the proposed system, the users can view the current damage assessment report of the school, create new reports, and upload reports. Figure 9.10 shows the sample screenshot of the rapid damage assessment report.

Process improvement on RADAS-PESBA summary report. The proposed project study also provides numerical and graphical representation of the rapid damage assessment reports. It speeds up the initial assessment of damages per school division and improves data analysis. Figure 9.11 shows the sample screenshots.

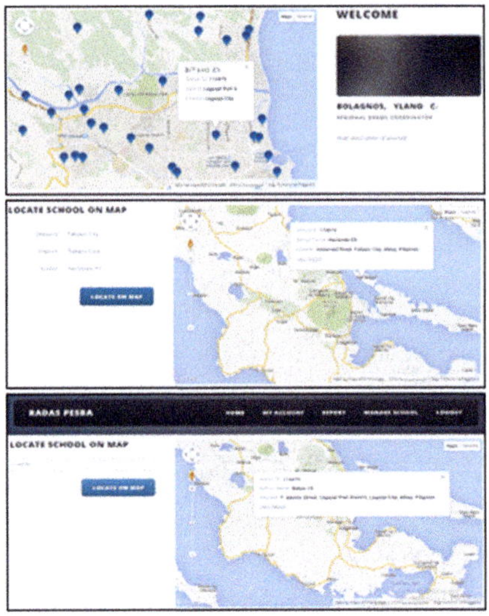

Figure 9.9 School info—GIS, school name, school id

RAPID DAMAGE ASSESSMENT SYSTEM
FOR PUBLIC ELEMENTARY SCHOOL BUILDING IN ALBAY
DAMAGE REPORT

Damage ID:	201502273200			
Disaster:	Flood			
Date Occurred:	2015-02-04			
Damage Building:	TEEP SB			
School ID:	114606			
School Name:	Agnas ES			
Date - Time (Created):	2015-03-06 10:15:14			
Brief Description:	jhgfds			
Evaluated By:	Obiasca, Arlin B.			
	Minor Damage	Major Damage	Totally Damage	TOTAL
No. of Damaged Rooms:	0	1	0	1
Damage Percentage:	0 %	26 %	0 %	-
Cost (PHP):	0	165,880	0	165,880

Figure 9.10 RADAS-PESBA report

9.3.6 System evaluation

The researcher used the ISO 9126 characteristic and sub-characteristics (ISO 1991) [26] to evaluate this study in terms of (a) functionality, (b) reliability, (c) usability,

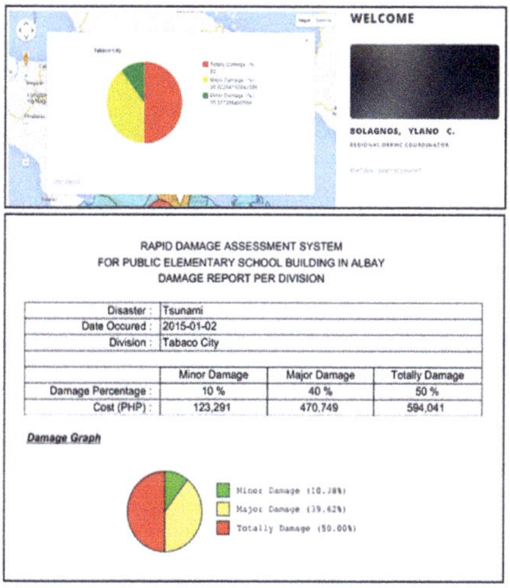

Figure 9.11 RADAS-PESBA summary report

and (d) efficiency, as shown in Table 9.2. The 5-point Likert rating scale was also used to provide the descriptive rating of the proposed study: 5 (4.50–5.00), highly acceptable; 4 (3.50–4.49), moderately acceptable; 3 (2.50–3.49), acceptable; 2 (1.50–2.49), less acceptable; and 1 (1.00–1.49), not acceptable.

For the evaluation of the proposed rapid damage assessment system, it was evaluated by 120 randomized focal persons from different public elementary schools of Albay, school divisions, as well as the focal person of regional office and APSEMO. The result shows that functionality has a weighted mean of 4.55, highly acceptable; reliability has 4.10 weighted mean, moderately acceptable; usability has 4.40 weighted mean, moderately acceptable; and efficiency has 4.15 weighted mean, moderately acceptable. The total weighted mean of system evaluation is 4.30, which is moderately acceptable.

9.4 Conclusions

This project study was intended to design and develop a rapid damage assessment system for elementary schools of Albay. Based on the findings, conclusions were derived: (1) the current procedures in conducting rapid damage assessment in the public elementary school building of Albay are properly implemented as perceived by the key informant respondents; (2) the features of the proposed rapid damage assessment system are efficient and effective in determining the type and extent of damages of disaster to Albay public elementary school building; and (3) the report formats generated and produced by the project study are reliable and useful to the end users.

Table 9.2 ISO 9126 characteristics and sub-characteristics

Characteristics	Sub-characteristics	Explanation	Weighted mean	Interpretation
Functionality	Suitability	Can software perform the tasks required?	4.40	Moderately acceptable
	Accurateness	Is the result as expected?	4.70	Highly acceptable
		WM	**4.55**	**Highly acceptable**
Reliability	Fault tolerance	Is the software capable of handling errors?	4.00	Moderately acceptable
	Recoverability	Can the software resume working and restore lost data after failure?	4.20	Moderately acceptable
		WM	**4.10**	**Moderately acceptable**
Usability	Understandability	Does the user comprehend how to use the system easily?	4.30	Moderately acceptable
	Learnability	Can the user learn to use the system easily?	4.80	Highly acceptable
	Operability	Can the user use the system without much effort?	4.20	Moderately acceptable
	Attractiveness	Does the interface look good?	4.30	Moderately acceptable
		WM	**4.40**	**Moderately acceptable**
Efficiency	Time behavior	How quickly does the system respond?	4.10	Moderately acceptable
	Resource utilization	Does the system utilize resources efficiently?	4.20	Moderately acceptable
		WM	**4.15**	**Moderately acceptable**
		Total weighted mean	**4.30**	**Moderately acceptable**

The findings show that this new technique of assessing school damages is fast and efficient. The main advantage is the quickness and automation of the procedure. The proposed system could be implemented by DepEd and the procedure should be evaluated to determine if changes or additions need to be made and if a standardized rapid damage assessment procedure would be beneficial.

References

[1] Lindell M.K., and Prater C.S. Assessing community impacts of natural disasters. Natural Hazards Review, 4, 176 [online]. 2003. Available from http://www.tc.umn.edu/~blume013/Lindell_Prater2003.pdf [Accessed Feb 2015]

[2] Department of Education (DepEd). *Philippines Disaster Risk Reduction Resource Manual (Safer Schools Resource Manual).* Published by the Department of Education (DepEd) Philippines in partnership and with the support of the United Nations Children's Fund (UNICEF) Philippines [online]. 2008 [Accessed Jan 2015]

[3] Hyogo Framework. *Hyogo Framework for Action 2005–2015. Building the Resilience of Nations and Communities to Disasters. Geneva: United Nations International Strategy for Disaster Management.* Available from http://www.unisdr.org/2005/wcdr/intergover/official-doc/L-docs/Hyogo-framework-for-action-english.pdf [online]. 2007 [Accessed Jan 2015]

[4] International Federation of Red Cross and Red Crescent Societies. *Disaster Preparedness Training Programme. Participant Resource & Learning Module. Disaster Emergency Needs Assessment* [online]. 2000. pp. 10–11. Available from wwww.ifrc.org/global/publications/disasters/all.pdf [Accessed Feb 2015]

[5] Skavdal T. *Introduction to Disaster Assessment and Assessment Methodologies*. International Training Program on Total Disaster Risk Management [online]. 2003. Available from http://www.adrc.asia/publications/TDRM2003June /16.pdf [Accessed Jan 2015]

[6] Global Facility for Disaster Reduction and Recovery (GFDRR). *Disaster, Damage, Loss and Needs Assessment.* World Bank Office, Dhaka, Bangladesh [online]. 2008. Available from http://www.gfdrr.org/2008/disaster_ damage_loss_and_needs_assessment.pdf [Accessed Feb 2015]

[7] Florida Division of Emergency Management. *The Damage Assessment Process* [online]. November 15, 2008. Available from http://www.floridadisaster. org/brm/ Damage%20Assessment.htm [Accessed Jan 2015]

[8] National Fire Academy (NFA). *Rapid Damage Assessment. Executive Analysis of Fire Service Operations in Emergency Management.* [online]. November, 2008. Emmitsburg, Maryland. Available from http://www.usfa.fema.gov/pdf/efop/efo43456.pdf [Accessed Mar 2015]

[9] Massarra C.C. *Hurricane Damage Assessment Process for Residential Building*. Published Thesis for Louisiana State University and Agricultural and Mechanical College [online]. August 2012. Available from http://etd.lsu.edu/docs/available/etd-07092012-145208/unrestricted/Massarra_Thesis-1.pdf [Accessed Feb 2015]

[10] Reddick C. *Information Technology and Emergency Management: Preparedness and Planning in US States* [online]. 2011. Retrieved from http://www.ncbi.nlm.nih.gov/pubmed/20722692 [Accessed Jan 2015]

[11] Amir, E. E., Hegazy, O., Noureldien, M., and Ali, A. Integrating Web Services with Geospatial Data Mining Disaster Management. *Geoinformatica – An International Journal (GIIJ)*, Volume 2, Issue 1, 2011.
[12] Alkhateeb F., Maghayreh E., and Aljawarneh M. 'The use of location based services for very fast and precise accidents reporting & locating'. *International Conference on Intelligent Systems, Modelling and Simulation*, 978-0-7695-3973-7/10, 2010.
[13] Cova, T. J. *GIS in Emergency Management*. Vol. 2, pp. 845–858. [online] 1999, John Wiley & Sons, New York. Available from http://www.geos.ed.ac.uk/~gisteac/ gis_book_abridged/files/ch60.pdf [Accessed Feb 2015]
[14] Sheehan Matt. *Improving Disaster Assessment: Disconnected Mobile App Uses ArcGIS Online*. [online]. 2013. Available from http://www.esri.com/esri-news/arcuser/summer-2013/improving-disaster-assessment [Accessed Jan 2015]
[15] OAK Ridge. National Laboratory. *Mobile Damage Assessment Testbed Development: DARS360 Assessment, Testing, and Evaluation* [online]. December, 2012. Available from http://www.runmobile.com/wp-content/uploads/2013/01/RunMobile-Project-92000-Final-Report-19-December-2012.pdf [Accessed Mar 2015]
[16] United Nations Children's Fund (UNICEF) and the Department of Education Philippines (DepEd). *Workshop Orientation on Disaster Risk Reduction in Schools & Emergency Preparedness and Response Planning Training* [online]. October 16–18, 2007. Available from www.unicef.org/philippines/drr-training-brochure-educ cluster.pdf [Accessed Mar 2015]
[17] *Deped Eyes 'Quick Fix' to School Damage, Evacuation Centers* [online]. 2014. Available from http://newsinfo.inquirer.net/621019/deped-eyes-quick-fix-to-school-damage#ixzz3C5Zr4MUs [Accessed Mar 2015]
[18] UNICEF. *DRR Outcomes for Children in the Philippines*. UNICEF, New York [online]. June, 2012. Available from http://learningforpeace.unicef.org/wp-content/uploads/2012/08/EEPCT_DRR_CaseStudyA_Philippines_2012.pdf [Accessed Feb 2015]
[19] International Finance Corporation (IFC). *Disaster and Emergency Preparedness: Guidance for Schools* [online]. 2010. Available from http://www.ifc.org/disaster-preparedness-guidance-schools.pdf [Accessed Jan 2015]
[20] UNISDR. *School Emergency and Disaster Preparedness* [online]. Available from http://www.unisdr.org/files/15655_1msshguidenotesprefinal0313101.pdf [Accessed Jan 2015]
[21] Romero A., and Bongcales I. *Glenda Property Damage Exceeds P10B* [online]. July 23, 2014. Available from http://www.philstar.com/nation/2014/07/23/1349516/glenda-property-damage-exceeds-p10b [Accessed Jan 2015]
[22] UNICEF. *Philippines EEPCT Fourth Progress Report*. UNICEF, New York [online]. June, 2011. Available from http://learningforpeace.unicef.org/wp-content/uploads/2011EEPCT_Fourth_Progress_Report_2011.pdf [Accessed Mar 2015]

[23] Province of Albay. Region V Bicol. About Albay [online]. Available from http://albay.gov.ph/about/ [Access Nov 2019]

[24] Bueza, V., Fatalism in a Hostile Geography? The case of Albay in Pacific Jinx [online]. January 29, 2014. Available from https://vincibueza.wordpress.com/2014/01/29/fatalism-in-a-hostile-geography-the-case-of-albay-in-the-pacific-jinx/ [Access Nov 2019]

[25] IBM and Aked, Mark. *RUP in Brief* [online]. November 25, 2003. Available from http://www.ibm.com/developerworks/rational/library/1826.html#N100E4 [Accessed Jan 2015]

[26] Chua B.B., and Dyson L.E. *Applying the ISO9126 model to the evaluation of an eLearning system.* In Atkinson R., McBeath C., Jonas-Dwyer D. Phillips R. (eds.). *Beyond the comfort zone: Proceedings of the 21st ASCILITE Conference*. [online]. December 5–8, 2004. pp. 184–90. Retrieved from http://www.ascilite.org.au/conferences/perth04/procs/chua.html [Accessed April 2015]

Chapter 10

Technologies for emergency rollout of broadband public protection and disaster relief (BB-PPDR) communications in humanitarian crisis zones

David López-Bueno[1], *Nikolaos Bartzoudis*[1], *Oriol Font-Bach*[2], *Marius Caus*[1], *Pere Gilabert*[3] *and Gabriel Montoro*[3]

10.1 Introduction

Over the past 10 years, we have witnessed an increasing trend in migratory flows of people that try to flee from war zones, famine, extreme poverty, abrupt climate-changing conditions and all sorts of violence and persecutions. In many cases, immigrants and asylum seekers cross land and sea national borders under extreme or life-threatening conditions. Numerous remote and difficult-to-access areas have been transformed into emergency first-aid camps to temporarily host immigrants and asylum seekers before they are transferred to organised camps which in the case of European Union countries are called hotspots. Unfortunately, this situation has resulted in a sequence of large-scale humanitarian crisis outbreaks in different geographic sites around the globe.[*] As of today, the heterogeneous conditions fuelling migration flows persist and thus, it is expected that humanitarian crisis sites will continue to appear under emergency conditions. In such sites, national authorities or relevant operators use public protection and disaster relief (PPDR) radio communications to respond to emergencies. More specifically, PPDR communications are used by first responders including the police, fire and ambulance services, civil defence forces, border guards, armed forces and search and rescue missions. In this sense, PPDR communications constitute a vital service for saving lives. Apart from emergency sites hosting

[1]Centre Tecnològic de Telecomunicacions de Catalunya (CTTC/CERCA), Castelldefels, Spain
[2]Software Radio Systems Limited, Cork, Ireland
[3]Universitat Politècnica de Catalunya (UPC), Department of Communications and Signal Theory, Castelldefels, Spain
[*]Indicative worldwide data are offered in this link: https://migrationdataportal.org

temporarily refugees and immigrants, there are several other scenarios of the humanitarian crisis provoked by devastating natural disasters such as floods, earthquakes, wildfires, tropical cyclones and tsunamis. On such occasions, the installed PPDR communications might be rendered unavailable or with limited operability due to the catastrophic failure of the PPDR radio and network infrastructure equipment. In such cases, emergent deployments of PPDR communications need to take place.

Most of the current PPDR communication networks use narrowband wireless technologies, primarily serving mission-critical voice communications with limited data capabilities. The PPDR services are delivered through public safety networks (PSNs) which use different communication standards such as the terrestrial trunked radio (TETRA) and Project 25 (P25). The spectrum needed for such systems has been harmonised over the past years on a regional basis. The stakeholders involved in PPDR communications are lately promoting the urgent need to add broadband data capacity in existing PPDR communication networks, to serve new operational requirements and adapt to transversal societal changes and technology trends. In particular, high-definition (HD) images or videos from humanitarian crisis sites (e.g. taken from drone-mounted HD cameras), monitoring rescue missions of migration flows, or places where devastating natural disasters take place, are becoming increasingly critical for real-time situational awareness and intelligent field assessment. The effective content analysis of such videos helps to timely detect and track critical and life-threatening conditions, by extracting data and identifying people, objects and their attributes.

10.2 Spectrum harmonisation and coexistence challenges

The international telecommunication union World Radiocommunication Conference 2015 (WRC-15) harmonised the spectrum needed for broadband PPDR (BB-PPDR) by revising the Resolution 646 [1]. Towards this end, the frequency range 694–894 MHz was adopted by WRC-15 as the globally harmonised frequency range for BB-PPDR. This frequency range includes the most commonly used PPDR bands in 700 MHz (3GPP bands 14, 28 and 68) and 800 MHz (3GPP bands 5, 20 and 26). A relevant working group, denoted as System Architecture SA6, was created in the Third Generation Partnership Project (3GPPP) to develop several new BB-PPDR features [2]. These features are included in the international mobile telecommunication (IMT) technology [3], starting with the long-term evolution (LTE) release 13 and continue to evolve and mature in specifications of Releases 14, 15 and 16, and going into IMT-2020. Fifth-generation (5G) networks will provide, in the years to come, ultra-reliable low latency and high mobility communications serving video and data requirements of different PPDR agencies [4].

Deploying new BB-PPDR infrastructure in the medium term requires among other issues the management of radio spectrum resources in a coordinated manner, considering that PPDR users belong to multiple agencies (e.g. police, fire and rescue units among others). Geographical coverage and capacity are also crucial services in mission-critical BB-PPDR applications, such as real-time video communications and real-time data. As of 2019, only a few nationwide BB-PPDR networks have been deployed in the 700 MHz band; an example is a public safety LTE (PS-LTE) network deployed in the United States. PPDR deployments might coexist with other terrestrial mobile users depending on the frequency planning applied in each case. The detailed arrangements regarding the sharing of the spectrum vary from country to country. The interference provoked to BB-PPDR systems from non-PPDR users should be minimised as much as possible. In this respect, considering the national regulations, systems supporting PPDR may be required to use specific channel spacing between user equipment (UE) and base station (BS) transmit frequencies. In addition, a greater reuse factor is required because of potential interference from distant mobiles into cells designed for portable coverage.

Cross border frequency spectrum coordination is a cumbersome effort involving the regulator bodies and related stakeholders of two or more neighbouring countries. The transnational negotiations and coordination among different PPDR agencies typically take a long time to settle, sometimes leading to situations where spectrum coordination is not entirely or effectively achieved. The derived issues and obstacles of such a situation are highlighted and detailed in a related GSMA report [5] (e.g. figure 8, p. 28). For instance, in the case that the deployment of PS-LTE with the rest of LTE carriers and the digital terrestrial television (DTT)[†] transmissions is coordinated, interference between DVB-T and LTE could be relatively easily mitigated [5]. The risk of interference between PPDR and DTT can be reduced by a set of technical measures including a guard band between DTT and PPDR BSs and an appropriate limit of the corresponding PPDR BS out-of-band (OoB) emissions [6]. Other related considerations include the improvement of the BB-PPDR UE blocking capability to limit the risk of interference from DTT Channel 21. On the side of the BB-PPDR BS receiver, interference can be perceived by DTT transmitters using channel 48 and located in the vicinity. Another case is illustrated in [7], where LTE-based BB-PPDR systems operating in the 420–430 MHz band could cause harmful interference to radiolocation systems, implying the need to apply exclusion zones in case of adjacent channel operation. Finally, in the case of deployments reaching the border limits of neighbouring countries, cross-border interference may require more rigorous analysis because, on top of the previously mentioned indicative sources of interference, the potential absence of an established frequency spectrum coordination and emissions regulation could even provoke more complex scenarios of interferences. As an example the report in [6] highlights that cross-border coordination will be a hard problem to address, in the case where one

[†]Making use of the digital video broadcasting-terrestrial (DVB-T), DVB-T2 (release 2), or other DTT transmission standards

country deploys mobile/fixed communications network supplementary downlink (SDL) and the other deploys BB-PPDR, considering that the band 738–743 MHz will be used in the opposite direction, ultimately resulting in interference from the SDL BS to the BB-PPDR one over a long distance [8]. On top of that, the interference management issue gets even more complex with the inclusion of other narrowband transmissions of Internet of things (IoT) devices (i.e. either in-band or in guard bands).

10.3 Motivation

The spectrum resources, coverage, availability and capacity of BB-PPDR communications could not always be guaranteed in geographic zones witnessing humanitarian crisis outbreaks. The latter could be in cross-border sites that temporarily host migrant flows, or zones hit by major scale natural disasters. In the latter case, PPDR communications infrastructure might be damaged and the replacement transmitters may interfere with transmitters that survived the disaster. In such case, frequency allocations, power levels, operating waveforms and device configurations must be dynamically managed based on the assessment of the pertinent conditions. The presence and quality of BB-PPDR services can be of vital importance in such zones. To summarise, the emergency rollout of BB-PPDR radio communications serving critical field requirements in humanitarian crisis zones has to be conducted in the following three cases:

1. Complete lack of coverage in the affected area (known a priori).
2. Inefficient cross-border coordination of the spectrum resources in the bands of interest.
3. Downgraded coverage availability (or complete unavailability) of a BB-PPDR network infrastructure, due to catastrophic failure of network assets provoked by natural disasters.

In the case of the cross-border problematic, a rapid context-aware on-field assessment of the situation might be necessary (i.e. scan the spectrum of interest to identify interferers and jammers), to apply a coarse grain minimisation of interference sources. However, in all three cases and especially in the second one, the BB-PPDR radio transmission must also satisfy the following key performance indicators (KPI):

- A high degree of flexibility of the transmitted waveform able to be reconfigured according to the local spectrum landscape and coexist with adjacent or even in-band secondary or co-primary transmissions (living in in-band spectral holes).
- Given the previous requirement, the transmitted BB-PPDR waveform must satisfy a spectral mask where the OoB emissions are minimised.
- BB-PPDR also require a high spectral efficiency considering for example the on-site operational needs for HD video transmission.

Chapter motivation

Satisfying these KPIs in emergency BB-PPDR rollouts serving humanitarian crisis zones is the main motivation behind the work presented in this chapter. All four KPIs are contemplated in the context of specific use cases.

Particular focus is given on the last KPI related to the energy-efficient linearisation of the transmitters, which is critical for guarantying low OoB emissions and to exploit the inherent spectral efficiency of the proposed transmitted waveform.

- The high peak-to-average power ratio (PAPR) of the wideband multicarrier waveforms used in BB-PPDR (e.g. PS-LTE) limits the transmitter energy efficiency. The power amplifier (PA) needs to be operated with significant back-off, avoid the waveform peaks falling in the non-linear region and meet the signal quality requirements. Such operation is extremely inefficient and not convenient in an energy-constrained scenario (i.e. without electrical grid availability). The use of digital linearisation techniques contributes to optimally handle the efficiency versus linearity trade-off.

The mentioned KPIs are important for both variations of PPDR communications. In the first one, terminals establish communication without the help of a network using the direct mode operation (DMO), whereas the second uses a PPDR network with fixed radio planning over-regulated BB-PPDR bands to enable the communication between terminals. The two variations are typically used for different operational scenarios. For example, DMO is used in a local reduced range to enable the coordination of first responders, in those areas without coverage, or where a natural disaster renders the assets of a PSN non-operational. In a scenario of humanitarian crisis with an emergency rollout of BB-PPDR services, it is highly likely that both variations of PPDR communications need to be employed, replacing where necessary compromised radio access and backhaul equipment or extending coverage with temporary infrastructure equipment that might include a combination of drones, portable BSs, IoT devices and satellite connectivity.

10.4 Target use cases and waveform selection

The LTE extensions for device-to-device (D2D) communications, initially introduced in Release 12 of 3GPP, specifically address PPDR needs through the added proximity services (ProSe) and group communications system enabler. LTE devices with ProSe can also serve as a fall back alternative for PSNs that must function when cellular networks are not available or fail. In the PPDR context, D2D must function even without radio access network (RAN) support.

Currently, two variations of PS-LTE for BB-PPDR are used [9]. The first one encompasses a number of non-critical BB-PPDR services that could be deployed through existing LTE commercial networks making use of a mobile virtual network operator (MVNO). The second variation is based on LTE-based PSNs entirely owned by the PPDR agent using for instance licenses spectrum at the 700 MHz (3GPP bands 14, 28 and 68) and 800 MHz (3GPP bands 5, 20 and 26). This variation is one of the use cases contemplated in this chapter and bounds to the coexistence challenges highlighted in Section 10.3.

The use of reconfigurable radio systems specifically designed for emergency BB-PPDR deployments and able to operate in a licensed shared access (LSA) regime is another of the potential use cases where the assumptions and results presented in this chapter could apply too. Such systems may operate over a licensed frequency band of a mobile network operator, who temporarily leases part of the local spectrum resources, following the evolved LSA framework [10,11]. This scheme requires conformance with national regulations and potentially implies sharing the spectrum with one or more waveforms (e.g. LTE).

Due to the criticality of the communications in question, agencies and stakeholders of the PPDR ecosystem are reluctant to adopt the operation in license-exempt spectrum due to coexistence issues and thus, target exclusively licensed spectrum to guarantee quality of service. The use of licensed-exempt spectrum was introduced in release 13 of 3GPP under the term licensed assisted access through the listen before talk (LBT) procedure. LBT provides fair spectrum coexistence conditions with other wireless technologies operating in the same band, a fact however that has not yet managed to effectively attract the attention of the PPDR agencies.

Despite having a number of clear advantages, the LTE waveforms currently used in BB-PPDR also feature some shortcomings when considering the KPIs and use cases covered in this chapter (linked to emergency BB-PPDR infrastructure rollout). In more detail, the spectral mask of the LTE waveforms implies elevated OoB emissions. This means that in a spectrum coexistence scenario, appropriately sized guard bands must be added, at the cost of valuable spectral resources. Similarly, when applying spectral gaps within the LTE waveform to accommodate other secondary or co-primary transmissions, a significant spectrum waste must be assumed. In addition, the inclusion of the cyclic prefix (CP) reduces the spectral efficiency. Hence, looking ahead at the evolution of BB-PPDR waveforms and the future communication standards (beyond 5G), this chapter makes an introduction to the utility of an alternative type of waveform for the use cases of interest.

Over the past years, different flavours of filter-bank multi-carrier (FBMC) waveforms have been proposed for the 5G communication systems [12], due to their superior spectral efficiency and spectral shape characteristics (resulting in nearly optimal OoB emissions). The different FBMC waveform variations can be considered ideal for spectral cohabitation scenarios. A number of previous works expose the advantages of FBMC over LTE waveforms in spectral coexistence scenarios. For instance, in [13,14] experimental results of narrowband TETRAPOL transmissions in coexistence with flexible broadband FBMC or LTE waveforms operating at the same band are presented.

In this chapter, FBMC is assumed to be a flexible BB-PPDR waveform in coexistence with other secondary or co-primary waveforms. To preserve the superior characteristics of the FBMC waveform and preserve the radio equipment energy efficiency, it is crucial to compensate for the potential non-linearities introduced by the radio transceiver and the PA when dealing with such high-PAPR waveforms. Figure 10.1 depicts the focused use cases with the two variations of the PPDR communications (DMO and network-based), where indicative interference sources and spectrum coexistence, either in-band or adjacent, must be efficiently addressed.

10.5 An insight to the FBMC modulation

The coexistence scenarios described in Section 10.4 pose very stringent challenges on the multicarrier modulation format. To allow a dynamic spectrum allocation while minimising the mutual interference between BB-PPDR and secondary and co-primary services, the generated waveform must achieve a fine-grained control of the spectrum. Indeed, the key feature that the waveform design must account for is the channelisation of the available spectrum with very narrow transition bands.

To meet this requirement, it is common practice to partition the transmission bandwidth into narrowband subchannels or subcarriers. In this section, we indistinctly refer to subchannels and subcarriers. The rationale is to intentionally leave empty those subcarriers that are occupied by the users of the secondary and co-primary legacy services so that only those subcarriers that are localised in the spectral gaps will remain active. It becomes evident that some guard subcarriers must be placed to protect primary transmissions and reduce to the highest possible extent the leakage. It is important to highlight that this approach yields a substantial spectral efficiency degradation unless the signals conveyed on each subcarrier exhibit good frequency localisation.

10.5.1 Multicarrier schemes for dynamic spectrum allocation

In the light of the previous discussion and in line with the works available in the literature, it can be resolved that the most eligible multicarrier schemes are as follows: orthogonal frequency division multiplexing (OFDM), FBMC, generalised frequency division multiplexing (GFDM), filtered-OFDM (f-OFDM) and universal filtered multicarrier (UFMC).

OFDM has been the modulation of choice in the fourth generation of cellular standards and so far it has been the dominant technology in the fifth generation new radio (5G NR). The beauty of OFDM lies in its ability to turn the end-to-end communication system into a set of parallel flat fading channels with a reduced complexity, by using fast Fourier transform (FFT) and inverse FFT (IFFT) algorithms [15]. To achieve this behaviour, data symbols are mapped into an orthonormal basis that is formed by a set of subcarrier filters, which are obtained by frequency-shifting the rectangular pulse. To combat the detrimental effects induced

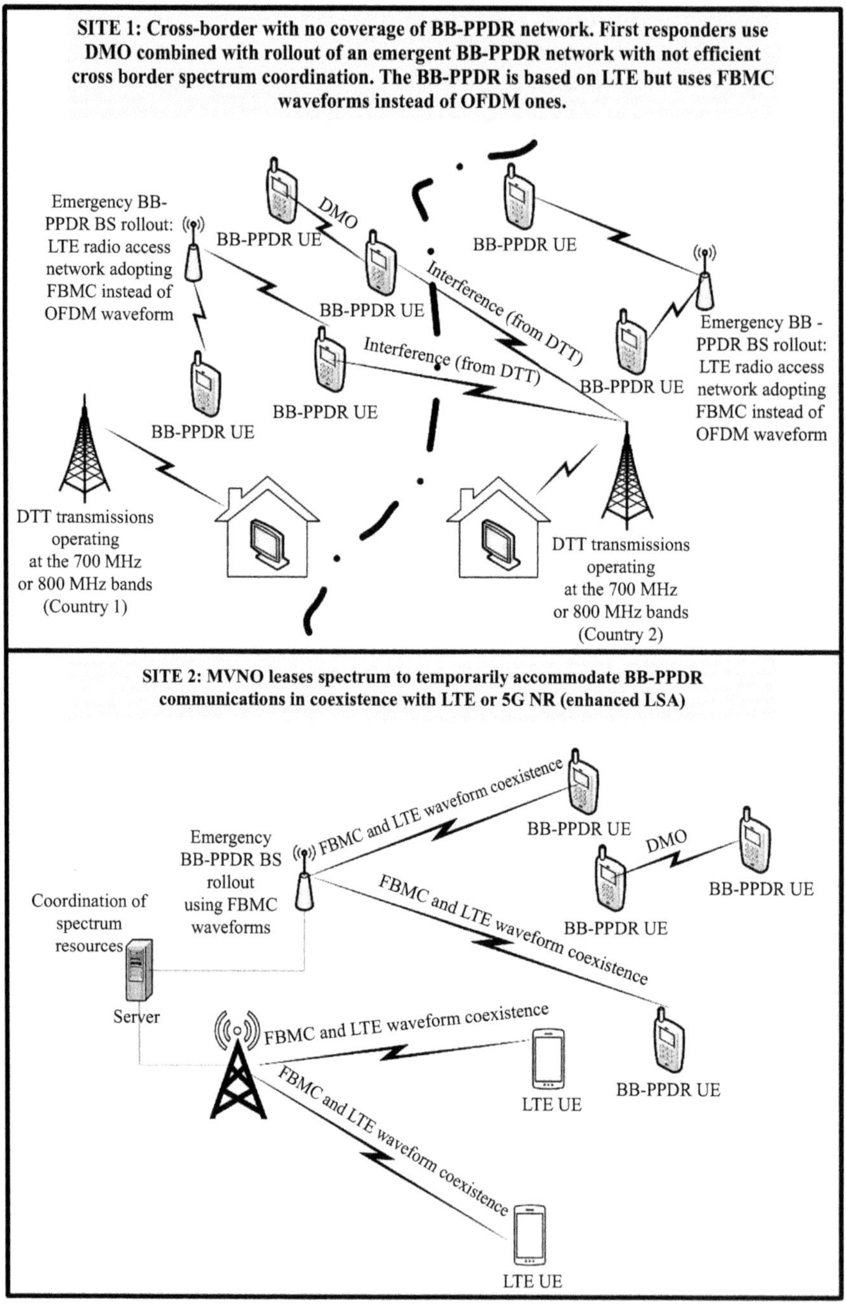

Figure 10.1 Potential spectral coexistence scenarios in emergency BB-PPDR rollouts. Envisioned from [5] (top) and [10] (bottom)

by the multipath fading a CP is transmitted ahead of each multicarrier symbol. Since the CP introduces redundancy, the penalty is the spectral efficiency degradation. Concerning the frequency response, it is known that the rectangular pulse exhibits large sidelobes, which calls for more advanced waveform designs.

A viable alternative to overcome the limitations of OFDM consists in filtering a group of subcarriers. Multicarrier schemes that lie within this category include f-OFDM [16] and UFMC [17]. To further reduce the sidelobe decay, adopting a subcarrier-wise filtering is deemed necessary. In this category, FBMC and GFDM are the most promising schemes. The so-called GFDM scheme benefits from the circular convolution to perform a block-based processing, while employing a CP ahead of each block to combat the channel frequency selectivity [18]. It must be mentioned that GFDM is a non-orthogonal multicarrier modulation, which means that the demodulated signals are always subject to considerable inter-carrier interference (ICI) even in the absence of multipath fading. The FBMC modulation belongs to the family of multicarrier schemes that shape the transmitted signal on a subcarrier basis and do not rely on the CP-assisted block transmission [19]. Hence, FBMC attains the maximum bandwidth efficiency. The main drawback of FBMC lies in preserving the orthogonality over frequency selective channels, which usually requires complex equalisation techniques [20].

The selection of the most suitable multicarrier scheme depends on the metric to be optimised. Table 10.1 gathers information about different multicarrier schemes and performs a comparison. It is worth recalling that the symbol density has been computed as $1/(T\Delta_f)$, where T is the symbol period and Δ_f is the subcarrier spacing. Desired features include maximum bandwidth efficiency, simple equalisation and good localisation in the frequency domain. These three metrics may trade off with each other, and thus there may not be a unified solution to solve all the issues and satisfy all the requirements. To cast some light into the selection of the most suitable scheme, thorough comparisons between different waveform contenders have been recently published. In this regard, the comparative performance assessment conducted in [12] reveals the superiority of FBMC in some pertinent use cases in front of GFDM, UFCM and OFDM. Leveraging on existing comparative performance assessments available in the literature, it can be concluded that if priority is

Table 10.1 Comparison among different multicarrier schemes

Scheme	CP	Orthogonal	Equalisation	Filtering	Symbol density
OFDM	Yes	Yes	Simple	No	<1 (complex symbol)
f-OFDM	Yes	No	Simple	Group of subcarriers	<1 (complex symbol)
UFMC	No	Yes	Simple	Group of subcarriers	<1 (complex symbol)
GFDM	Yes	No	Complex	Subcarrier basis	<1 (complex symbol)
FBMC	No	Yes (real domain)	Complex	Subcarrier basis	2 (real symbols)

given to the spectral efficiency and the frequency confinement, then it follows that the balance is clearly tipped towards FBMC. Therefore, a schematic view of FBMC and a concise overview of the system model will be provided hereafter. It is important to remark that the comparison between different multicarrier schemes has been presented concisely because it is not within the main scope of the chapter.

10.5.2 FBMC-based system model

The FBMC scheme presented in this section is based on the offset quadrature modulation (OQAM), usually referred to as FBMC/OQAM or, less commonly, as staggered multitone. In the sections following, FBMC will be used to refer to the FBMC/OQAM modulation. The idea behind FBMC is to perform a parallel transmission by employing time-frequency shifted versions of the prototype pulse. The transmission format is characterised by introducing a time offset on the real (imaginary) part of the QAM symbols on the even (odd) subcarriers. Hence, the component that is delayed depends on the parity of the subcarrier. This can be understood as a transmission of real-valued pulse amplitude modulation (PAM) data symbols at twice the symbol rate, which highlights that FBCM does not inherits the OFDM kernel. This creates a specific pattern where adjacent symbols in the time-frequency grid have a phase shift of $\pi/2$, creating a real-domain orthogonality. Hence, even in the absence of distortion, the received data symbol will be affected by a self-interference term that is in quadrature with the desired data. Accordingly, under ideal conditions the data can be perfectly recovered. In the presence of multipath fading, equalisation techniques must be performed to restore the orthogonality in the real domain. In the following, the basics of FBMC are provided. For the sake of brevity, some derivation steps are omitted. Further details are provided in [20].

The baseband modulated signal that is generated by the synthesis filter bank (SFB) can be expressed in discrete-time domain as

$$s[n] = \sum_{k\in\mathbb{Z}} \sum_{m\in\mathcal{M}_u} x_m[k] f_m\left[n - k\frac{M}{2}\right], \tag{10.1}$$

where $x_m[k]$ is the symbol transmitted on the frequency-time position (m, k) and $f_m[n]$ is the subcarrier pulse that shapes the symbols conveyed on the m-th subcarrier. Note that the set $\mathcal{M}_u$ gathers the indexes of the active subcarriers, which are not necessarily contiguous. It is considered that the transmission bandwidth is divided into M subcarriers. It is noteworthy to mention that the low- and the high-rate signals are indexed with variables k and n, respectively. For what concerns the data symbols, it is common to multiply the PAM data symbols $\{d_m[k]\}$ by the phase term:

$$\theta_m[k] = \begin{cases} 1 & m+k \text{ even} \\ j & m+k \text{ odd} \end{cases} \tag{10.2}$$

to get $x_m[k] = d_m[k]\theta_m[k]$, which is either real or pure imaginary. Another possibility is to set $\theta_m[k] = j^{m+k}$, which is equivalent to the previous definition aside from the sign.

It is important to remark that the class of FBMC systems that are tackled in this section is uniformly modulated. This means that the subcarrier pulses are obtained by frequency-shifting the prototype pulse $p[n]$ as follows:

$$f_m[n] = p[n]e^{j\frac{2\pi}{M}m(n-D)}. \tag{10.3}$$

Under the design guidelines of [19], the delay D is used to force the pulse to be causal, and thus, it depends on the length. Provided that the prototype pulse is a unit-energy finite impulse response filter of order $L-1$, then it follows that $D = (L-1)/2$. In general, the length of the pulse is close to a multiple of M, that is, $L \in \{KM-1, KM, KM+1\}$. The parameter K is called the overlapping factor because it controls the number of multicarrier symbols that are overlapped in the time domain. Typically, $K \in \{1, 2, 3, 4\}$ albeit other configurations are allowed. It becomes evident that the higher K is, the faster decays the pulse in the frequency domain. As an example, we have depicted in Figure 10.2 the design proposed in [21] with $K = 4$. It must be mentioned that the frequency is normalised to the subcarrier spacing and the time is normalised to the symbol period. At the expenses of increasing the duration of the pulse in FBMC, the spectral confinement is improved with respect to OFDM.

In the presence of multipath fading, the received signal is given by

$$r[n] = s[n] \star h[n] + w[n], \tag{10.4}$$

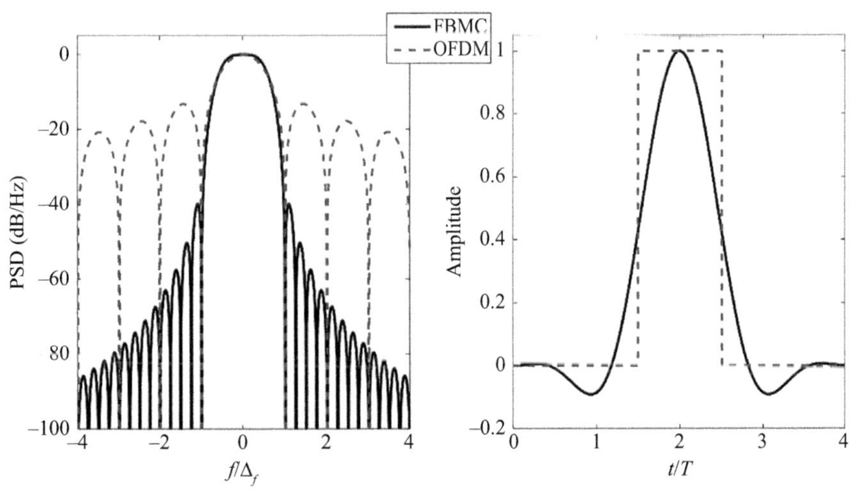

Figure 10.2 Normalised power spectral density and time response of FBMC and OFDM

where $\star$ is the linear convolution, $h[n]$ denotes the channel impulse response and $w[n]$ denotes the additive noise that contaminates the reception. To recover the data symbols, the receiver executes the operations performed by the transmitter in a reversed order. That is, the analysis filter bank (AFB) is composed of a bank of filters that are matched to the filters used in the SFB. The idea is illustrated in Figure 10.3.

The direct implementation of FBMC entails a complexity that may be hardly affordable. Fortunately, in the literature, one can find more efficient implementations either based on polyphase networks [19] or frequency domain processing [22]. For the sake of the analytical tractability, we stick to the direct implementation. In this sense, coming back to the receive side, an estimate on the (q,k) frequency-time position can be computed by feeding the received signal into the AFB, namely:

$$y_q[k] = \theta_q^*[k] \times \sum_{n\in\mathbb{Z}} r[n] f_q^*\left[n - k\frac{M}{2}\right]. \tag{10.5}$$

After several derivation steps, it can be demonstrated that the signal demodulated on the q-th subcarrier can be compactly formulated as

$$y_q[k] = \theta_q^*[k] \times \sum_{m\in\Omega_q} \Gamma_{qm}[k] \star x_m[k] + \eta_q[k]. \tag{10.6}$$

The set Ω_q gathers the indexes of the subcarriers that contribute to ICI. In most of the FBMC systems ICI mainly comes from the adjacent subcarriers, that is, $\Omega_q = \{q-1, q, q+1\}$. The transmultiplexer (TMUX) response and the filtered noise are, respectively, formulated as

$$\begin{aligned} \Gamma_{qm}[k] &= \left(f_m[n] \star h[n] \star f_q^*[-n]\right)_{\downarrow M/2} = \left(f_m[n] \star h[n] \star f_q^*[-n]\right)_{n=kM/2} \\ \eta_q[k] &= \theta_q^*[k] \times \left(w[n] \star f_q^*[-n]\right)_{\downarrow M/2} = \theta_q^*[k] \times \left(w[n] \star f_q^*[-n]\right)_{n=\frac{kM}{2}}. \end{aligned} \tag{10.7}$$

The operation $(\cdot)_{\downarrow x}$ performs a decimation by a factor x. In low-frequency selective channels, the channel coherence bandwidth is usually wider than the subcarrier spacing. This is tantamount to saying that the channel frequency

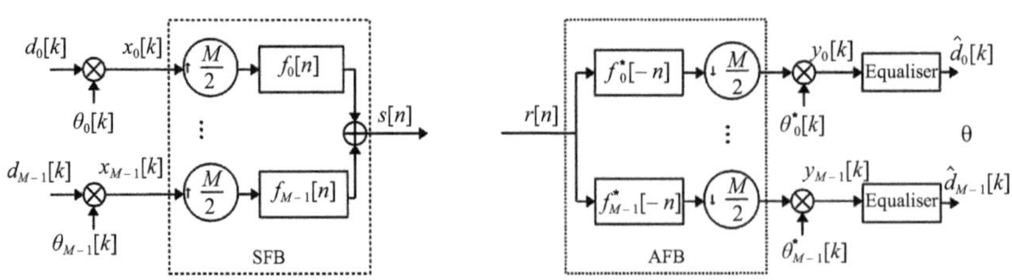

Figure 10.3 Block diagram of the FBMC/OQAM transceiver

response is flat in the passband region of subcarrier signals. In such a case, the TMUX can be drastically simplified as

$$\Gamma_{qm}[k] = H_q\left(f_m[n] \star f_q^*[-n]\right)_{\downarrow M/2} = H_q \sum_n f_m[n] f_q^*\left[n - \frac{kM}{2}\right] = H_q \alpha_{qm}[k], \tag{10.8}$$

where H_q represents the channel frequency response evaluated on the radial frequency $2\pi q/M$. Notice that the term $\alpha_{qm}[k]$ is characterised by the prototype pulse and the subcarrier indexes. In essence, $\alpha_{qm}[k]$ represents the interplay between the q-th and the m-th subcarrier signals under ideal conditions, that is, in the absence of multipath fading. Accordingly, the global communication system reads as

$$y_q[k] = H_q \times \sum_{m \in \Omega_q} \sum_{l \in \mathbb{Z}} \theta_q^*[k] \alpha_{qm}[l] \theta_m[k-l] d_m[k-l] + \eta_q[k]. \tag{10.9}$$

Under the perfect reconstruction (PR) property, the functions $\{\theta_m[k] f_m[n - kM/2]\}$ constitute an orthonormal basis in the real domain. Hence:

$$\Re\left\{\theta_q^*[k] \alpha_{qm}[l] \theta_m[k-l]\right\} = \delta_{q,m}\delta_{l,0} \tag{10.10}$$

must be satisfied. It can be readily verified that $\alpha_{qm}[l]$ must be either real or pure imaginary to fulfil (10.10). Exploiting the PR property, the data symbols can be perfectly recovered except for the noise if the channel is equalised on a per-subcarrier basis according to the zero-forcing criterion. In other words, a single-tap equaliser followed by a stage that extracts the real component suffices to get rid of the interference, that is

$$\begin{aligned} \hat{d}_q[k] &= \Re\left\{\frac{y_q[k]}{H_q}\right\} = d_q[k] \\ &+ \sum_{(m,l) \neq (q,0)} \Re\left\{\theta_q^*[k] \alpha_{qm}[l] \theta_m[k-l]\right\} d_m[k-l] + \Re\left\{\frac{\eta_q[k]}{H_q}\right\} \end{aligned} \tag{10.11}$$

If the power spectral density of the noise at the input of the receiver is N_0, then it follows that $E\left\{\left|\eta_q[k]\right|^2\right\} = N_0$, where $E\{.\}$ is the statistical expectation [20]. From this result, it can be inferred that the quality of the symbol estimates will be similar to that achieved in OFDM.

In the most general case, no assumption can be made about the channel frequency selectivity, and thus, the TMUX response cannot be simplified. Hence, if we focus the attention on highly frequency selective channels, the frequency response is not flat anymore on the frequency range spanned by one subcarrier signal. Then, it is crucial to adopt more complex equalisation techniques to counteract the channel and invoke the PR property. Some alternatives to single-tap equalisation have been reviewed in [20]. The most prominent solutions involve multi-tap filtering [23] and complex architectures based on multi-stage processing [24]. The signal processing developments to overcome the channel conditions have been omitted because the primary objective of this section is to give a concise overview of FBMC. Building upon this model in Section 10.6, we will analyse the spectral coexistence of narrowband and broadband private mobile radio (PMR) waveforms and we will describe PA digital linearisation techniques to make the most efficient use of spectrum resources.

10.6 Digital linearisation techniques for enhanced and efficient LTE-FBMC spectral coexistence and experimental validation

In the previous sections, the benefits of using a broadband FBMC waveform for the PPDR transmissions have been demonstrated. If the peaks of the waveform fall in the PA non-linear region in transmission, the properties of the FBMC BB-PPDR system will be lost both in terms of spectral efficiency and spectral containment and those primary transmissions coexisting with the FBMC BB-PPDR system will be distorted. If the PA is operated at a significant power level back-off to keep the waveform peaks in the linear region, the PA energy efficiency will be very low, more complex heat dissipation or cooling structures will be required and the PA reliability could be further constrained. The use of digital linearisation techniques such as the crest factor reduction (CFR) and digital predistortion (DPD) that optimise the linearity-versus-efficiency trade-off is required to preserve the FBMC waveform properties and have an efficient and reliable BB-PPDR communication system.

CFR reduces the peaks of the modulated waveform to a satisfactory level: due to the potentially added in-band and OoB distortion, CFR does not necessarily introduce system gain; but, thanks to the reduction in the peaks, it is possible to operate the PA at higher average power and therefore closer to its saturation point, where it is most efficient. CFR is combined with DPD that relies on digitally cascading a non-linear system before the PA, which provides an inverse response to the PA, that provokes a linear response at the output of the cascaded system, including the PA. To accomplish that, complex non-linear models are used to characterise the PA behaviour taking into account bandwidth-related memory effects which are usually present in wideband RF PAs operated under changing modulated waveforms or whose performance may change over time and temperature and require adaptive DPD correction. Combining CFR and DPD techniques may have a positive impact on the radio equipment due to the following: first, as the transmit system linearity is improved, this technique allows using cost-effective and lower power consuming PA solutions with less expensive board, enclosure and heat sink designs and simplified power supply and cooling subsystems. Second, the system performance can be enhanced extending the modulation order (i.e. increasing transmission capacity) or the reach and robustness against interference of the radio equipment, all of which maximises the system efficiency. Third, closed-loop adaptive DPD may enable self-calibration and correction of the transmitter increasing its operating stability.

Another relevant consideration to be taken into account is that digital linearisation techniques, thanks to favouring the spectral containment and in-band distortion reduction in the transmitter, are also an enabler not only for enhanced cell size and coverage scenarios, but new licensed/unlicensed intra-band or adjacent inter-band spectral coexistence scenarios and multi-cell deployments. This section provides experimental validation not only of the beneficial impact of the digital linearisation techniques in the FBMC BB-PPDR system (both in terms of performance and energy efficiency), but also of the coexistence between the BB-PPDR system and LTE co-primary signals in a scenario like that shown in Figure 10.1 (bottom). In such scenario, one or several LTE MVNOs lease spectrum to temporarily accommodate the FBMC BB-PPDR communications that coexist with the rest of LTE waveforms. The FBMC system will be validated considering LTE-like requirements, as specified in [25], regarding the adjacent channel power ratio (ACPR) and the error vector magnitude (EVM). The ACPR requirement is set at −45 dBc and the EVM requirement is set at 18.5 per cent, 13.5 per cent, 9 per cent and 4.5 per cent considering the QPSK, 16-QAM, 64-QAM and 256-QAM subcarrier modulations, respectively. The details of the test setup are found in Section 10.6.3.

10.6.1 Linearisation of PAs operated under FBMC waveforms

The PAPR reduction techniques may be divided into three groups according to the taxonomy shown in [26]. In the first group, the signal distortion techniques reduce the PAPR by distorting the OFDM signal before the PA at the expense of some tolerable increase in the in-band and OoB distortion. In this group, we find the clipping and filtering, peak windowing, companding and peak cancellation techniques. The second group includes the multiple signalling and probabilistic techniques that either modify the OFDM waveform (i.e. by introducing phase shifts, adding peak reduction subcarriers or changing the symbol constellation points) or generate multiple permutations of the OFDM signal and transmit the one with the minimum PAPR. In this group, we find the selective mapping, partial transmit sequences, tone injection, tone reservation, active constellation shaping and the interleaved OFDM, among others. In the third group, some coding schemes like linear block coding, turbo coding or Golay complementary sequences are provided both for error detection and correction and to perform PAPR reduction. The PAPR reduction solutions are typically analysed in terms of implementation complexity, bit rate loss, induced distortion and in terms of power increase. The signal distortion techniques are good solutions in terms of implementation complexity, preserving the bit rate, and not requiring a power increase but, on the contrary, the EVM and bit error rate (BER) may increase. The multiple signalling and probabilistic techniques are of higher complexity but do not provoke distortion. The bit rate loss and power increase will depend on the specific technique. For instance, while in the active constellation extension the symbol constellation modification requires higher power but does not imply bit rate loss, the partial transmit sequence does not require power increase but does imply a bit rate loss due to the need of side information (i.e. sent to inform the receiver of what modifications have been implemented at the transmitter). Many of the above-introduced solutions cannot be directly applied to FBMC given the fact that, unlike in OFDM, the time domain symbols overlap and cannot be treated separately. The solutions found in the literature are mainly based on clipping, tone reservation or active constellation extension. In [27], different clipping and windowing methods are applied to both OFDM and FBMC and their usability is validated for the latter. In [28], an iterative clipping-based solution is compared by simulation with tone reservation and active constellation extension and the combination of the two when applied to FBMC. The authors in [29] compare the clipping and companding techniques concluding that u-law companding could be the best solution for the reduction of PAPR in FBCM/OQAM systems. Some novel adaptations of the traditional schemes have been proposed to reduce their implementation complexity and add extra PAPR reduction. For example, in [30], the authors propose an algorithm based on the active constellation extension for FBMC.

Regarding DPD, in [31], the authors evaluate the degradation of the FBMC performance due to the transceiver RF impairments and its impact on the DPD module. To do that, they consider the IQ gain and phase imbalances and gain variations in the forward and feedback paths, the number of bits in the digital-to-analogue conversion (DAC) and analogue-to-digital conversion (ADC) chains and fractional delay errors when aligning the DPD observation signal. An orthogonal polynomial memory series is utilised for the DPD but neither information on the measurement setup to get the PA excitation signals nor the waveform bandwidth is provided. In any case, the article shows simulation results that are worth considering. In [32], an overlapped recursive error-correcting predistortion is proposed for FBMC based on Rapp PA AM/AM modelling, but such model appears to be limited for real applications with both amplitude and phase distortion and memory effects. The authors in [33] present an interesting benchmark of OFDM and FBMC DPD performance in terms of EVM and ACPR for different PA devices and operating input power levels by considering a Volterra DDR-based predistorter. The results show how the non-linearities destroy the advantages of using FBMC instead of OFDM in terms of ACPR when the DPD is not employed. The DPD results show how the ACPR advantages of FBMC are restored but also show that the FBMC signals tend to be more sensitive to the PA non-linear distortion than OFDM (i.e. in terms of EVM) which justifies the use of a DPD engine to boost the overall performance. The authors in [34] also propose two adaptive DPD schemes for FBMC and evaluate the performance in simulation. In this case, both AM/AM distortion and AM/PM distortion are considered by using a PA Saleh model but the extent of the results is limited considering again that they lack from experimental measurement data and that the model will not be the best fit when targeting linearisation of PAs under wideband modulated signals showing memory effects. Two DPD schemes are employed for either joint and separate AM/AM and AM/PM distortion compensation. The non-linear modelling is based in feed-forward neural network that models the inverse response of the PA in the feedback path and that will then be copied in the forward path given the indirect learning architecture presented. Once more, the results show that the PA distortion degrades more significantly the FBMC signals when compared to OFDM, due to the higher sensitivity of the former to phase distortion. For this reason, the joint approach is outperformed by the separate one at the cost of increasing the system complexity.

10.6.2 Proposed digital linearisation architecture and principles

In this work, we have considered the high-level digital linearisation architecture shown in Figure 10.4.

The CFR technique to reduce the PAPR of the FBMC modulated signal is the peak cancellation (PC) scheme described in [35], due to the reasonable implementation complexity and tuning capabilities. As shown in the block diagram in Figure 10.5, it is based in the clipping and filtering techniques.

In a first stage, a peak signal $p[n]$ is generated as

$$p[n] = s[n] - s[n]c[n] \tag{10.12}$$

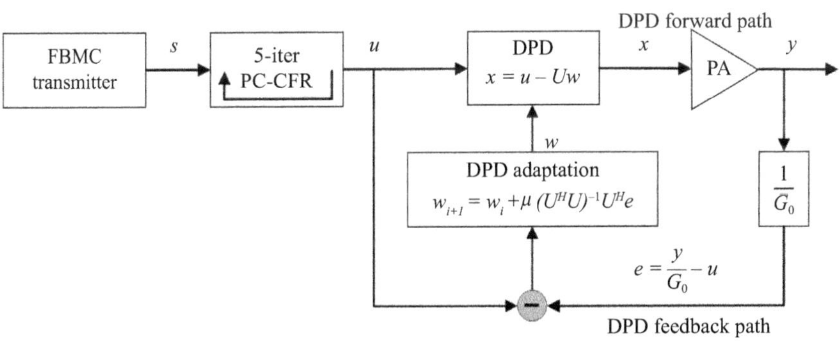

Figure 10.4 CFR and direct learning DPD digital linearisation architecture proposed for a BB-PPDR BS

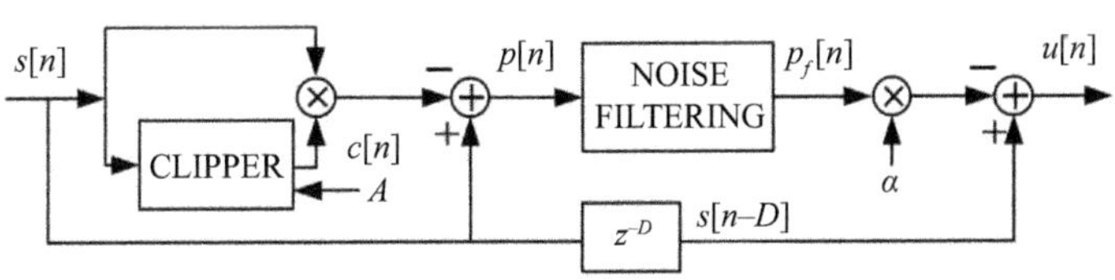

Figure 10.5 Block diagram of the PC-CFR technique

$s[n]$ being the input signal and $c[n]$ the clipping signal defined as

$$c[n] = \begin{cases} \frac{A}{|s[n]|} if |s[n]| > A \\ 1 \ if \ |s[n]| \leq A \end{cases} \tag{10.13}$$

with A being a clipping threshold. In a second stage, the peak signal $p[n]$ is filtered (noise shaping), obtaining $p_f[n]$. Finally, the reduced PAPR output signal is calculated by subtracting a weighted version of the filtered peak signal from the original input signal (after proper time alignment):

$$u[n] = s[n - D] - \alpha p_f[n] \tag{10.14}$$

with α being the subtraction parameter. To guarantee the PAPR reduction factor for a given set of threshold and subtraction parameters, the output signal is fed back to the input and the process is iterated four times more. By benchmarking in advance the PAPR reduction factor and the quality of the signal for all the different combinations of threshold and subtraction parameters, it is therefore possible to tune the PC-CFR trading-off between induced distortion and PAPR reduction to meet the requirements imposed by the application scenario.

The DPD can be defined by the parameter estimation architecture, the PA behavioural model being used, and the parameter estimation method. Regarding the first, a joint direct learning architecture is employed to compensate for the AM/AM and AM/PM distortion. In the direct learning architecture, the distortion introduced by the PA is estimated and then subtracted from the original signal to compensate for the PA distortion. In an indirect learning architecture, the output of the PA is used as input of a postdistorter that models the inverse response of the PA in the feedback path and replicates this structure in the forward path. The direct learning architecture is less prone to suffer from noisy PA output observations than the simpler and widely used indirect learning one; performs better when the PA non-linearity is severe; and can benefit from reduced ADC sampling rate techniques and better control of the regularisation process (aiming at reducing the non-linear modelling order), as demonstrated in [36]. For the PA behavioural model, we have chosen the generalised memory polynomial (GMP) model since for this scenario it has outperformed the other alternatives also considered (i.e. MP, DDR-Volterra). The direct learning scheme employs the linear least squares (LS) solution as estimation method since the DPD function is non-linear but it is, however, linear in the parameters. The equations appearing in Figure 10.4 are now further detailed:

The input–output relationship at the DPD block in the forward path is

$$x[n] = u[n] - d[n] \tag{10.15}$$

where $x[n]$ is the signal at the output of the DPD block, $u[n]$ is the input signal, and $d[n]$ is the distortion signal which is described in this paper through the GMP behavioural model defined in [37]:

$$\begin{aligned} d[n] \quad = & \sum_{i=0}^{N_a-1}\sum_{p=0}^{P_a-1}\alpha_{pi}\cdot u\left[n-\tau_i^a\right]\left|u\left[n-\tau_i^a\right]\right|^p \\ & +\sum_{j=1}^{M_b}\sum_{i=0}^{N_b-1}\sum_{p=1}^{P_b-1}\beta_{pij}\cdot u\left[n-\tau_i^b\right]\left|u\left[n-\tau_i^b-\tau_j^b\right]\right|^p \\ & +\sum_{j=1}^{M_c}\sum_{i=0}^{N_c-1}\sum_{p=1}^{P_c-1}\gamma_{pij}\cdot u\left[n-\tau_i^c\right]\left|u\left[n-\tau_i^c+\tau_j^c\right]\right|^p. \end{aligned} \tag{10.16}$$

The first term is the MP expression having N_aP_a DPD coefficients; the second and third terms introduce both negative and positive cross-term delays to the MP model, having $M_bN_b(P_b-1)$ and $M_cN_c(P_c-1)$ coefficients, respectively, $N_{a|b|c}$ and $M_{b|c}$ relate to memory depths while $P_{a|b|c}$ are polynomial orders. The sum of all these coefficients is the behavioural model order. The memory delays being applied to the signal at each of the three GMP terms are introduced in $\tau_{i|j}^{a|b|c}$. In a more compact notation, (10.15) can be rewritten as

$$x[n] = u[n] - \varphi^T[n]\boldsymbol{w}[n] \tag{10.17}$$

where $\boldsymbol{w}[n] = (w_1[n], \ldots, w_i[n], \ldots, w_M[n])^T$ is a vector of coefficients at time n with dimensions $M \times 1$, M is the order of the behavioural model, and $\varphi^T[n] = (\varphi_1[n], \ldots, \varphi_2 i[n], \ldots, \varphi_M[n])$ is the vector containing the basis functions $\varphi_i[n]$, with $i = 1, \ldots, M$. The mapping between the GMP specific coefficients (α_{pi}, β_{pij}, γ_{pij}) in (10.16) and the general purpose DPD coefficients $w_i[n]$ is straightforward. If (10.15) is expressed in matrix notation, we have:

$$\boldsymbol{x} = \boldsymbol{u} - \boldsymbol{U}\boldsymbol{w} \tag{10.18}$$

where $\boldsymbol{x} = (x[0], \ldots, x[n], \ldots, x[N-1])^T$ and $\boldsymbol{u} = (u[0], \ldots, u[n], \ldots, u[N-1])^T$, with $n = 0, \ldots, N-1$, are the predistorted and input vectors, respectively, and $\boldsymbol{U} = (\varphi[0], \ldots, \varphi[n], \ldots, \varphi[N-1])^T$ is the $N \times M$ data matrix, with N being the number of samples and M being the number of basis functions (order of the model). With a closed-loop error minimisation technique, the coefficients can be extracted iteratively by finding the LS solution. At the n-th iteration (and considering buffers of N data samples), the coefficients are obtained as

$$\boldsymbol{w}[n+1] = \boldsymbol{w}[n] + \mu\left(\boldsymbol{U}^{\mathbf{H}}\boldsymbol{U}\right)^{-1}\boldsymbol{U}^{\mathbf{H}}\boldsymbol{e} \tag{10.19}$$

where $\mu(0 \le \mu \le 1)$ is the weighting factor and $\boldsymbol{e} = (e[0], \ldots, e[n], \ldots, e[N-1])^T$ is the $N \times 1$ vector of the error defined as

$$\boldsymbol{e} = \frac{\boldsymbol{y}}{G_0} - \boldsymbol{u} \tag{10.20}$$

where G_0 determines the desired linear gain of the PA, and $\boldsymbol{u}$ and $\boldsymbol{y}$ are the $N \times 1$ vectors of the transmitter input and the PA output, respectively.

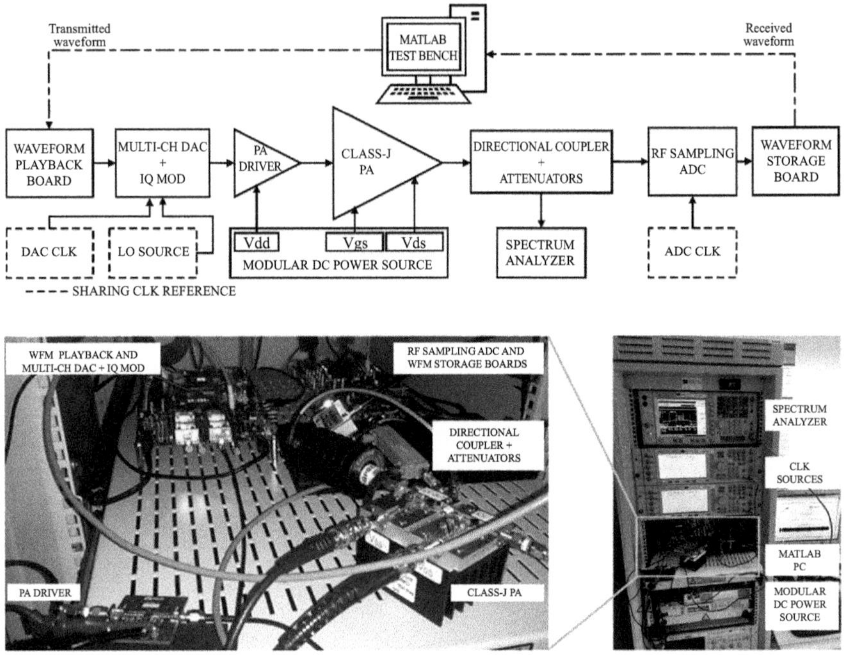

Figure 10.6 Digital linearisation test bench

In [38], the authors show how the forward path of this architecture can be implemented in hardware by focusing on the most relevant basis functions. The DPD follows a look-up table (LUT) interpolation implementation approach and is designed for linearisation of highly efficient concurrent dual-band envelope tracking PAs. The same authors propose in [39,40] dynamic estimation techniques for the feedback path that only take the minimum necessary components to reach the targeted linearity levels at every iteration.

10.6.3 CFR tuning, DPD linearisation and experimental evaluation

To tune fast prototype and evaluate the digital linearisation techniques, the test setup in Figure 10.6 was used. The test bench employs 20 MHz bandwidth downlink signals which are modulated with fast convolution FBMC and that follow the structure described in [14]. In the tests, 64-QAM modulated FBMC signals were employed. These waveforms feature one-by-one subcarrier deactivation to optimise the spectrum occupation and allow in-band spectral coexistence with co-primary systems operating. Two spectrum holes have been created by subcarrier deactivation at $\pm$4.5 MHz (from centre frequency) to allow coexistence with two LTE signals with 5 MHz and 1.4 MHz bandwidth, respectively. The FBMC signals follow a frame and timing structure similar to that used in LTE (LTE FDD Rel. 9) for which the original baseband sample rate for 20 MHz bandwidth is 30.72 MHz and the signals are then oversampled by five at 153.6 MHz (maximum DPD signal bandwidth) to allow for OoB compensation when predistorting. These waveforms feature around 10.5-dB PAPR at baseband. When doing linearisation experiments it is meaningful to know the waveform PAPR CDF for a given configuration not only to properly dimension the non-linear phenomena but to evaluate the DPD robustness. Waveform analysis and design techniques have been employed to synthesise two-subframe length test waveforms (with 307,200 samples considering the DPD oversampling factor) being PAPR statistically representative of the above-mentioned configuration with 10-subframe per frame structure. The MATLAB®-controlled digital linearisation test bench employs some commercial boards for waveform playback and data capture, DAC, in-phase and quadrature-phase (IQ) modulation, and ADC for direct RF sampling. The DPD signal is in these experiments IQ modulated to the RF frequency to feed a class-J PA based on a gallium nitride high-electron-mobility transistor biased at 28 V drain-to-source voltage and −3.3 V gate-to-source voltage. The RF operating frequency is located at 875 MHz which is near the centre of the 3GPP band 26 (downlink) also employed for PPDR. The waveform at the PA output is then attenuated and RF sampled at 2457.6 MSa/s, to be further decimated and time aligned with the signal that was originally sent.

10.6.3.1 CFR simulations and measurement results

When using the PC-CFR algorithm in a digital front-end, the OoB distortion can be partially FIR filtered out. However, effectively filtering out the in-band distortion produced by the PC-CFR scheme in those subcarriers being deactivated for spectral coexistence can be hardly implementable and the FBMC subcarrier deactivation configuration may not be known in advance and change over time to adapt dynamically to the environment. To show how the PC-CFR algorithm impacts the in-band signal quality, we have benchmarked the normalised mean square error (NMSE) in a MATLAB simulator by sweeping the 5-stage PC-CFR threshold and weighting (alpha) parameters as shown in Figure 10.7 (top). For a given desired PAPR reduction shown in steps of 0.5 dB, the plot circles show the parameter subset (a threshold, α-subtraction factor) that maximises the NMSE. Considering the previous PC-CFR optimal parameter subsets, the EVM has been calculated to show how the CFR technique impacts the in-band signal quality (the higher the PAPR reduction the worse the EVM value that is obtained). Finally, Figure 10.7 (bottom) shows how the FBMC in-band distortion is increased with the PAPR reduction value (the OoB distortion, which would be equivalent to that generated in-band in terms of spectral density, has been ideally filtered by a 20 MHz bandpass filter), and thus how the FBMC waveform spectral containment is progressively diminished. This effect will increase the interference to the coexisting primary systems. With prior knowledge and coordination between heterogeneous systems (see scenario in Figure 10.1, bottom), the FBMC BB-PPDR transmitter could adaptively trade-off the PAPR reduction (which may impact the transmitter efficiency as it will be demonstrated) with the interference provoked to those co-primary signals while guaranteeing the EVM of the two heterogeneous systems.

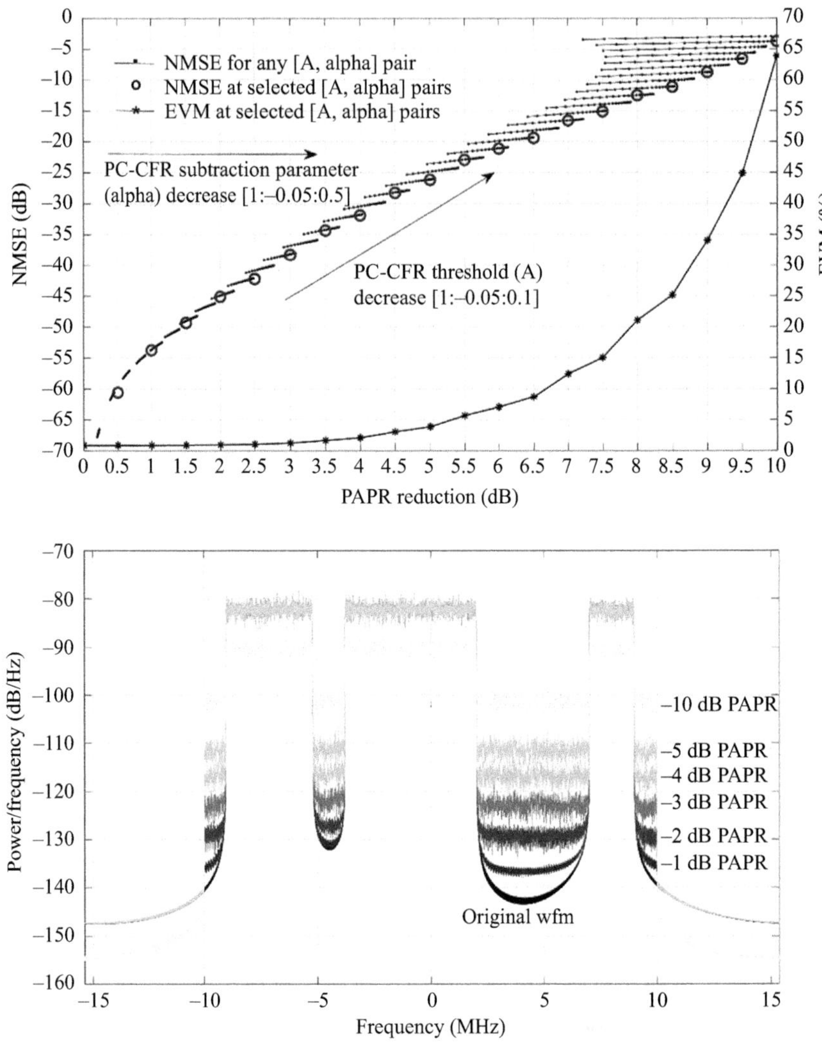

Figure 10.7 NMSE and EVM versus PC-CFR parameters (top), and FBMC waveform power spectral density versus PAPR reduction (bottom) simulation results

Figure 10.8 shows the spectra plots at the output of the PA (at +33 dBm mean output power) taken with the experimental test setup when the CFR is applied to the original signal (no DPD is applied yet). By analysing the third-order intermodulation distortion (IMD3), which is equivalent to the adjacent channel power rejection, and the fifth-order intermodulation distortion (IMD5), being equivalent to the alternate channel power rejection, the improvement is negligible for 3 dB PAPR reduction and between 2.5 dB and 5 dB (IMD3) for 7 dB PAPR reduction.

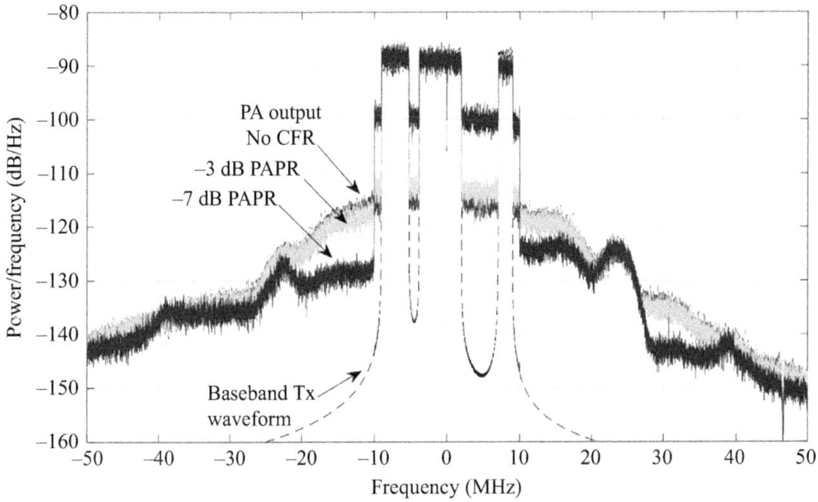

Figure 10.8 PA output spectra plots when only PC-CFR is applied to the FBMC waveform

The PC-CFR technique does not provide any relevant improvement when used in stand-alone mode in this test setup because the induced in-band distortion is too critical if we want to reduce the ACPR. If we consider the 7 dB PAPR reduction, the original EVM of 8.8 per cent at the output of the PA (no CFR applied) reaches an unaffordable value of 27.4 per cent and the distortion in the spectral holes is increased by about 14 dB on top of that generated by the PA distortion.

Despite the fact that the previous results show that using the PC-CFR technique as stand-alone digital linearisation scheme is not providing any relevant benefit in the scenario being evaluated, the rationale to include such a block follows: the predistorter may be assisted by the CFR block as a protection mechanism to minimise the effects of having signals with larger PAPR than those with the highest probability of occurrence and that are typically used for training the DPD or calculating a specific set of coefficients in non-adaptive DPD architectures. The low probability of occurrence but highest PAPR transmissions can feature up to 2–3 dB higher PAPR than the mean statistical value. Preventing the peaks from going hard into compression allows keeping under control the ACPR values and thus increasing the output power to have further system gain and power efficiency. In adaptive DPD architectures, the CFR mechanism facilitates compensating the response of a highly compressed PA (i.e. by the predistorter) by easing the parameter identification and the convergence of the DPD engine. It helps to keep under control the PAPR values of the predistorted signal being produced (always some dBs higher than the nominal ones as the compressed AM/AM curve needs to be inverted) which otherwise could affect the quantisation performance at DAC devices or impose hard restrictions on the mean power at the output of such devices to avoid DAC saturation.

10.6.3.2 DPD behavioural modelling simulation and CFR+DPD measurements

In this work, the DPD engine follows the direct learning architecture. To model the non-linearities of the PA, the test bench in Figure 10.6 is used to get the PA output waveforms that will be compared to the originally sent ones after time alignment and gain compensation. The GMP model in (10.16) has been employed for behavioural modelling in simulation. The chosen configuration leads to about 100 DPD coefficients which are a considerable amount but it is also a reasonable figure considering a PA showing very strong non-linearities due to operating under hard compression (no CFR is applied yet). The complexity of the DPD engine may be further reduced (i.e. by a factor from four to eight in the forward path and from six to ten in the feedback path) by applying the regularisation and dimensionality reduction techniques described in [40], where a very similar test setup is employed with four 20 MHz bandwidth carrier-aggregated FBMC signals (80 MHz overall bandwidth). By considering the GMP structure in the predistorter, the coefficients will be estimated iteratively according to the method shown in Section 10.6.2. Eight iterations (Tx-Rx and DPD processing loops) are used to reach the results found for different output power experiments (considering no DPD, DPD and CFR+DPD) that are shown in Table 10.2. At every experiment, the worst case IMD5, IMD3 and in-band channel power rejection (IBPR) values are shown together with the EVM and the PA drain efficiency. Figure 10.9 shows the spectra plots and constellation before and after linearisation for the +33 dBm (top) and +35 dBm (bottom) output power levels and the corresponding symbol constellations.

The results show the benefits of using the digital linearisation techniques. If the same PA is kept and only back-off was considered to meet the ACPR requirements, the mean output power would be reduced by more than 13 dB and the PA would be operated very inefficiently. When considering a nominal operating output power of +33 dBm, the IMD3 and the IBPR can be improved by about 18 dB and the EVM can be reduced to more than 7 per cent points, both on account of the DPD. When the output power is increased to +34 dBm, when only applying DPD the IMD3 (or ACPR) requirement of −45 dBc is no longer met because the predistorter cannot compensate the PA hard compression when inverting its response. If CFR is applied before DPD, the results show how it is feasible to reach at least a 2 dB higher output power level while meeting the ACPR and EVM requirements, by clipping the signal with some controlled in-band distortion. Thanks to increasing the output power, when comparing back-off operation with the nominal one at +33 dBm when applying DPD, the drain efficiency is enhanced by 31 per cent points. As a result of applying CFR+DPD, the drain efficiency can be enhanced by at least 12 per cent points more at the cost of increasing the in-band distortion. For this reason, the BS needs to be aware of the presence of co-primary signals and the interference received by the UEs to trade-off between the FBMC transmitter efficiency and the allowable interference to the co-primary system. A more detailed analysis in this regard is provided in the next section but the results indicate that

Table 10.2 Digital linearisation experimental results at different power levels

Case	Pout (dBm)	Worst IMD5 (dBc)	Worst IMD3 (dBc)	Worst IBPR (dBc)	EVM (%)	Drain Eff. (%)
No DPD+ back-off	20	−51.85	−45.74	−42.44	6.75	4
No DPD	33	−45.67	−30.83	−26.36	8.55	35
DPD	33	−55.57	−48.01	−44.98	1.12	35
No DPD	34	−46.72	−29.48	−24.38	11.41	41
DPD	34	−53.28	−43.16	−42.38	1.30	41
No DPD	35	−46.28	−28.35	−22.76	12.32	47
CFR[1]+DPD (−2.6 dB)	35	−53.76	−45.06	−35.26	3.10	47
CFR[2]+DPD (−4 dB)	35	−56.11	−53.13	−21.70	5.27	47

[1]2.6 dB PAPR reduction
[2]4.1 dB PAPR reduction

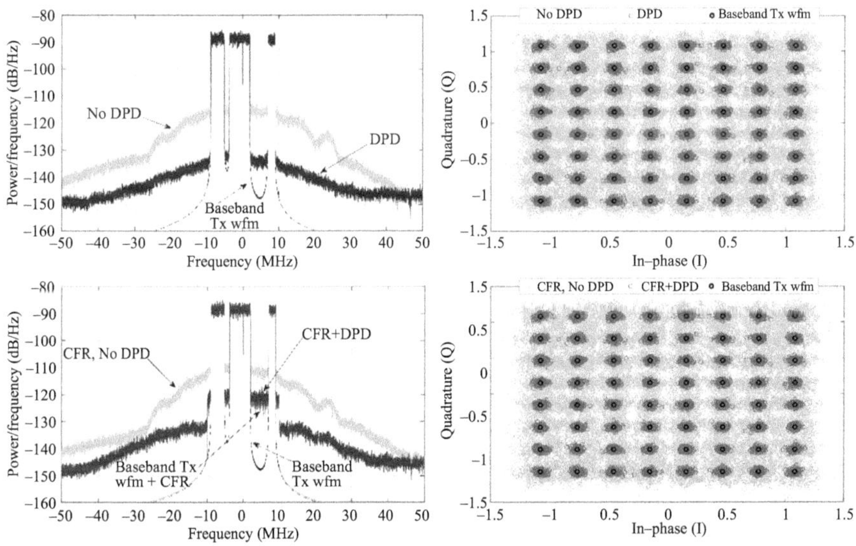

Figure 10.9 Spectra plots and constellations of the linearisation experiments at +33 dBm (top) and +35 dBm (bottom) mean output power levels

with CFR+DPD the PA drain efficiency could overpass the 50 per cent easily; The PA output power could be increased beyond 36 dBm if the interference to the in-band channels requirement was alleviated or the in-band co-primary channels were freed (in which case the FBMC waveform could also occupy the whole 20 MHz channel).

A comparison with the works that are found in the literature on class-J PA DPD can be hard to conduct given the different waveforms, PA devices and biasing conditions, and DPD engine complexity levels. The most recent work is found in [41]. The authors employ the same PA device and similar output power levels and compare the performance of their DPD technique by using a 5G-NR 20 MHz signal with some other previous works that use LTE waveforms. In this work, similar DPD results are reached in terms of ACPR and EVM, but at setting the gate-to-source voltage at a much lower value meaning that the PA efficiency is significantly better in our case. Moreover, the PAPR of the waveforms employed in our experiments is about 1.5 dB higher and the subcarrier spacing half of that found in the 5G-NR 20 MHz signal. By combining CFR and DPD this work can outperform the benchmarking in [41] in terms of ACPR (while keeping EVM at the required levels) at the cost of a higher system complexity (i.e. FBMC vs 5G-NR, CFR and direct learning DPD vs indirect learning DPD). The proposed solution, however, will operate more efficiently the PA and employ a waveform allowing better spectral coexistence than OFDM.

10.6.4 Impact of the digital linearisation in spectral coexistence

In this section, we emulate a coexistence scenario where the 1.4 MHz bandwidth and 5 MHz bandwidth LTE signals that coexist with the FBMC BB-PPDR transmission are received by different users. These users will filter out the content of the FBMC signal but they will receive a certain amount of FBMC to LTE interference in-band. The amount of interference is induced in these experiments by sweeping the power difference between the LTE signals (i.e. appearing in the deactivated subcarriers) and the broadband FBMC one from 0 dB to −25 dB. Figure 10.10 (top) illustrates this concept for a power difference swept between the LTE signals (which are set at same output power level) and the FBMC one. In this evaluation, we have considered the FBMC system experimental data taken with and without PA digital linearisation to demonstrate its beneficial impact from the waveform coexistence point of view.

The two LTE signals are 256-QAM modulated and the EVM is evaluated under the different LTE versus FBMC power ratios. As expected, the results for the 5 MHz bandwidth signal will be worse than those of the 1.4 MHz bandwidth signal because the in-band noise level will be higher in the first case. In Figure 10.10 (bottom), the benefits of using digital linearisation towards spectral coexistence are demonstrated. If we observe the results for the 5 MHz bandwidth LTE signal (worst case EVM), when comparing the traces with and without DPD and the LTE-FBMC power difference at which the 4.5 per cent EVM threshold is met, about 18.5 dB of signal-to-interference-plus-noise ratio (SINR) is gained. Nearly 6 dB will be gained when using CFR and DPD (i.e. for a 2.6 dB PAPR reduction setting). The 12.5 dB degradation in terms of robustness against interference when applying CFR and DPD (compared to applying only DPD) is coherent with the data shown in

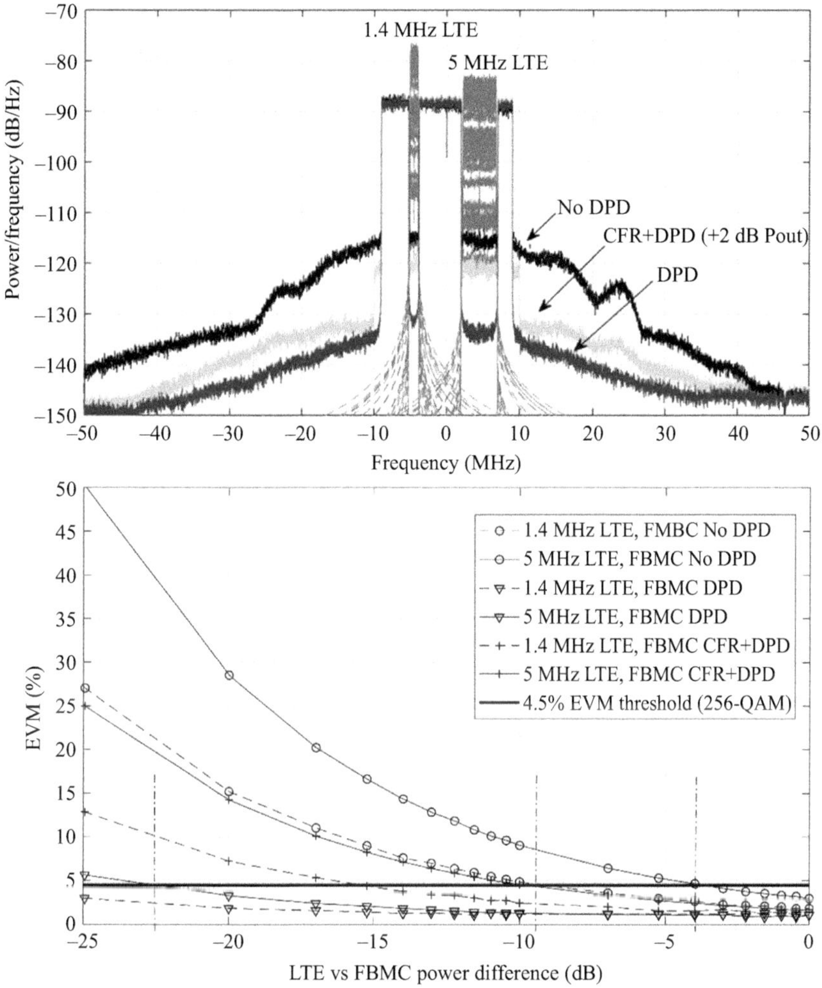

Figure 10.10 Waveform coexistence spectra plots (top) and EVM (bottom) for different LTE versus FBMC power ratios and digital linearisation experiments

Table 10.2. In this example, applying CFR may contribute to increase the FBMC transmitter output power and PA drain efficiency by 2 dB and 12 per cent points, respectively. Moreover, in a coordinated scenario (as in Figure 10.1 (bottom)), the amount of FBMC in-band induced CFR distortion could be adaptively controlled according to the LTE user's SINR information being shared with the FBMC BS. In such scenario, we consider the LTE-compatible frame structure of the FBMC system physical layer and assume that the rest of the protocol stack remains LTE both at the BS and the UE sides of the BB-PPDR radio access network. By adapting the PAPR reduction value accordingly, it is possible to meet both the FBMC EVM and ACPR requirements and tune the in-band distortion level to a level that maximises the FBMC transmitter efficiency and at the same time allows the LTE signals meeting the most stringent EVM requirement at 256-QAM subcarrier modulation.

10.7 Conclusion

The PPDR communications are used by first responders in humanitarian crisis zones constituting a vital service for saving lives. The international telecommunication regulation organisations are fostering the use of broadband services to expand the usability of PPDR communications in emergencies, and promoting more efficient and agile spectrum usage in interference-prone scenarios like those potentially appearing in the borders of neighbouring countries. The BB-PPRDR radios require highly performing and flexible waveform technologies to accommodate the local spectrum landscape and coexist with adjacent or in-band secondary or co-primary transmissions. Therefore, both maximising the spectral efficiency demanded by the broadband services and meeting the spectral emission requirements to avoid interference to the coexisting signals are needed. To meet the spectral efficiency requirements, multi-carrier modulations that generate high-PAPR waveforms are employed. Such waveforms put stringent requirements on the RF transceivers and the PA, which cannot be efficiently operated as significant back-off needs to be taken to meet the linearity requirements. Highly energy-efficient radios are also a must in emergency rollouts where the access to the energy supply can be constrained.

To allow for dynamic spectrum allocation while minimising the mutual interference between heterogeneous systems, the generated waveform must achieve a fine-grained control of the spectrum and each subcarrier must exhibit good frequency localisation. In this chapter, the use of a flexible FBMC waveform, featuring subcarrier deactivation to allow both OoB and in-band spectral coexistence, together with digital linearisation technologies is proposed. When compared to OFDM, FBMC features superior spectral efficiency and spectral containment. However, the PA non-linearities can minimise the advantages of the FBMC waveform which may be more sensitive than OFDM in this regard. To have a cost-effective and energy efficient radio equipment, FBMC is combined with CFR and DPD optimising the RF transceiver linearity versus efficiency trade-off. The benefit of the herein proposed solution for BB-PPDR has been experimentally validated with a test setup that includes the FBMC baseband modulator, a digital linearisation architecture based on a PC-CFR and an adaptive direct learning DPD architecture, and the necessary hardware to synthesise, amplify and capture the signal at RF. The relevant improvements in terms of EVM, ACPR, power efficiency and spectral coexistence between LTE and FBMC have been demonstrated.

Acknowledgements

This work was supported in part by the Spanish Government (Ministerio de Ciencia, Innovación y Universidades) and the Fondo Europeo de Desarrollo Regional (FEDER) under grants RTI2018-099841-B-I00, TEC2017-88373 and TEC2017-83343-C4-2-R and in part by the Generalitat de Catalunya under grants 2017 SGR 891, 2017 SGR 1479 and 2017 SGR 813.

The authors would like to thank Dr J.A. García, Dr M.N. Ruiz, Dr J.R. Pérez-Cisneros, and D. Vegas from the Universidad de Cantabria for the donation of the class-J PA used in Section 10.6.3, and A. Mars and M. Cañal for their contribution to the results in Section 10.6.4.

References

[1] International Telecommunication Union (ITU). *Final Acts WRC-15 World Radiocommunication Conference*. ITU-R Conference Publications, Geneva, 2015

[2] 3rd Generation Partnership Project (3GPP). *Technical Specification Group Services and System Aspects; Mission Critical Communication Interworking with Land Mobile Radio Systems; Stage 2 (Release 16)*. 3GPP TS 23.283 V16.3.0 Technical Specification, Sophia Antipolis, Jun 2019.

[3] International Telecommunication Union (ITU). *The Use of International Mobile Telecommunications (IMT) for Broadband Public Protection and Disaster Relief (PPDR) Applications. M Series Mobile, Radiodetermination, Amateur and Related Satellite Services*. Report ITU-R M22.91-1, Geneva, Nov 2016.

[4] 3rd Generation Partnership Project (3GPP). *Technical Specification Group Services and System Aspects; Common Functional Architecture to Support Mission Critical Services; Stage 2 (Release 16)*. 3GPP TS 23.280 V16.3.0 Technical Specification, Sophia Antipolis, Jun 2019.

[5] Global System for Mobile Communications Association (GSMA). Securing the Digital Dividend Across the Entire ASEAN, A Report on the Status of the Implementation of the APT700 Band for ATRC. GSMA Executive Summary and Overview, London, Aug 2018.

[6] European Conference of Postal and Telecommunications Administrations (CEPT), European Electronic Communications Committee (ECC). *Harmonised Conditions and Spectrum Bands for the Implementation of Future European Broadband Public Protection and Disaster Relief (BB-PPDR) Systems*. ECC Report 218, Copenhagen, Oct 2015.

[7] European Conference of Postal and Telecommunications Administrations (CEPT), European Electronic Communications Committee (ECC). *Compatibility Studies Regarding Broadband PPDR and Other Radio Applications in 410-430 MHz and 450-470 MHz and Adjacent Bands*. ECC Report 240, Copenhagen, Sept 2015.

[8] European Conference of Postal and Telecommunications Administrations (CEPT), European Electronic Communications Committee (ECC). *Compatibility and Sharing Studies for BB PPDR Systems Operating in the 700 MHz Range*. ECC Report 239, Copenhagen, Sept 2015.

[9] Jarwan A., Sabbah A., Ibnkahla M., and Issa O. 'LTE-based public safety networks: a survey'. *IEEE Communications Surveys & Tutorials*, 2nd quarter 2019, vol. 21, no. 2, pp. 1165–1187.

[10] European Telecommunications Standards Institute (ETSI). *Reconfigurable Radio Systems (RRS); Evolved Licensed Shared Access (eLSA); Part 1: System Requirements*. ETSI TS 103 652-1 V1.1.1 Technical Specification, Sophia Antipolis, Feb 2019.

[11] European Telecommunications Standards Institute (ETSI). *Reconfigurable Radio Systems (RRS); Feasibility Study on Temporary Spectrum Access for Local High-Quality Wireless Networks*. ETSI TR 103 588 V1.1.1 Technical Report, Sophia Antipolis, Feb 2018.

[12] Guerzaguet R., Bartzoudis N., and Gomes Baltar L. *et al.* 'The 5G candidate waveform race: a comparison of complexity and performance'. *EURASIP Journal on Wireless Communications and Networking*, Jan 2017, art. no. 13.

[13] Font-Bach O., Bartzoudis N., and Mestre X. *et al.* 'When SDR meets a 5G candidate waveform: providing efficient use of fragmented spectrum and interference protection for PMR networks'. *IEEE Wireless Communications*, 2015, vol. 22, no. 6, pp. 56–66.

[14] Font-Bach O., Bartzoudis N., and López-Bueno D. *et al.* 'Real-time implementation and experimental validation of a DL FBMC system'. *Orthogonal Waveforms and Filter Banks for Future Communication Systems*, Elsevier - Academic Press, Jul 2017, pp. 513–546.

[15] Farhang-Boroujeny B. 'OFDM versus filter bank multicarrier'. *IEEE Signal Processing Magazine*, 2011, vol. 28, no. 3, pp. 92–112.

[16] Zhang X., Jia M., Chen L., Ma J., and Qiu J. *'Filtered-OFDM – Enabler for flexible waveform in the 5th generation cellular networks'*. IEEE Globecom; San Diego (CA), USA, 2015.

[17] Vakilian V., Wild T., Schaich F., ten Brink S., and Frigon J.-F. 'Universal-filtered multi-carrier technique for wireless systems beyond LTE'. *IEEE Globecom Workshops*; Atlanta (GA), USA, 2013.

[18] Fettweis G., Krondorf M., and Bittner S. 'GFDM – generalized frequency division multiplexing'. *Proceedings of the IEEE 69th Vehicular Technology Conference*; Barcelona, Spain, 2009, pp. 1–4.

[19] Siohan P., Siclet C., and Lacaille N. 'Analysis and design of OFDM/OQAM systems based on filterbank theory'. *IEEE Transactions on Signal Processing*, 2002, vol. 50, no. 5, pp. 1170–1183.

[20] Pérez-Neira A. I., Caus M., and Zakaria R. *et al.* 'MIMO signal processing in offset-QAM based filter bank multicarrier systems'. *IEEE Transactions on Signal Processing*, 2016, vol. 64, no. 21, pp. 5733–5762.

[21] Bellanger M. G. 'Specification and design of a prototype filter for filter bank based multicarrier transmission'. *Proceedings of the 2001 IEEE*

International Conference on Acoustics, Speech, and Signal Processing; Salt Lake City (UT), USA, 2001, pp. 2417–2420.

[22] Bellanger M. 'FS-FBMC: an alternative scheme for filter bank based multicarrier transmission'. *Proceedings of the 5th International Symposium on Communications, Control and Signal Processing*; Rome, Italy, 2012, pp. 1–4.

[23] Waldhauser D.S., Baltar L.G., and Nossek J.A. 'MMSE subcarrier equalization for filter bank based multicarrier systems'. *Proceedings of the IEEE 9th Workshop on Signal Processing Advances in Wireless Communications*; Recife, Brazil, 2008, pp. 525–529.

[24] Mestre X., Majoral M., and Pfletschinger S. 'An asymptotic approach to parallel equalization of filter bank based multicarrier signals'. *IEEE Transactions on Signal Processing*, 2103, vol. 61, no. 14, pp. 3592–3606.

[25] European Telecommunications Standards Institute (ETSI). *LTE; Evolved Universal Terrestrial Radio Access (E-UTRA); Base Station (BS) Conformance Testing*. 3GPP TS 36.141 V12.5.0 R12 Technical Specification, Sophia Antipolis, Oct 2014.

[26] Rahmatallah Y., and Mohan S. 'Peak-to-average power ratio reduction in OFDM systems: a survey and taxonomy'. *IEEE Communications Surveys & Tutorials*, 4th quarter 2013, vol. 15, no. 4, pp. 1567–1592.

[27] Rahim M. U., Hidalgo Stitz T., and Renfors M. 'Analysis of clipping-based PAPR-reduction in multicarrier systems'. *Proceedings of the IEEE 69th Vehicular Technology Conference*; Barcelona, Spain, 2009, pp. 1–5.

[28] Kollar Z., Varga L., and Czimer K. 'Clipping-based iterative PAPR-reduction techniques for FBMC'. *Proceedings of the 17th International OFDM Workshop 2012 (InOWo'12)*; Essen, Germany, 2012, pp. 1–7.

[29] Varghese N., Chunkath J., and Sheeba V. S. 'Peak-to-average power ratio reduction in FBMC-OQAM System'. *Proceedings of the Fourth International Conference on Advances in Computing and Communications*; Cochin, India, 2014, pp. 286–290.

[30] Van der Neut N., Maharaj B. T., de Lange F. H., González G., Gregorio F., and Cousseau J. 'PAPR reduction in FBMC systems using a smart gradient-project active constellation extension method'. *Proceedings of the 21st International Conference on Telecommunications (ICT)*; Lisbon, Portugal, 2014, pp. 134–139.

[31] Gotthans T., Maršálek R., Gotthans J., and Baudoin G. 'Influence of filter-bank RF transceiver chain imperfections on digital predistortion performance'. *Proceedings of the International Symposium on Wireless Communication System (ISWCS) 2016*; Poznan, Poland, 2016, pp. 623–627.

[32] Bulusu S. S., Shaiek H., and Roviras D. 'PA linearization of FBMC-OQAM signals with overlapped recursive error correcting predistortion'. *Proceedings of the International Symposium on Wireless Communication System (ISWCS) 2016*, 20–23 Sept 2016, pp. 533–537.

[33] Gotthans T., Marsalek R., Blumenstein J., and Baudoin G. 'Experimental evaluation of digital predistortion with FBMC and of DM signals'.

Proceedings of the 16th Annual Wireless and Microwave Technology Conference (WAMICON); Cocoa Beach (FL), USA, Apr 2015, pp. 1–3.
[34] Zayani R., Medjahdi Y., Bouhadda H., Shaiek H., Roviras D., and Bouallegue R. 'Adaptive predistortion techniques for non-linearly amplified FBMC-OQAM signals'. *Proceedings of the IEEE Vehicular Technology Conference*; Seoul, Korea, May 2014, pp. 1–5.
[35] Kim W.-J., Cho K.-J., Stapleton S.P., and Kim J.-H. 'An efficient crest factor reduction technique for wideband applications'. *Analog Integrated Circuits and Signal Processing*, 2007, vol. 51, no. 1, pp. 19–26.
[36] Braithwaite R. N. 'A comparison of indirect learning and closed loop estimators used in digital predistortion of power amplifiers'. *Proceedings of the 2015 IEEE MTT-S International Microwave Symposium*; Phoenix (AZ), USA, 2015, pp. 1–4.
[37] Morgan D. R., Ma Z., Kim J., Zierdt M. G., and Pastalan J. 'A generalized memory polynomial model for digital predistortion of RF power amplifiers'. *IEEE Transactions on Signal Processing*, 2006, vol. 54, no. 10, pp. 3852–3860.
[38] Pham Q. A., López-Bueno D., Wang T., Montoro G., and Gilabert P. L. 'Multi-dimensional LUT-based digital predistorter for concurrent dual-band envelope tracking power amplifier linearization'. Proceedings of the 2018 IEEE Topical Conference on RF/Microwave Power Amplifiers for Radio and Wireless Applications (PAWR); Anaheim (CA), USA, 2018, pp. 47–50.
[39] López–Bueno D., Pham Q. A., Montoro G., and Gilabert P. L. 'Independent digital predistortion parameters estimation using adaptive principal component analysis'. *IEEE Transactions on Microwave Theory and Techniques*, 2018, vol. 66, no. 12, pp. 5771–5779.
[40] Pham Q. A., Montoro G., López-Bueno D., and Gilabert P. L. 'Dynamic selection and estimation of the digital predistorter parameters for power amplifier linearization'. *IEEE Transactions on Microwave Theory and Techniques*, 2019, vol. 67, no. 10. pp. 3996–4004.
[41] Madero-Ayora M. J., Crespo-Cadenas C., and Becerra-González J. A. 'Procedimiento de linealización de amplificadores de alta eficiencia operando cerca de saturación'. *Proceedings of XXXIV Simposium Nacional de la Unión Científica Internacional de Radio (URSI)*; Sevilla, Spain, Sept 2019.

Chapter 11

Connecting the disconnected: a combination of DTN, CDN, TCP/IP, and OpenFlow approach

Khondokar Habibul Kabir[1]

Information technology plays a vital role in the development of every aspect of our lives. Internet, one of these many technologies, is the most utilised entity for accessing the ocean of information. However, many remote regions of the world are still deprived of this information available today. The prime reasons behind this deprivation are inadequate infrastructure facilities and remoteness of the concerned places. Delay-tolerant networks (DTNs) support communication in an unorthodox way for the intermittent networks, where direct links always may not be available. A network architecture by combining DTNs, content delivery network using TCP/IP, and OpenFlow with the help of wireless connectivity enabled devices can be one of the solutions to mitigate this problem. The network scheme aims to be implemented in Third-World countries where sometimes technological amenities are scarce.

11.1 Introduction

At this age of technology, most of the places in the world are interconnected via the Internet. The people of developed countries have flawless, fast, and reliable connectivity. However, unfortunately, the scenario is not quite the same in most of the developing countries or the Third-World countries. As a result, the rural people here are significantly deprived of the ocean of information available on today's Internet; whereas, extensive use of information technology could be an excellent catalyst for rural development.

The people, living in such disconnected scenarios, do not even get their fundamental amenities, let alone information, and technologies. So, they are somewhat detached from modern technologies. For example, according to the Bangladesh Telecommunication Regulatory Commission (November 2018), 65% of rural communities in Bangladesh do not have Internet facilities to get the necessary information [1].

[1]Department of Electrical and Electronic Engineering (EEE), Islamic University of Technology (IUT), Gazipur, Bangladesh

According to World Internet Statistics (2017) [2] in Eritrea, Fiji, Niger, Chad, and Congo, only 1.3%, 1.5%, 2%, 2.6%, and 3.8% of the countries' respective population are under Internet coverage. Besides, more than 94% of the total population in Western Sahara, Sierra Leone, Central African Republic, Madagascar, Somalia, Guinea-Bissau, and Burundi are deprived of Internet connectivity [2].

Besides, the regions where conventional networks are not possible such as communities on hilly mountain areas or communities on islands, for example islands in the Maldives, Comoros, Indonesia, Fiji, and the Philippines, are also deprived of part of the modern network coverage and hence are separated from the pool of information superhighway. Moreover, intra-rural community communication, that is communication among neighbouring communities, for information exchange, also has a significant impact on the journey to rural development. The utilisation of information from neighbouring communities brings remarkable benefits in rural socio-economic development, which have high potentials in rural community empowerment.

The proposed model overcomes the crisis of information and enhances the development of the concerned places. The people, in those areas, do not have access to the online newspaper, e-books, stock rates, current market prices, etc., and all other latest events happening around the world. People of every line of works need updated information related to their respective jobs. A rural farmer needs to know the current market price of daily essentials in the city. He may need to know the appropriate time and state-of-the-art rules of sowing, irrigating, fertilising, and harvesting to get maximum crop yield. A rural fisherman needs to know the fish market price for his business purpose, while a student at a remote village has the right to access the e-books and other educational materials available online just like an urban student.

Besides these, most of the rural people are not often technically sound even to operate today's state-of-the-art communication technologies. This limitation creates one more challenge to ensure information availability among rural people. Thus, the prime objective of this proposed network platform is to bring the disconnected and technically less sound rural people under the network coverage, letting the deprived rural community access desired information in time.

Unsoundness in the technical domain is one of the significant barriers before the target clients to utilise the Internet facility. Therefore, providing the Internet facility at anyhow is not enough for the target clients. Instead, a separate entity should be available before them to collect and serve their demand for information in a convenient way. Most of the infra-structure-less areas in the least developed countries do not have enough infrastructure to provide Internet facilities, and hence this is our cordial effort to put up the desire information towards the target clients by even tolerating long delays. Yet, it is better to achieve the desired content even experiencing a long delay than getting nothing. Note here that in this proposed network real-time data communication such as voice over or playing the video may not well match.

11.2 Network model for connecting the disconnected

The prime reasons behind this deprivation are inadequate infrastructure facilities and remoteness of the concerned places that cause a delay in data communication because of discontinuation and disruption.

From this motivation, the proposed model uses a combination of delay-tolerant networking (DTN) [3] to make the disconnected rural community connect with global resources. To implement the proposed model, a smart application package is essential to collect and serve the demands of target clients in a convenient way which involves user-friendly interaction for technically less sound communities.

DTN is a unique approach to computer network architecture that enables communication in heterogeneous networks and allows data transfer in situations where traditional networks are obsolete, that is where physical end-to-end connectivity is not present. To communicate over long distances through time delays, DTN becomes the standard method nowadays. The fundamental technique of DTN referred to as the bundle protocol (BP) [4] is similar to the transmission control protocol/Internet protocol (TCP/IP) [5] that is the building block of the Internet. The big difference between BP and IP is that, while IP assumes a more or less smooth pathway for packets going from start to endpoint, whereas, BP has the capabilities to store the data and allows data communication through disconnections, disruptions, discontinuation, glitches, etc. The DTN architecture creates an overlay networking combining an in-network or node-level storage. The bundle in the DTN architecture is referred to as an application message. This bundle is stored in node-level storage on DTN gateway (or nodes) for an arbitrary length of time. It waits for finding the forwarded path to reach the destination. This differs from the IP model, where packets must be forwarded immediately, otherwise, dropped.

In the proposed DTN scenario, the network model is an example of the inter-cluster DTN referred to as ferry-assisted multi-cluster DTN, which is shown in Figure 11.1. It consists of several isolated *clusters* (e.g. towns and villages) where each consists of several *nodes* (e.g. houses/clients). In each cluster, nodes can directly communicate only with other nodes through Wi-Fi within the transmission range. To collect data bundles from one cluster to distantly located another cluster, we apply the ferry-initiated message ferry scheme [6–13]. The *message ferry* refers to as *movable server* in the proposed scenario, collects the client's requested data from the primary *admin server* in the town with Internet connectivity, and then transfers it to the villages (without Internet connectivity) by travelling in a round trip. Within a cluster, the movable server (i.e. the ferry) and several nodes (i.e. the clients) connect each other through Wi-Fi and synchronise their data. In some cases, in a cluster, the movable server connects one *local server* (special node), where that specific node is responsible for storing the data and delivering them to other clients/nodes within the cluster. Here, ad hoc network is used to connect the Wi-Fi-enabled wireless devices. This networking technique allows practically establishing connectivity within a closed remote boundary using wireless-equipped devices. In the proposed network platform, the Wi-Fi-enabled wireless-equipped

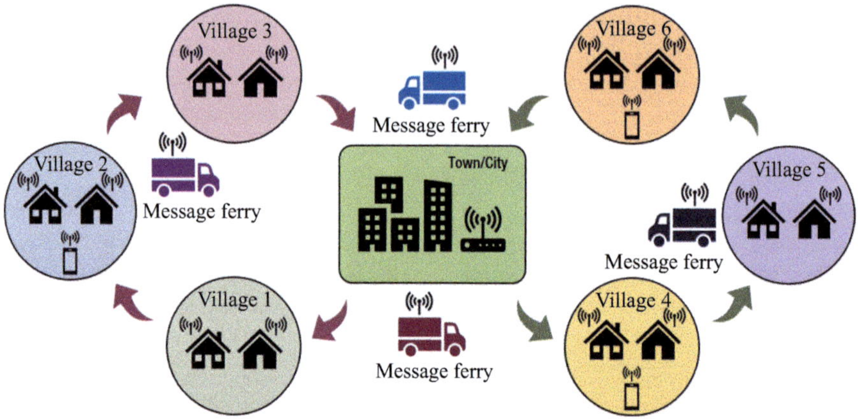

Figure 11.1 Ferry-assisted multi-cluster delay-tolerant network

computing devices are used for its higher data transfer speed, ability to connect to multiple users, and relatively better radio signals. The proposed DTN scenario differs from the standard DTN scenario used for challenged networks, where the network must rely on the opportunistic connectivity. On the contrary, in the proposed ferry-assisted DTN scenario, the locations of clusters (i.e. distant villages) are permanent, and the message ferry (movable vehicle equipped with wireless device) travels round trips in the predefined path to communicate nodes in each cluster. Here, nodes know the path and data delivery point of the message ferry as well as where and when to collect data from them. Although the ad hoc network connects the physically closed message ferry and one node (or server), it establishes the reliable end-to-end connectivity between them. Then, the message ferry moves to another cluster and again establishes the reliable end-to-end connectivity between the physically closed node in that cluster. Thus, when we consider data communication between the source (i.e. admin server) to the destination (i.e. node in a distant cluster), the overall network does not have any end-to-end connectivity; however, during data synchronisation between message ferry and each specific node, there establishes reliable end-to-end connectivity. The data exchange between physically closed message ferry, and each node is accomplished through an ad hoc network facilitated by TCP/IP. Thus, we can utilise all the advantages of TCP/IP. Therefore, instead of using BP, we use TCP/IP. It is advantageous in the current scenario because most of the state-of-the-art devices use TCP/IP coherently, which is easier to customise for the ad hoc network. However, the proposed platform is fully adaptable with BP, and the proposed platform can be customisable and convertible for BP at any time.

In the proposed scenario, for better data delivery performance, the technique of content delivery networking (CDN) [14] is also applied, where the movable server (i.e. message ferry) acts as a CDN server. For example, the CDN server copies the pages of a website caching the contents of the page to a network of servers located geographically different places. When a user requests that website, that is part of a content delivery network, the CDN server delivers the cached contents by redirecting the request from the server of the originating site to a server in the CDN that is closest to the user. When the connection is available, the CDN delivers any cached content to the originating server that has not been previously cached.

The proposed network platform adopts CDN technique [14] to improve the data delivery latency, in combination with TCP/IP [5] and OpenFlow* protocol suite for data transmission mechanism with the help of wireless connectivity enabled devices in an ad hoc manner. OpenFlow approach defines the route of network packets across switches over a network, that is it facilitates network management between routers and switches of different models and different vendors.

The primary contributions are as follows. First, a sustainable network model and corresponding related software entities are proposed to connect the disconnected rural communities within themselves and with the global resources. Second, a bundle of potential applications of information technology in the rural development sector to improve the socioeconomic prospects is proposed adopting the network model.

11.3 Related work

Many unorthodox and unique alternative schemes are proposed to connect the disconnected communities [15,16]. Among them many are tested or in the process of being tested, for example, Bytewalla [15,17], DakNet [18], 7DS [19], SNC [20], SWIM [21], and Motopost [22]. DakNet [18] uses Linux operating system-based Laptops with Wi-Fi located inside the bus which travels among rural villages. Here, along with the bus route, users can download and upload the e-mails for transferring to the Internet or other villages. On a similar kind of bus network, a system of throwboxes [23,24] was implemented to extend the capacity of the DTN. Besides, KioskNet [25] creates a rural Internet kiosks network which provides data transferring services in disconnected villages using vehicles with Laptops. KioskNet exchanges the data between the kiosks or Internet gateways. On the other hand, cinema-in-a-backpack [26] delivers movies (educational and entertainment) in remote villages like DakNet [18]. Besides, TACO-DTN [27] disseminates a content-based information system composed of fixed and mobile infostations. It allows mobile users to subscribe to media contents for a period of time. Also, campus bus networks designed to serve the commuters inside the university campus are proposed in [28–30]. It may be noted that a significant number of disconnected rural communities are getting connectivity by various telecommunication companies. Although they are offering many Internet schemes before them, still it is expensive and unaffordable for most disconnected communities.

*OpenFlow is a communications protocol, managed by Open Networking Foundation (ONF).

In most of the related works, they use BP, which limits the mass adaptation with existing devices available there. Moreover, for better data delivery, so far from our knowledge, none of them considers the smart management of content, delivery offers by CDN, and/or per individual user account management. Furthermore, very few of them specifically address potential user applications, their applicability, and prospect related to the village people associated with the system. The proposed network platform incorporates all the above facilities by focusing on the technically less sound community.

11.4 Details of network entities

The proposed network establishes connectivity by implementing the combined concept of DTN and CDN, where TCP/IP suite and OpenFlow are utilised for data transmission mechanism. It employs wireless-equipped smart devices (for example, smartphones, tablets, etc.), which work in store-carry and content delivery fashion. However, as discussed earlier, unlike traditional DTN, end-to-end connectivity between any two communicating devices is possible in the proposed network. Hence, the proposed network allows data transfer among wireless-equipped devices employing TCP/IP, a standard transport layer protocol. This also improves compatibility with other platforms. Note that the applications can be flexibly extended through a software update, to support any other protocol, for example bundle protocol.

To ensure effective data communication facility among the protocol layers, with the help of the host operating system, the TCP layer incorporates with the IP layer and medium access control layer. *AsthaNet* adopts OpenFlow that allows switches from different vendors to be remotely managed by a single open protocol. OpenFlow provides software-defined networking. In OpenFlow, all control can be flexibly configured through software by separating the programming of switches and routers from their hardware (no hardware configuration needs to be done).

Note that since we are considering a challenged network situation or alternative to the Internet system, real-time communication is compromised in the proposed network platform.

The basic idea of the data transfer mechanism is illustrated in Figure 11.2 and explained by the following points.

1. Two wireless-equipped devices create wireless ad hoc networks and transfer data among them when devices are within the transmission range.
2. No data transfer occurs when the devices are out of the range.
3. The devices can again transfer data among themselves when a third wireless device is placed between these out of range devices.
4. When a wireless device is placed inside a house, or a vehicle or a building, then that house, vehicle, or the building can transfer data among themselves.
5. While one of the connecting devices is placed inside a moving object, connectivity range among the disconnected devices increases to a great extent.

The proposed network comprises basic two entities: (i) end user and (ii) server.

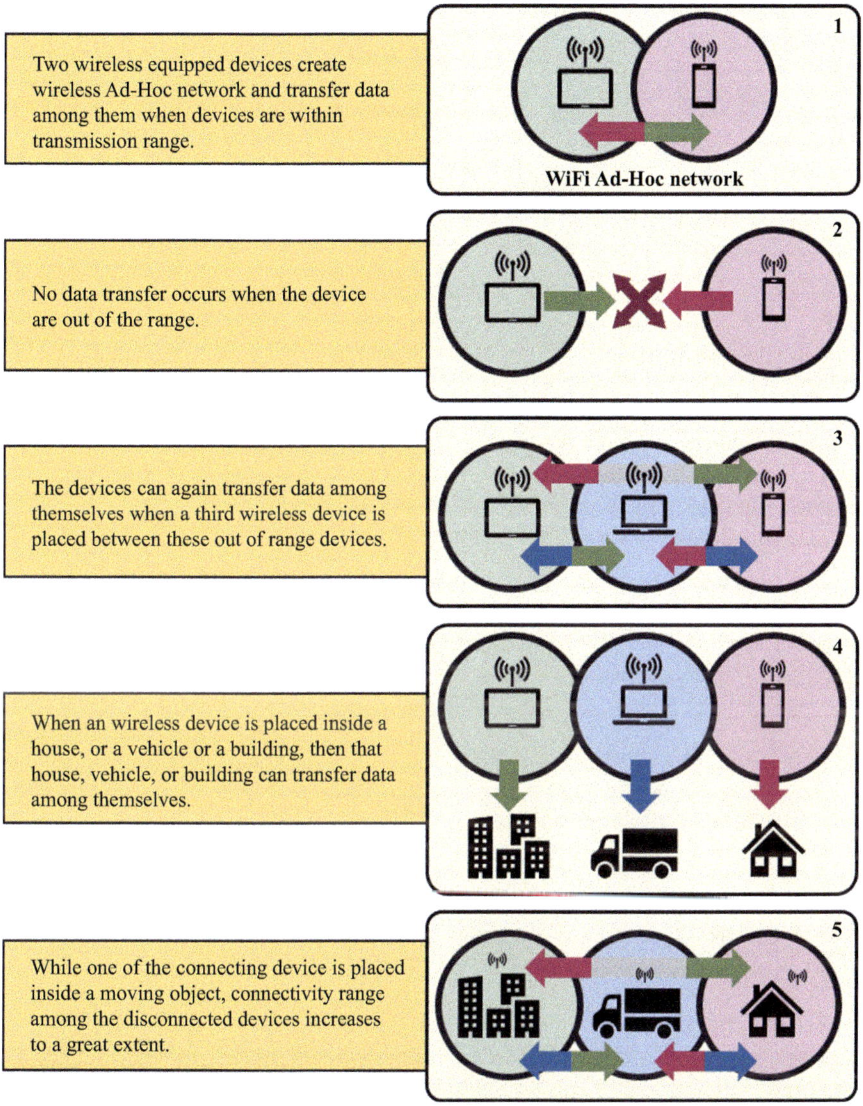

Figure 11.2 Fundamental idea of the proposed networking

11.4.1 End user

End user is referred to as an individual from a disconnected community possessing a wireless device (e.g. smartphone, tablet, etc.) as shown in Figure 11.3. Here, ad hoc network is used to connect the Wi-Fi-enabled wireless devices to each other within the transmission range. Wi-Fi technology is considered because of its higher data transfer speed, ability to connect multiple users, and relatively better radio

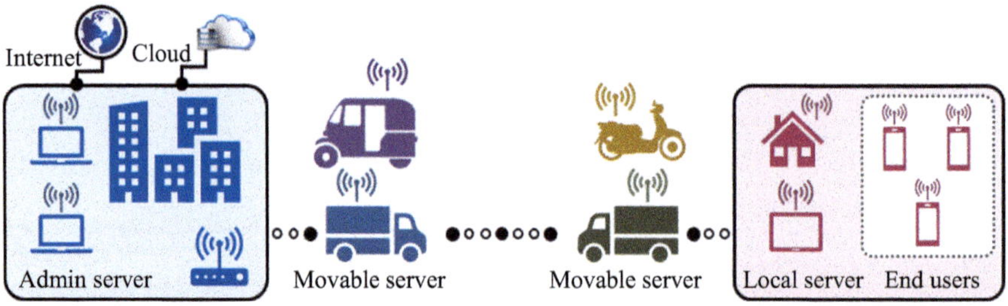

Figure 11.3 Different entities and servers of the proposed network platform

signals. Besides, for the clients, the Wi-Fi-equipped small smart devices, that is, smartphone, tablet, etc., with android operating system become the best choice due to the cheaper price, better processing specification, better software customisation option, openness for an upgrade, easy user interface, and other additional attractive features. Their prices are becoming more reasonable and more affordable to general people day by day.

11.4.2 Server

Wireless network interface card equipped computing device with moderate computing and storage capabilities acts as a server in the proposed network. It is responsible for one or more of the following tasks.

- Collecting user queries and data.
- Carrying user queries and data.
- Processing user queries and data.
- Delivering user-desired data.

Thus, as shown in Figure 11.3, the proposed network requires the following three categories of servers: (i) admin server, (ii) movable server, and (iii) local server.

11.4.2.1 Admin server

A computing device directly connected with a high-speed Internet connection and possessing high storages and computing capabilities acts as *admin server*. It is installed at the nearest urban area where a high-speed Internet connection is available. Availability of bulk of sequence of data and global Internet facilities allow the admin server to process the user queries and to present meaningful information for end users. However, such processing requires a few manual operations accomplished by one or more human operators. Therefore, the proposed network requires employing one or more human entities to coordinate the processing of user queries employing the Internet facility. However, processed data are delivered to the movable server by establishing an ad hoc network while it comes within its transmission range. It launches a reliable TCP connection while establishing the ad hoc networking between them. It is to be noted that, admin server plays the role of a client device while it receives row data or user queries from the movable server.

11.4.2.2 Movable server

Server carried by the message ferry acts as a *movable server* in the proposed network. Different motor or nonmotor vehicles in any disconnected communities are utilised in the proposed network. These vehicles are used to carry the movable server across the disconnected villages. Any traditional vehicle used in intra-village transportation can also be utilised for this purpose as shown in Figure 11.3. Thus, the utilisation of available resources from disconnected communities are adopted. However, the cost of power consumption by the movable server can be minimised by fitting a solar panel on the vehicle's top. Therefore, uninterrupted data delivery can be facilitated in case of power failure or any disaster scenario.

Movable server roams around the disconnected villages for collecting user queries and delivering processed information. It collects furnished data from the admin server and delivers those to appropriate local servers. Moreover, the collection of row data or user queries from local servers and the delivery of those to the admin server also belong to its responsibilities. While passing through any local server or admin server, it waits for a significant amount of time to deliver and collect respective information. Once a reliable TCP connection is established between these connected devices, data transfer goes on until it goes out of their mutual transmission range. However, the movable server also behaves like a client while it collects information from the admin server or local server. Note here that if required secure communications can be provided by transport layer security or secure sockets layer to secure all communications between the server and clients and to provide privacy and data integrity between them.

11.4.2.3 Local server

A server owning moderate storage and minimal computing capabilities take the role of a local server in the proposed network. It is placed in a general meet point of a certain class of disconnected community people. It serves the end users directly by delivering the desired information and collecting user queries. It establishes a reliable TCP connection with the movable server, while the movable server comes within its transmission range. Once the connection is established, the local server acts as a client with respect to the movable server. Thence, it collects processed data and delivers user queries to the movable server. Figure 11.4 depicts the role of these servers.

End users collect and store desire information from the nearest local server by reaching within the transmission range of the local server. Hence, an ad hoc network is established employing TCP connection. However, a local server also projects a few important information on its display board or wider screen. Thus, people from disconnected communities get essential and general information without having their own devices even.

Let us have a practical scene of the proposed network where rural businessmen are willing to know the market price of rice and farmers desire to know the perfect

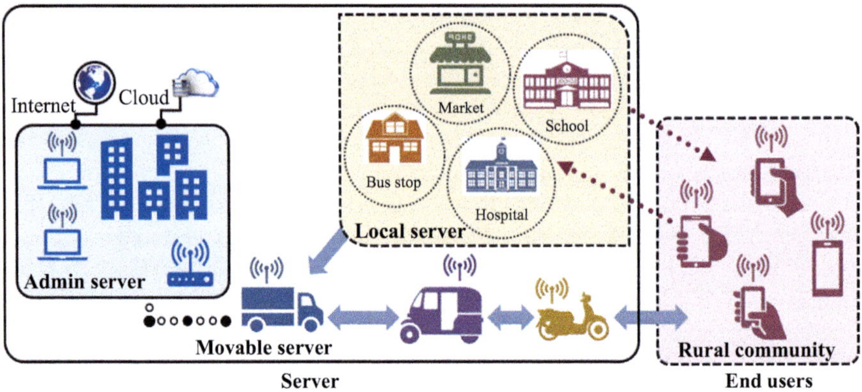

Figure 11.4 Role of different servers

time of employing different insecticides and place their respective queries to the nearest local server. These queries will be processed by the admin server and will be available at that local server according to the mentioned principle. However, these businessmen have to collect inquired information from that local server along with those farmers whose demands belong to a different domain of interest. The performance of such a data delivery procedure at a local server puts down in a highly dense community as data from all domains are placed together. Thus, the efficiency of the whole process degrades, as dealings of all categories of data together create complexity in local server and inflexibility for end users.

To elevate the network performance, the proposed network introduces the use of multiple local servers in a single disconnected community, where each server is dedicated to a particular domain of interest. The network proposes the following categories of local servers.

- Local server at the school.
- Local server at the hospital.
- Local server at the sports centre.
- Local server at the business centre.
- Local server at the entertainment centre.
- Local server for miscellaneous purposes.

11.5 Data flow mechanism

The movable server travels to all the local servers and uploads user queries as well as row data and delivers already processed data to different local servers based on the category of data. The proposed network uses identification keywords to distinguish between the category types. Processed data are transferred to the appropriate local servers depending on the matched keyword. Figure 11.5 describes all the local servers. People desire for particular information collects from the

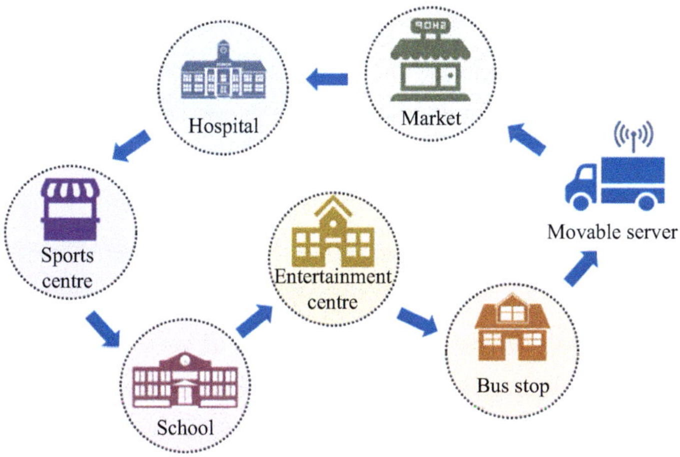

Figure 11.5 Data flow of the proposed network model

respective local server. Hence user density is lessened, and message exchange takes place with higher speed and reliability.

11.5.1 Multi-user data management

Once a connection is established, the data management system compares the contents present at both storage devices and synchronises data with the updated and latest one and deletes old data. Through this process, the storage reuse can be ensured for the overall network.

As information is not delivered to the end users immediately, storage management in both devices of *server* and *end user* is another important issue. A priority queue mechanism manages this aspect efficiently. Data management system categorises data sorted by private data (personalised requested data for registered clients) or public (general information for all clients). If private data, then the data are ensured restriction from unauthorised access and deleted by the administrative server from memory after delivery to ensure privacy and reuse of memory. However, frequently asked popular public information contents remain in the memory of the offline local servers for future use until a predefined period which can be instantly accessed without any online Internet connection.

11.5.2 Content delivery mechanism

The proposed network includes an efficient content delivery procedure which synchronises the information contents among the communicative devices. Once a connection is established, a content delivery unit at the server-side compares the contents present at both storage devices. Thence, it synchronises both storage devices with updated and latest data and information. However, for smarter and faster data delivery, the proposed scheme applies the concept of caching and redirecting the requested contents from the closest server instead of the admin server if possible.

11.5.3 Storage management

Storage management in all these servers is an important issue because the information is not delivered to intermediate recipient immediately. A priority queue mechanism manages this aspect efficiently. Moreover, processed queries for individuals are kept secret to restrict unauthorised access. A server deletes the successfully delivered resources from its storage device. However, frequently asked category of data is preserved for future usages till a fixed period is elapsed. The priority queue mechanism also handles this requirement. Thus, a server can instantly serve a few queries and works as an online server.

11.6 Software entities

Each device of the proposed network requires an efficient software entity. However, the behaviour of the software entity may be different at different devices. The network model demands two unique application software in executing all the functionalities of *servers* and *end users*. One of these applications is for *end users* and is referred to as UserApp. Another software application is installed on different *servers* and is referred to as ProxyServerApp. The software flow of the proposed network platform is shown in Figure 11.6. Each of these application software has distinct features and system architecture. As the network model does not demand the implementation of BP explicitly, both applications employ traditional TCP/IP in enabling data communication procedure. Efficiency, flexibility, and portability are the additional properties of this application package.

11.6.1 Software entity at server

Establishing reliable end-to-end connectivity, uploading and delivering concerned information are the key responsibilities of software entity at the server is referred to as ProxyServerApp. Minimising power consumption and implementing different autonomous features are the important issues which need to be considered while

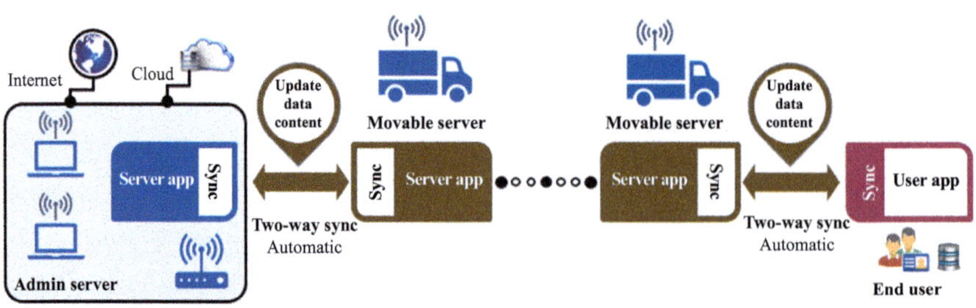

Figure 11.6 Software flow of the proposed network model

developing this software entity. Every single action of this software entity is completely autonomous.

11.6.1.1 ProxyServerApp

ProxyServerApp is mostly a content delivery solution provider or CDN Server. Besides providing the basic networking features, it ensures an efficient content delivery procedure among *servers* and *end users*. Storing, carrying, and delivering information for *end users* are the basic working paradigm of it. Implementation of centralised user control in content delivery networking is also coordinated by ProxyServerApp.

Autonomousness is another key feature of the ProxyServerApp. File transfer, user login, service gain, client monitoring and performance analysing, etc. are handled by fully automated processes. Redirection of downloading authority for certain download requests from *administrative server* to a certain CDN node is also coordinated by it; whereas, this certain CDN node is a *local server* which already cached the requested content earlier and closest to the requesting *end user*. Such an approach greatly reduces the data delivery delay.

11.6.1.2 Software architecture of ProxyServerApp

A conceptual model of ProxyServerApp which defines the structure, behaviour, and more views of this system is discussed here. A set of structural components together facilitates all the features of this application software. The kernel component (KCmp), communication component (CCmp), station management component (SMCmp), repository management component (RMCmp), registration component, and interface component (ICmp) are the building components of ProxyServerApp. The kernel is the heart of this application and coordinates the activities of all the remaining components. Any component can interact with its co-related components whenever the coordination required. An illustration of the mutual relationship among those components is shown in Figure 11.7.

CCmp ensures the communication paradigm by utilising underlying TCP/IP. Connection established with any other *server* or *end user* is handled by CCmp. Information contents are placed before CCmp for delivery based on *end user's* service profile and its transaction history. Upon connection establishment, the detailed profile of connected devices is collected by SMCmp and are kept in log files.

To store the collected information and request contents of end users, ProxyServerApp uses a resource repository (RRps) and a request repository (RqRps), respectively. RMCmp executes the resource distribution policy among multiple CDN clients. Synchronising the information contents between the CDN server and CDN client is a key task performed by this component. RMCmp compares the information contents available at both storage devices and synchronises both storage devices with the updated and latest one.

Redirecting downloading authority for certain user requests is coordinated by this component. This component also maintains priority queues for its storage management and preserving frequently asked information contents.

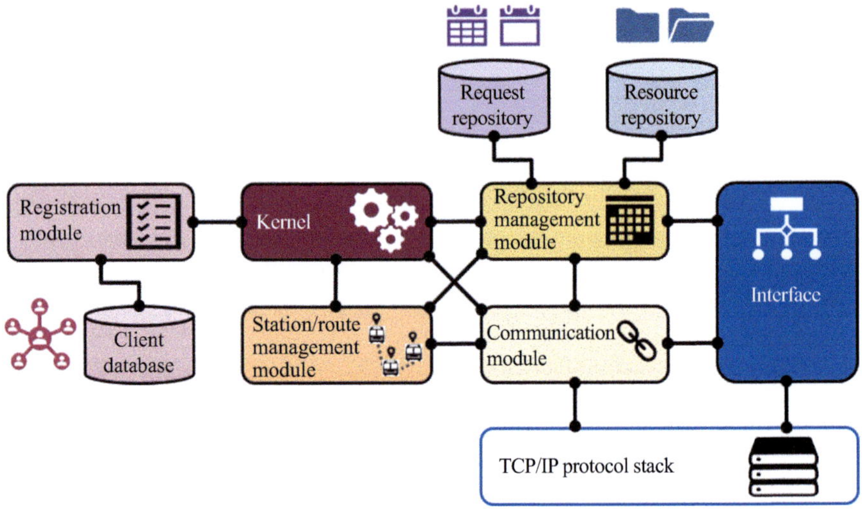

Figure 11.7 Software architecture of ProxyServerApp

Moreover, RMCmp also facilitates three scheduling policies, each with a particular service target for distributing contents among CDN clients. If less number of partial delivery in a single trip is expected, then the shortest remaining time scheduling is activated. A content delivery among multiple CDN clients ruled by fixed priority pre-emptive scheduling or the earliest deadline first ensures the quality of service for which delivery deadline of content and priority of data must be mentioned. Round-robin scheduling is an option for ensuring fairness among CDN clients.

RMCmp and SMCmp together enable remote configuration of user's service profiles and resource contents. The concept of modularisation service is also implemented by these two components.

Similar to RRps and RqRps, the ProxyServerApp includes a client repository (CRps) for preserving CDN client profiles on it. The registration module is responsible for service module registration and updating CRps accordingly, which relates to the client database. The basic interfacing of ProxyServerApp is provided by ICmp.

11.6.2 Software entity at end user

For successful coordination and interaction between a client's device and network navvy, an efficient, reliable, and flexible software entity at the client device is necessary. This software entity is specially developed, focusing on a user's actual needs and state-of-the-art software development technologies. Along with those specialisations, this software entity is specially designed with friendly (user interface (UI) which requires natural and minimum user interactions.

The simplest software to establish a reliable connection with a particular local server, collecting desire resources from them and presenting those in a suitable format before the end user are the tasks of this software entity, which is referred to as UserApp. Minimum and user-friendly interactions are the most important issues relating to the design of UserApp.

11.6.2.1 UserApp

UserApp primarily emulates the job of a CDN client in *end user*. Establishing a reliable TCP connection with the designated *server*, collecting the keen resources from them, and presenting those in perfect UI are the basic tasks executed by it. Delivering the user queries to the designated *server* is also accomplished by this application software. Minimising power consumption in a battery-powered *end user's* device is another issue which is also considered here.

11.6.2.2 Software architecture of UserApp

A number of linked-up structural components together form this application software. Discovery component (DCmp), CCmp, resource management component (RsMCmp), request management component, registration helper component (RHCmp), and ICmp are controlled by a KCmp. Besides these components, it has two data repositories called received resource repository (RRRps) and RqRps. The co-relationship among those components is depicted in Figure 11.8.

Discovering the *servers* and establishing TCP connections are handled by DCmp. Node scanning consumes equal energy that requires in transmission or reception. Hence, to minimise the power consumption of the *end user's* device,

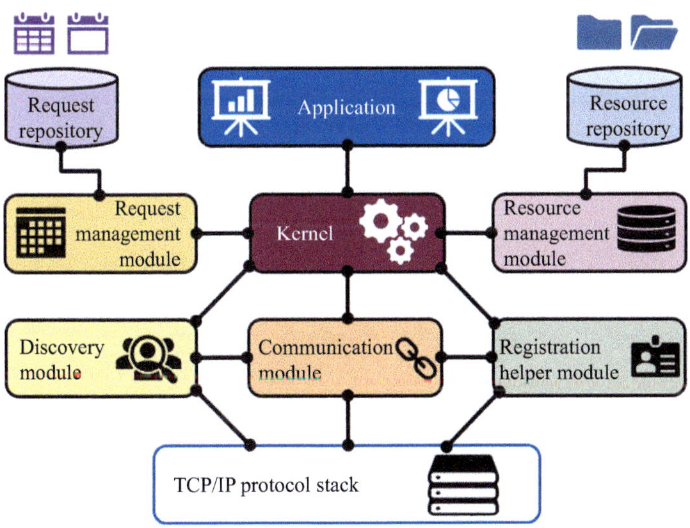

Figure 11.8 Software architecture of UserApp

UserApp adopts an adaptive node scanning mechanism by DCmp. On the other hand, CCmp is responsible for facilitating the TCP/IP communication service.

RsMCmp decides what to receive and store in RRRps. This component controls redundant receiving while any *server* performs resource synchronisation with it. It also acts as a helper component in establishing remote configuration by any *server*. On the contrary, RMCmp manages all user queries, whereas RqRp is utilised to store the queries in an organised way. User registration prompts is provided by RHCmp. However, the most important component of this UserApp is ICmp, which presents information contents user interface from RRRps. The desired visualisation and user interface of UserApp is discussed in the following subsection.

11.6.2.3 Visualisation and user interface

Appealing visualisation and user interface is crucial for interactive information retrieval in any system. As most of our target users are not sound in technological and literal interpretation. Hence, appropriate visualisation and appealing UI of UserApp are the utmost requirements in boosting its acceptance level.

The use of local language in UserApp UI is one of the key requirements in achieving so. Extensive experiments claim that visual representation can communicate much more rapidly and effectively than any other method. Hence, the use of related logos and appropriate graphics in the interface is captivating and appealing, especially when well designed. For easy access to different options in UserApp, the different features of the UserApp are placed in a grid view layout. Grid view layout gives a wholesome view of the available options. As a result, the users do not need to search for other options by clicking some buttons.

Touch-screen facility in using UserApp offers some significant advantages for our target users. Ease of use is one of the major benefits of it over other input methods. A touch-screen interface reduces training time for *end users* in looking up information or placing the request by themselves.

The availability of the voice request feature beside text request provides better user interaction. This feature is useful even for technically sound people, especially in their hurry. Voice notification on different events can make our *end users* more interactive. Hence, it announces pressed options when the user explores UserApp. It announces the incoming messages from recorded data or using text-to-speech technology.

11.7 Adoption of the proposed network model for supporting community needs

The proposed network can be implemented in various aspects of disconnected communities. However, most of the following aspects do not require real-time service of desire information. Hence, the proposed network concept is mostly applicable instead of the real-time Internet. Elaborate descriptions are present in the following subsections.

11.7.1 Serving rural youth

11.7.1.1 Disseminating etiquettes in youth

Proposed network can be utilised for boosting social ethics and spiritual value among rural youth. Documentaries on social and spiritual etiquettes can be projected or delivered to rural youth by availing those in a local server at the sports centre. These resources can be downloaded at the administrative server and delivered to all local servers at sports centres. On the availability of sufficient information, a number of training can be conducted on electronics, solar home systems, mobile phone repair, fisheries, looms, carpentry, breeding cows and goats, milk production, growing seedlings, setting up nursery, small businesses, tailoring, potteries, pickles, chalk, pure drinking water, etc.

11.7.1.2 Enlarging employment opportunities

The proposed network can also play a role to extend the employment opportunities to the disconnected youth community. They can get access to the information pool relating to job opportunities. Moreover, online application form for college and university admission and job can be downloaded and provided to end users. Local servers at the sports centre can be utilised in this purpose.

11.7.2 Serving rural women

To Promote and strengthen the rural womenfolk, the proposed network can be incorporated by providing appropriate audio-visual contents or educational documentaries to raise consciousness in the following major areas: health, childcare, sanitation, cleanliness, proper cooking methods, handicrafts, making pickles, and similar other activities. Local servers for miscellaneous purposes can be utilised for this purpose. The availability of such resources at a local server can allow it to project those on its widescreen or deliver those to end users.

11.7.3 Promoting health and Medicare

People of the village and rural areas mostly know little about health. As a result, they are easy prey to different malicious and seasonal diseases. However, a little concern can prevent these things.

11.7.3.1 Availing health care

Rural hospital authority sometimes may feel the necessity to consult with a specialised doctor regarding critical patients or exceptional medical cases. Hospital authority can upload the clinical health status of critical patients in its local server. However, the availability of remote diagnostic testing facilities in the hospital can make the total process very efficient. It might include a specific vital sign, such as blood glucose or heart ECG or a variety of indicators for distant patients. Once the health status is uploaded, the movable server can deliver those to the destined local server or admin server. Although the experts are out of the proposed network scope, the admin server can deliver the content employing a global Internet facility.

However, the experts can then check the symptoms and prescribe accordingly, which are delivered in a reverse way.

11.7.3.2 Raising healthcare awareness

The symptoms of the common diseases and their cures can be downloaded by the administrative server and given on a local server at the hospital. Those can be projected on a widescreen available at the local server even.

11.7.4 Elevating educational facilities

Education resources can be collected by admin servers and delivered to all local servers at different schools. Gradual collection of such resources enriches the resource pool at every Local server at school. However, local resources (e.g. lecture material prepared by local resource personnel) can be distributed among local servers at schools to enrich the resources pool further.

11.7.4.1 Offline e-library

The electronic copy of different books can be made available to disconnected students employing the proposed model. All textbooks of primary and secondary levels and books of different writers on different topics can be uploaded on local servers at different schools. These may be downloaded on demand and will not be necessary anymore to collect from the admin server.

11.7.5 Promoting business and commerce

Accessibility of appropriate and up-to-date information by disconnected business communities works as a great catalyst in maximising their scope and prospect. The availability of state-of-the-art information helps the disconnected rural businessmen to become more concerned and skilled. A disconnected businessman can avoid unexpected losses and cheat whenever he/she is informed of the latest state of commodities in neighbouring village and city markets. This information is an asset to promote the financial state of rural businessmen to a great extent.

The remarkable progress of business and commerce in disconnected communities can be achieved with minimal cost by employing the proposed network. Each local server at the Business centre can gather business and commerce related row data from respective individuals. However, once all the row data are grouped and processed at the admin server, summarised information is prepared. This summary report represents the overall business-like picture of disconnected communities. The availability of such a picture before disconnected businessmen has an immensely positive impact on their business.

11.7.6 Lifting agricultural development

Most of the developing countries mostly rely on their agricultural sector. This is the main driving force of these countries. However, they are victims of nature and no more or less, lack of proper knowledge. Although the agriculture department is working relentlessly to provide them with the proper help, still farmers are sometimes victims of drought, heavy rain, hail storms, etc. Most of the time, they lack the knowledge of modern farming and they do it the traditional way. So, the output is less. For this, the instruction and videos of modern farming can be a saviour. They can learn how to protect the crops from different threats like pests. They can also learn how things are done in the technologically advanced world. The proposed model provides a basic structure to implement a number of promising applications, which benefit the agricultural development in a disconnected community, such as (i) arising agricultural awareness, (ii) providing remote treatment to livestock, etc.

11.7.7 Elevating rural communication

11.7.7.1 Interconnection within villages

If an end user wants to send any text or recorded voice message, he/she can send it to the local server at the Entertainment centre. However, to facilitate this, each local servers require to maintain a list containing end users currently associated with it. Thus, the cost of Internet or traditional connectivity can be bypassed.

11.7.7.2 Offline community network

In today's virtual world, we socialise with other people via different social networks such as Facebook, Twitter, and LinkedIn. However, for the regions without infrastructure, this facility is hard to get. Local servers at the Entertainment centre can be utilised to facilitate such a feature. Comments, pictures, and even videos can be shared by employing the proposed network, not instantly, but with a delay. Similarly, e-mails can also be shared.

11.7.7.3 Convey information

National important news, news for the local community, and news bulletins can be conveyed by the movable server and stored on a local server at Entertainment centre. Notifications can be sent in advance for the national and other religious holidays.

11.7.7.4 Weather forecast

Weather forecast is a crucial event, especially for the crop farmers, fishermen at sea, and the coastal area populations. So, this thing must be delivered in real time. However, for the intermittent network, we have no other choice. So, for the regions where the conventional network is not present like islands and hilly areas, the proposed network concept can be used as an alternative.

Local server for miscellaneous purposes can announce this information once it receives an updated forecast collected from the Internet by the admin server.

11.8 Requirements for implementation

For the implementation of the proposed system in intermittent communication scenarios, some special and some general facilities must be ensured. Some of these facilities are as follows:

(i) Training the volunteers: A group of enthusiastic and energetic young boys and girls must be selected first from among the local community, who would be given proper training by the skilled trainers so that they attain computer-literacy and become ready to impart training to others. Later, this group would take up the responsibility of training and coaching the various local groups of people to continue or sustain the process.
(ii) Proper equipment: Vehicles, mobile devices, technical support, power supply, etc.
(iii) Funding and space allocation: Proper funding for the project and adequate space for setting up different local servers.
(iv) Security facilities: Proper hardware and skilled manpower for preventing cybercrime, and blocking harmful and inappropriate content while requesting for a search.

11.9 Conclusion

In this proposed network platform, a combination of DTN and CDN is implemented among devices using wireless connectivity, where data communication uses TCP/IP+. The platform supports a sustainable network model to interconnect the rural villages which are detached from modern technologies and give them autonomous solutions so that the village people do not require any costly Internet connection or infrastructure to connect within themselves and with the global Internet. The proposed network connects these disconnected communities by utilising minimum technical tools and existing rural infrastructure. Moreover, the use of the proposed model is not only a cost-effective solution but also a profitable and sustainable project for investors. This proposal emphasises on the idea of a DTN and CDN network that interconnects a village without an Internet facility. However, routing protocol, security protocol, data acquisitions, calculations, etc. are yet to be done. The future goal will be implementing this proposal in real-world scenarios.

References

[1] Bangladesh Telecommunication Regulatory Commission. Accessed: 2019-01-25. Available from: http://www.btrc.gov.bd/content/ internet-subscribers-bangladesh-november-2018.
[2] World Internet Usage, Population Statistics. Accessed: 2018-05-25. Available from: http://www.internetworldstats.com.
[3] Cerf V, Burleigh S, Hooke A, *et al.* SPARTA, Inc. Delay-Tolerant Networking Architecture, Network Working Group, 2007, Request for Comments: 4838. Available from: https://tools. ietf.org/html/rfc4838.
[4] Scott K, and Burleigh S. NASA Jet Propulsion Laboratory. Bundle Protocol Specification, Network Working Group, 2007, Request for Comments: 5050. Available from: https://tools.ietf.org/html/rfc5050.

[5] Jorgensen JW. Transmission control protocol/internet protocol (TCP/IP) packet-centric wireless point to multi-point (PtMP) transmission system architecture. Google Patents 2005, US Patent 6,862,622.

[6] Zhao W, Ammar M, and Zegura E. A message ferrying approach for data delivery in sparse mobile ad hoc networks. Proceedings of the 5th ACM International Symposium on Mobile Ad Hoc Networking and Computing; Roppongi, Tokyo, Japan, ACM 2004. pp. 187–98.

[7] Chuah MC, and Yang P. A message ferrying scheme with differentiated services. Proceedings of the Military Communications Conference; MILCOM 2005, IEEE, Atlantic City, NJ, USA; 2005. pp. 1521–7.

[8] Sasabe M, Kabir KH, and Takine T. Nonlinear integer programming formulation for quasi-optimal grouping of clusters in ferry-assisted DTNs. IEICE Transactions on Communications. 2013; E96-B(8):2076–2083.

[9] Kabir KH, Sasabe M, and Takine T. Integer programming formulation for grouping clusters in ferry-assisted DTNs. Proceedings of the 2nd International Workshop on Complex Information Flows (CIF) of 26th IEEE International Conference on Advanced Information Networking and Applications (AINA); Fukuoka, Japan; 2012.

[10] Kabir KH, Sasabe M, and Takine T. Optimal visiting order of isolated clusters in DTNs to minimize the total mean delivery time of bundles. Numerical Algebra, Control and Optimization of American Institute of Mathematical Science (AIMS). 2011;1(4):563–576.

[11] Kabir KH, Sasabe M, and Takine T. Self-organized data aggregation among self- ish nodes in an isolated cluster. Proceedings of the 5th International ICST Conference on Bio-Inspired Models of Network, Information and Computing Systems; Springer, Boston, MA; 2010.

[12] Kabir KH, Sasabe M, and Takine T. Evolutionary game theoretic approach to self-organized data aggregation in delay tolerant networks. IEICE Transactions on Communications. 2010;E93-B(3):490–500.

[13] Kabir KH, Sasabe M, and Takine T. Design and analysis of self-organized data aggregation using evolutionary game theory in delay tolerant networks. Proceedings of the 3rd IEEE WoWMoM Workshop on Autonomic and Opportunistic Communications; Kos, Greece; 2009.

[14] Lazar I, and Terrill W. Exploring content delivery networking. IT Professional. 2001;3(4):47–49.

[15] Yanggratoke R, Azfar A, Marval MJP, and Ahmed S. Delay tolerant network on android phones: implementation issues and performance measurements. Journal of Communications. 2011;6(6):477–484.

[16] Raj VS, and Chezian RM. DELAY—disruption tolerant network (DTN), its network characteristics and core applications. International Journal of Computer Science and Mobile Computing, 2013;02(09):256–262.

[17] Ntareme H, and Domancich S. Security and performance aspects of Bytewalla: a delay tolerant network on smartphones. Proceedings of the IEEE 7th International Conference on Wireless and Mobile Computing, Networking and Communications (WiMob); Wuhan, China; 2011. pp. 449–454.

[18] Pentland AS, Fletcher R, and Hasson A. Daknet: rethinking connectivity in developing nations. Computer. 2004;37(1):78–83.

[19] Papadopouli M, and Schulzrinne H. Performance of data dissemination among mobile devices. En cours de soumission; 2001.

[20] Doria A, Uden M, and Pandey D. Providing connectivity to the saami nomadic community. Proceedings of the International Conference on Open Collaborative Design for Sustainable Innovation; Development by Design, Bangalore, India 2002;1(2):3.

[21] Small T, and Haas ZJ. The shared wireless infostation model: a new ad hoc networking paradigm (or where there is a whale, there is a way). Proceedings of the 4th ACM International Symposium on Mobile Ad Hoc Networking & Computing; ACM, Annapolis, MA, USA; 2003. pp. 233–244.

[22] Naidu S, Chintada S, Sen M, and Raghavan S. Challenges in deploying a delay tolerant network. Proceedings of the Third ACM Workshop on Challenged Networks; ACM, San Francisco, California, USA; 2008. pp. 65–72.

[23] Zhao W, Chen Y, Amma M, Corner M, Levine B, and Zegura E. Capacity enhancement using throwboxes in DTNs. Proceedings of the Mobile Adhoc and Sensor Systems (MASS), 2006 IEEE International Conference on IEEE; Vancouver, British Columbia, Canada; 2006. pp. 31–40.

[24] Banerjee N, Corner MD, and Levine BN. Design and field experimentation of an energy-efficient architecture for DTN throwboxes. Transactions on Networking (TON). 2010;18(2):554–567.

[25] Guo S, Falaki MH, Oliver EA, *et al.* Very low-cost internet access using KioskNet. ACM SIGCOMM Computer Communication Review; ACM New York, NY, USA; 2007;37(5):95–100.

[26] Galati A, Bourchas T, Siby S, Frey S, Olivares M, and Mangold S. Mobile-enabled delay tolerant networking in rural developing regions. Proceedings of the Global Humanitarian Technology Conference (GHTC); IEEE, San Jose, CA, USA 2014. pp. 699–705.

[27] Sollazzo G, Musolesi M, and Mascolo C. TACO-DTN: a time-aware content-based dissemination system for delay tolerant networks. Proceedings of the 1st International MobiSys Workshop on Mobile Opportunistic Networking; ACM, San Juan, Puerto Rico, USA; 2007. pp. 83–90.

[28] Burgess J, Gallagher B, Jensen D, and Levine BN. MaxProp: routing for vehicle-based disruption-tolerant networks. INFOCOM 2006; IEEE, Barcelona, Spain; 2006. Vol. 6, pp. 1–11.

[29] Zhang X, Kurose J, Levine BN, Towsley D, and Zhang H. Study of a bus-based disruption-tolerant network: mobility modeling and impact on routing. Proceedings of the 13th Annual ACM International Conference on Mobile Computing and Networking; ACM, Montreal, Quebec, Canada; 2007. pp. 195–206.

[30] Balasubramanian A, Levine B, and Venkataramani A. DTN routing as a resource allocation problem. ACM SIGCOMM Computer Communication Review. 2007;37(4):373–384.

Chapter 12

A framework for developing a smart and adaptive environment for aging population

Nirmalya Thakur[1] *and Chia Y. Han*[1]

One of the distinct features of this century has been the population of older adults which has been on a constant rise. At present, the worldwide population of elderly people is about 962 million and their population is anticipated to rise even further by the year 2100 and reach 3.1 billion. Elderly people have several needs and requirements due to physical disabilities, cognitive issues, weakened memory and disorganized behavior that they face with their increasing age. The extent of these limitations also differs according to the varying diversities in elderly, which include age, gender, background, experience, skills, knowledge and so on. To sustain these dynamic needs of elderly population, it is essential to develop and implement advanced health-care solutions that can incorporate the fundamental features of a personal care system—which involve care anywhere, care networking and care customization. There has been a rapid increase in the cost of caregiver solutions incurred as a result of addressing these varying needs of elderly. This looming crisis will affect many areas of society and its valuable resources, such as manpower in care giving, and economy.

Thus, it is also equally essential to develop economically viable solutions to ensure smooth functioning of the world. Therefore, the work talked about in this chapter discusses a framework for developing an intelligent technology-based adaptive solution that can foster the biological, psychological, behavioral, physical, mental and emotional well-being and contribute toward active aging of elderly. This multifunctional framework takes a holistic approach in this regard, first by its ability to sustain independent living in smart homes, second by its ability to predict falls, as a typical example of potentially highly harmful event in terms of physical well-being, from a human-centered personalized care perspective and finally by ensuring social welfare for elderly communities in smart cities. This framework is envisioned to pave the way for the future of adaptive behavior intervention technologies in health care and social welfare that will be able to analyze, assess, model, profile and provide efficient personalized behavior interventions to enhance

[1]Department of Electrical Engineering and Computer Science, University of Cincinnati, Cincinnati, USA

user experiences and foster elderly care, while addressing the varying diversities of the elderly population.

12.1 Introduction

The integration of activity-centric computing in the context of human-centered design has the potential to improve the quality of life in the future of smart homes through smart and intelligent assistive technology-based services. The constantly increasing population of elderly people is associated with several health-care needs and requirements that raise the global economy's strain. Recent years have seen a rapid increase in the number of people with dementia worldwide, and their count is expected to almost double by 2030 [1] leading to an expected 76 million dementia-affected older adults globally. Roughly $604 billion in expenses were spent in caring for individuals with dementia in the health-care industry in 2010 alone, and this amount is rising at an alarming rate [1]. The lack of caregivers is also a significant problem in this respect, according to latest World Health Organization studies. The present number of caregivers worldwide is only 7.2 million and their population is anticipated to rise to about 12.9 million by 2035, which is considerably lower relative to the predicted large-scale population growth of the elderly [2] as mentioned earlier.

Cognitive impairments, reduced memory, behavior anomaly and even muscle disorders [3] are some of the main problems that older individuals experience with aging, causing them to experience difficulties in performing their regular routine tasks—this leads to the growing need of caregivers. It is therefore essential for the next generation of technology-based devices to be able to assist older individuals to perform their regular routine tasks; this serves as one of the major motivations for this work. Falls are common in elderly people. Falls can have serious consequences that could lead to various kinds of injuries and even death. Multiple falls have been observed to be one of the primary reasons for independent elderly people to move to various health-care institutions for long-term care [4]. Also, studies [5,6] have shown that once an elderly person experiences a fall, they are afraid to fall again. This syndrome of post-fall anxiety in elderly is quite often associated with self-imposed restrictions on carrying out activities specifically in the context of their home environment [4]. Loss of trust in the capacity to safely ambulate can lead to further operational decline, anxiety, emotions of helplessness and social isolation [7].

With increasing age, fall-related mortality rises considerably, particularly in individuals aged 70 years or more. Around 1% of fallers who suffer from a knee injury have a death rate of 20%–30% within 1 year of the fall [8]. In older people, the propensity for fall-related death is owing to an elevated incidence of clinical illnesses (e.g., osteoporosis) and age-related physiological alterations (e.g., decreased defensive reflexes), which cause even a moderate fall to have grave impacts on their health. Although most falls do not result in severe injuries, between 5% and 10% of elderly people who fall each year suffer severe injuries, such as a fracture, neck trauma or severe laceration [9], which comes from the way of them performing their daily routine tasks in an independent manner, either due to the injuries sustained from the falls or due to the fear of falling or a combination of both.

Therefore, it is essential for the future of smart and assistive environments to be able to predict falls even before the fall takes place, by an analysis of the users motion data and nature of user interactions with the given environment parameters. This serves as the second motivation for the development of this framework.

This work is presented as follows: Section 12.2 comprises the literature review and Section 12.3 provides an overview of activity theory and activities of daily living (ADLs) that form the basis for the development of this framework. Section 12.4 gives an overview of one of the related works in this field—complex activity (CA) recognition algorithm (CARALGO) that has been used and extended for this work. Section 12.5 discusses about implementation, details and functionalities of this framework. Discussion of the results is presented in Section 12.6; that is followed by conclusion and an overview of the scope for future work.

12.2 Literature review

In this section an overview is presented about the recent works [10–30] in the fields of human–computer interaction, activity-centric computing, Internet of Things (IoT) and assistive technology in the context of fostering independent living of elderly people and contributing toward their active aging, specifically within the premises of a smart home environment.

Activity recognition studies can be widely divided into two classifications—knowledge-based approaches and data-driven approaches. Knowledge-based approaches are mostly used to infer about activities using ontologies. Azkune *et al.* [16] suggested a multilayered model for analyzing activities. The model comprised various layers connected with the functions of acquiring information from wireless sensors, comprehending semantics associated with the underlining actions and tasks, and stating a knowledge-based explanation of the inferred activity. In the work done by Nevatia *et al.* [17] a formal language for activity analysis was proposed. The approach primarily used image recognition principles to study video data and infer the underlining activity being performed. The data-driven approaches for activity recognition have mostly focused on using machine learning and data analysis methods. In the work done by Kasteren *et al.* [18], hidden Markov models (HMMs) were used for activity recognition. Each state of the HMM was used for inferring different activities. An Artificial Neural Networks (ANNs) based activity recognition approach was proposed by Skocir *et al.* [19], which primarily used infrared sensors to detect enter-and-exit events in different rooms. Doryab and Bardram [20] developed a context-aware task recommender system to increase the performance of medical practitioners in a hospital setting. The system used a knowledge base and performed these task recommendations based on user behavior.

As far as assistive technology is concerned, a sensor-driven indoor navigation system for elderly people with disabilities was proposed by Abascal *et al.* [21]. The sensors provided the system information about the various context parameters, which was further analyzed in real time to help the user navigate the same. Yared *et al.* [22] developed a smart cooking environment to prevent the occurrence of accidents in the kitchen for elderly people. Their framework could detect any discrepancies from usual working conditions in terms of user interactions and use reasoning principles to detect an accident in the kitchen.

There have also been several works focused on elderly fall detection. Smartphones have been effectively used in designing fall detection approaches, by using the data from their inbuilt accelerometers and gyroscopes [23]. Zigbee controllers and Bluetooth-based solutions have also been used to design fall detection models. Benocci *et al.* [24] used Zigbee controllers to design a wireless body area network for fall detection. It also used a wearable accelerometer sensor for transmitting the motion data that was analyzed by their system for detecting various kinds of falls. Wang *et al.* [25] developed a fall detection system that used electrocardiogram (ECG) signals to detect the occurrence of a fall. Cheng *et al.* [26] developed a system that used both accelerometer data and ECG data to detect elderly falls in the premises of a smart home environment. Chen *et al.* [27] used a host of wireless sensors to set up a framework that could analyze the user's motion and detect falls in the given context. Felisberto *et al.* [28] used a combination of accelerometers, gyroscopes and magnetometers to analyze multiple dynamics of motion data to develop a system that could infer the occurrence of falls. Rescio *et al.* [29] developed a field programmable gate array based wearable device that could be worn by elderly people. This device continuously transmitted data at a frequency of 100 ms, and this information was analyzed by the model to detect instances of a fall. Lim *et al.* [30] developed a microcontroller-based wearable device that used the data from a three-axis accelerometer to perform fall detection with an accuracy of 96.25%.

Most of the work related to activity recognition and task recommendation, as described here, has primarily focused on modeling an "average user" and has not considered the personalized needs and requirements of an "actual user." Also, the fall detection approaches have mostly focused on detecting the fall after the user has fallen. Falls can have several health-related impacts even leading to death as described in Section 12.1. Thus, it is important for the next generation of intelligent assistive agents to be able to predict a fall even before the fall occurs. Such predictions would be useful to develop behavior intervention assistants that could prevent falls or reduce its impacts. Such a system would also help one to alert caregivers, medical practitioners or family members of the elderly people, for immediate care and support as might be needed if the user falls. Thus, this framework discusses an approach for implementing a "personalized" touch in the next generation of task recommender systems and presents an approach to predict elderly falls even before they occur.

12.3 Activity theory and activities of daily living

12.3.1 Overview of activity theory

In a broad manner, activity may be defined as a specific relationship between a subject (an actor) and an object (an entity existing in the world). Activity is commonly represented as "S ↔ O." which represents the relationship and interaction between a subject and an object [31]. This relationship between the subject and object is governed by two specific features: (a) subjects act on objects because they have specific needs that they seek to accomplish and (b) subjects and objects mutually determine each other in a given context. For instance, if a person must solve a mathematics problem, the person needs to have skills in mathematics. Similarly, the difficulty of the problem would determine the extent to which the person would use his or her skills to solve the given problem. According to the traditional concept of activity theory [31], an activity is a hierarchical structure that consists of three levels in the hierarchy as shown in Figure 12.1.

The uppermost layer is the activity that is associated with a specific motive that causes the subject to act on a specific object. The total process of the activity occurs in a step-by-step manner with each step leading to the motive. Activity theory defines these steps as actions, and the respective objects at which these actions are performed are called goals. Accomplishment of goals associated with a given activity leads to the task of successfully completing the activity.

12.3.2 Overview of activities of daily living

ADLs refer to the daily routine tasks and self-care activities that are performed mostly in the context of the premises of one's home. There are multiple definitions to the exact list of activities that comprise ADLs but in a generic manner. ADLs can be broadly classified into five major categories that include activities related to maintaining hygiene, dressing, eating, continence and moving around as per the requirements and context. The ability of an individual to carry out these activities on their own or to seek help from caregivers gives an indication of the level of independence of the person [32]. Elderly people due to their increasing age that comes with associated disabilities are quite often

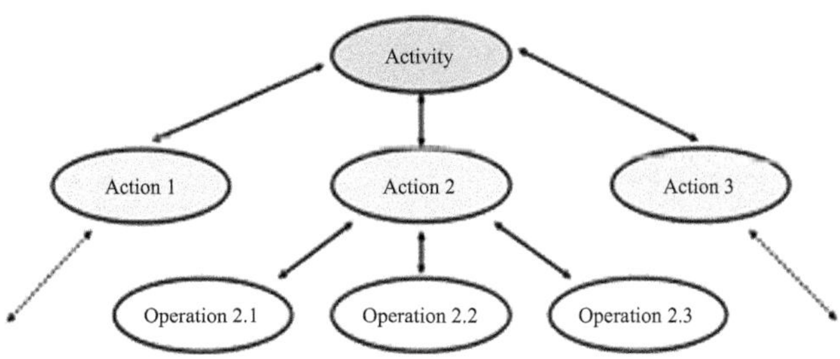

Figure 12.1 The various hierarchical levels comprising an activity

unable to accomplish these tasks that increase the need and relevance to provide technology-based solutions for their aid. Based on the varying needs and motives behind a user interacting with an object or certain objects in the given context, there may be the following broad ways in which activities may be broadly performed [33]:

1. Sequential activity: This refers to the user doing one activity at a time. When the user finishes off with the given activity at hand, the user initiates with the next activity.
2. Concurrent activity: This refers to the situation when the user would be doing more than one activity at a given time instance, although these activities may or may not start or end at the same time.
3. Interleaved activity: Real-life activities are mostly interleaved. This refers to the user starting off an activity in the middle of another activity and returning to the former after some time which may or may not be after completing the new activity that the user started.
4. False start: This refers to a situation like an interleaved activity, but the only difference is the user does not return to the initial activity to complete it. This may be because the other activity required more attention or could also be that the initial activity was no longer relevant.
5. Social interactions: This refers to activities that are performed by individuals in a social context. Such activities involve more than one person working toward the same goal with similar needs and motives.

12.4 Overview of complex activity recognition algorithm

Saguna *et al.* [34] proposed CARALGO for activity recognition based on the small tasks and actions that make up a CA. These small tasks are called atomic activities, and the environment parameters on which these tasks are performed are called context attributes.

In a broad sense, a CA may be defined as a tuple:

$$\text{CA} = (\text{Ati}, \text{Cti}, \text{AtS}, \text{CtS}, \text{AtE}, \text{CtE}, \gamma\text{At}, \rho\text{Ct}, \Delta T, \text{TS}, \text{TE})$$

where Ati stands for all the atomic activities associated with the CA; Cti, all the context attributes associated with the CA; AtS, start atomic activities; CtS, start context attributes; AtE, end atomic activities; CtE, end context attributes; γAt, core atomic activities; ρCt, core context attributes; TS, start time of CA; TE, end time of CA; and $\Delta T = \text{TE} - \text{TS}$ = duration of CA.

The atomic activities that mark the beginning of the CA are referred to as start atomic activities (AtS) and their associated context attributes are referred to as start context attributes (CtS). Similarly, those atomic activities that mark the end of the CA are referred to as the end atomic activities (AtE), and their associated context attributes are referred to as end context attributes (CtE). The atomic activities that are most relevant for the completion of a given CA are referred to as core atomic activities (γAt), and their associated context attributes are referred to as core context attributes (ρCt). The time at which a given CA starts is represented by TS, and the time when the given CA ends is represented by TE. The difference between the end time and start time is ΔT, which gives the time duration for the given CA. According to CARALGO, the weights associated with the CA and its associated atomic activities and context attributes can be determined by probabilistic analysis.

12.5 Proposed framework

12.5.1 ADL-based task recommendation system

This section discusses the functionality of this framework to recommend tasks in the context of ADLs. Its main characteristics are as follows:

1. In the context of any CA, it can analyze the tasks or actions that are performed by a user as well the context parameters associated with these.
2. The framework allows the modeling of the exhaustive nature of user interactions in terms of user behavior, by considering universal diversity, in the context of any given CA in an IoT setting.
3. This modeling of the exhaustive nature of user interactions allows for developing a knowledge base or database for intelligent systems to analyze user interactions with better accuracy and improved inferences.
4. The tool considers the impact of context parameters in the context of CAs for developing this knowledge base; thus, it may be applied to any context, and its implementation does not fail with the user moving to a different environment with a different context setting.
5. The model discusses a mathematical approach that allows the intelligent system to analyze user behavior in the context of any given CA and map discrepancies to assist the user refocus on the given activity. In this context, discrepancies may be three-fold: (1) the user gets distracted by some aspects of the environment and reduces focus on the current activity; (2) the user stops the current activity and focuses on something else; (3) the user feels confused and is not sure of how to proceed with the given activity.

By comparing the user interactions in terms of user behavior with the knowledge base for that specific CA in the given setting, the model is able to infer about these discrepancies and then a task recommender system based on the same would be able to assist the user refocus on the given activity at hand and therefore improve the user experience.

In addition to CARALGO [34], another work [35] has been used to propose this context-driven CA framework. This work [35] was done by Biggs, and it is the binomial theorem which is popular in the branch of probability theory, permutation and combination in mathematics. As per this theorem [35], the number of ways of selecting *k* items from an itemset containing *n* entities, assuming *k* to be less than or equal to n, without replacement, can be presented as follows:

$$nCk = \left(\frac{n!}{(n-k)! \times k!}\right) \tag{12.1}$$

where n is the total number of entities in the given itemset; k, the number of entities being chosen from this itemset; C stands for the notation of combination in probability theory of mathematics and ! denotes the factorial of a number.

Using mathematical induction, it can be proved that the total number of ways of selecting k items from an itemset of n entities without replacement where the value of k can be greater than equal to zero and less than equal to n is numerically equal to 2^n. According to CARALGO [34], every CA is associated with specific atomic activities and context attributes, based on the given contextual parameters. Owing to universal diversity, it is possible that every user performing this CA may have his/her own way of approaching the same which may involve performing any number of atomic activities associated with this CA. This may or may not lead to successful completion of the activity as it depends on the specific atomic activities that were performed by the user. The minimum condition to successfully accomplish any CA involves successfully completing the core atomic activities associated with the CA in the given context. These core atomic activities are associated with core context attributes that are also vital for the activity completion [34]. To create the mathematical foundation for the development of this ADL-based user interaction analysis framework, the following is being considered.

For a given CA let x_t, y_t, p_t and q_t stand for the total count of atomic activities, context attributes, core atomic activities and core context attributes, respectively. So, the exhaustive set of ways by which this CA may be performed according to (12.1) is as follows:

$$x_tC0 + x_tC1 + x_tC2 + \cdots + x_tCx_t = 2^{x_t} \tag{12.2}$$

where x_tC0 represents the situation when the user performs 0 atomic activities related to the given CA; x_tC1 represents the situation when the user performs 1 atomic activity related to the given CA; x_tCn represents the situation when the user performs n atomic activities, related to the given CA, such that $0<n<x_t$; x_tCx_t represents the situation when the user performs all x_t atomic activities related to the given CA.

Taking all these instances into consideration, the exhaustive number of ways by which a given CA may be performed is 2^{x_t}, where x_t is the number of atomic activities. Although performing only the core atomic activities by considering their associated context attributes is the minimum condition to successfully complete a given activity, any user may choose to perform any number of activities in addition to these core atomic activities, which are necessary to be analyzed. To mathematically model this situation, the following is considered:

$$D_t = (x_t - p_t) \tag{12.3}$$

where D_t are those atomic activities related to the given CA, which do not form the core atomic activities. These are being referred to as "other atomic activities."

Let r_t represent the number of atomic activities that a user performs from this list of atomic activities $(x_t - p_t)$. Clearly r_t may assume any value ranging from 0 to $(x_t - p_t)$ which is formulated, according to specific situations, as follows:

$$r_t = \begin{bmatrix} 0, \text{ when just the core atomic activities were performed} \\ > 0 \quad \text{and} \quad < (x_t - p_t), \text{when any number of the} \\ \qquad \text{“other atomic activities” were performed} \\ (x_t - p_t), \text{when all the “other atomic activities” were performed} \end{bmatrix} \tag{12.4}$$

Using (12.1), this can be mathematically modeled as follows:

$$(x_t - p_t)C0 + (x_t - p_t)C1 + (x_t - p_t)C2 + \cdots + (x_t - p_t)C(x_t - p_t)$$
$$= 2(x_t - p_t) \tag{12.5}$$

It is also essential to mathematically model those instances which would not lead to successful completion of the given CA. This may be mathematically modeled as follows:

From (12.2):

$$x_tC0 + x_tC1 + x_tC2 + \cdots + x_tCx_t = 2^{x_t}$$

From (12.5):

$$(x_t - p_t)C0 + (x_t - p_t)C1 + (x_t - p_t)C2 + \cdots + (x_t - p_t)C(x_t - p_t)$$
$$= 2^{(x_t - p_t)}$$

Let W_t be the total number of ways by which the process of the CA may be initiated but may not be successfully completed.

Then by comparing (12.2) and (12.4):

$$W_t = 2^{x_t} - 2(x_t - p_t) = 2^{(x_t - p_t)} \cdot (2^{p_t} - 1) \tag{12.6}$$

Thus, to summarize the mathematical foundation of this ADL-based user interaction analysis framework:

$$U_t = x_tC0 + x_tC1 + x_tC2 + \cdots + x_tCx_t = 2^{x_t}$$
$$V_t = (x_t - p_t)C0 + (x_t - p_t)C1 + (x_t - p_t)C2 + \cdots + (x_t - p_t)C(x_t - p_t)$$
$$= 2^{(x_t - p_t)}$$
$$W_t = 2x_t - 2(x_t - p_t) = 2^{(x_t - p_t)} \cdot (2^{p_t} - 1) \tag{12.7}$$

Based on this mathematical foundation, this work proposes an ADL-based task recommendation system (ATRS). Development of this proposed ATRS comprises the following steps:

1. Create a database of macro- and micro-actions related to ADLs in the given context.
2. Determine interesting characteristics of CAs in terms of their context parameters. This involves analyzing the appliance usage patterns in the context of a given CA.
3. Use CARALGO to analyze atomic activities and their associated context attributes, start atomic activities, start context attributes, end atomic activities, end context attributes, core atomic activities and core context attributes for different instances of all the CAs.
4. For each CA instance, compare the list of atomic activities and their associated context attributes with the activity definition in the knowledge base and perform the following:
    - (i) If user performs all atomic activities with respect to their associated context attributes,
        - (a) no activity recommendation.
    - (ii) If user performs just the core atomic activities with respect to their associated core context attributes,
        - (a) recommend "other atomic activities" and their associated context attributes.
    - (iii) If user does not perform one or more of the core atomic activities and their associated core context attributes,
        - (a) recommend the remaining core atomic activities with respect to their core context attributes,
        - (b) then recommend the "other atomic activities" and their associated context attributes.
    - (iv) Develop a supervised learning model that can analyze the information from the database and recommend atomic activities with their associated context attributes as mentioned earlier.

To evaluate the efficacy of ATRS, it was tested on two different datasets which comprised of different ADLs performed in smart homes. These two datasets were based on the UK Domestic Appliance Level Electricity dataset, also known as UK DALE dataset [36] and a dataset of ADLs developed by Ordóñez *et al.* [37].

12.5.1.1 Modeling an "average user" using the ADL-based task recommendation system

To uphold the relevance of ATRS, a dataset was developed based on the UK DALE dataset [36]. The UK DALE dataset consists of information about contextual parameters related to an activity, in terms of appliance usage patterns, for different appliances in a smart home. As a part of the study [36] which led to the development of this dataset, appliance usage patterns for a total of 109 appliances were recorded in 5 different smart homes over a time resolution of 6 s for a period of 3 years from 2012 to 2015 in Southern England. A related work done by Yassine *et al.* [38] on this dataset helps us to relate these appliance usage patterns to CAs. Some of the major CAs that are regularly featured on this dataset are Watching TV, Using Laptop, Using Subwoofer, Using Washing Machine, Cooking in Kitchen, Using Microwave and Using Toaster. Multiple instances and varying patterns of these CAs were analyzed to create general definitions for these CAs as per this framework. Then these definitions were used to develop the ATRS, and its performance accuracy was tested. Multiple instances and varying patterns of these CAs were analyzed from the dataset. Figure 12.2 shows different instances of these activities.

Two specific activity states, according to CARALGO, are analyzed in this approach. For each time instant, "1" represents the fact that the activity was performed and "0" represents the fact that the activity was not performed. Analysis of one of the CAs, CA of Watching TV (WT), is presented next (Table 12.1).

As observed from Table 12.2, there may be 128 different ways by which the user can perform this activity in the given context. This includes those activity instances when the activity was successfully completed as well as those activity instances when the activity was not successfully completed by the user. Table 12.2 also shows that there may be four different ways by which this CA may be performed, and the user can still meet with the end goal of completing the given activity and meeting the desired outcome. These four different ways can be explained by (12.4), as discussed earlier. These ways mostly refer to the user performing the core atomic activities with respect to their context attributes and then performing any number of the "other atomic activities" associated with the CA. For the remaining 124 ways of performing this activity, refer to those ways by which

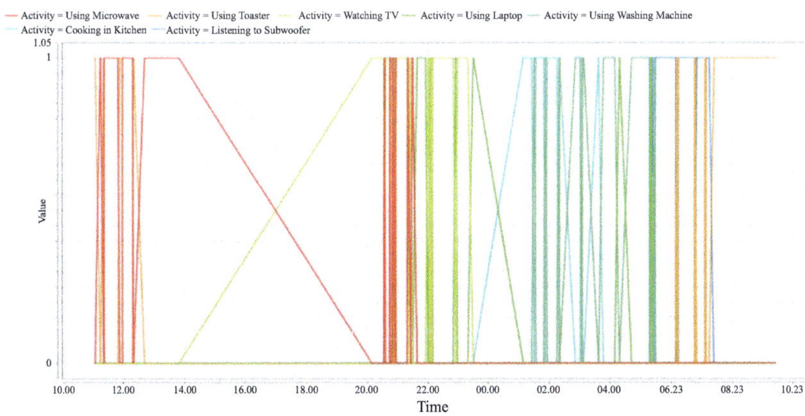

Figure 12.2 Multiple occurrences of the complex activities—Watching TV, Using Laptop, Using Subwoofer, Using Washing Machine, Cooking in Kitchen, Using Microwave and Using Toaster from this dataset

Table 12.1 CARALGO study—Watching TV (WT)

	Activity name: WT (0.67)
Atomic activities (weights)	At1: getting up (0.15); At2: moving near TV (0.15); At3: powering it on (0.25); At4: getting the remote control (0.15); At5: sitting (0.08); At6: changing to desired channel (0.12); At7: adjusting display and audio (0.10)
Context attributes (weights)	Ct1: lights on (0.15); Ct2: TV location (0.15); Ct3: TV present (0.25); Ct4: remote control present (0.15); Ct5: place to sit (0.08); Ct6: channel available (0.12); Ct7: settings functional (0.10)
Core γAt and ρCt	At1, At2, At3, At4, At5 and Ct1, Ct2, Ct3, Ct4, Ct5
Start AtS and CtS	At1, At2 and Ct1, Ct2
End AtE and CtE	At5, At6, At7 and Ct5, Ct6, Ct7

Table 12.2 Analysis by ATRS—Watching TV (WT)

x_t, total atomic activities, Ati	7
y_t, total context attributes, Cti	7
p_t, total core atomic activities, γAt	5
q_t, total core context attributes, ρCt	5
U_t, total ways the complex activity can be performed	128
V_t, total ways by which the activity end goal would be met	4
W_t, total ways by which the activity end goal would never be met	124

the activity may be started, and the user may perform one or more atomic activities related to it but never be successful in meeting the end goal and the desired outcome. These instances can be mostly performed by the user because of one or more of the following reasons:

1. The user was confused during the process of the CA and did not know how to proceed further toward completing the given activity.
2. The user might have been distracted by some attributes of the context surrounding him or her.
3. The user might have started this activity but moved on to another activity and never returned to complete this one.
4. The user did not find the current activity any more relevant as per his or her needs in the given context at that specific time instant.

The essence of developing this ATRS is to recognize all such instances and recommend tasks as per the situation, to the user to enable them to complete the given activity. Similarly, the other CAs were analyzed, and different supervised learning models based on this information and approach were developed in RapidMiner [39]. A total of 448 samples with 64 different occurrences of each of the seven CAs, Watching TV, Using Laptop, Using Subwoofer, Using Washing Machine, Cooking in Kitchen, Using Microwave and Using Toaster, were selected

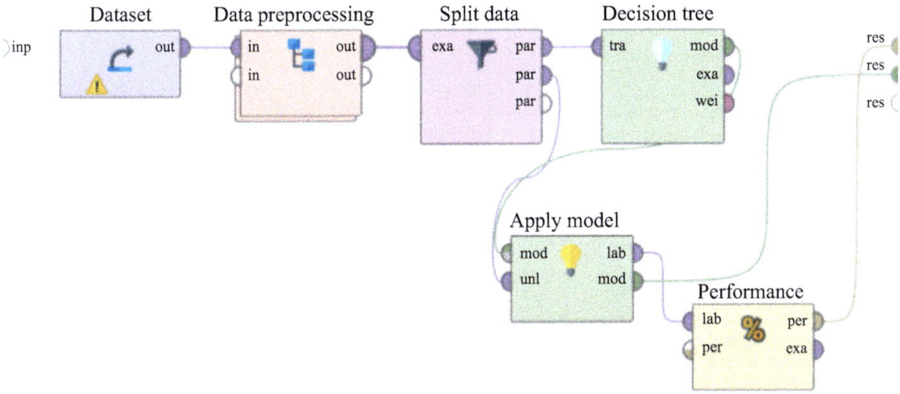

Figure 12.3 Implementation of the ATRS for an "average user" using the Random Forest learning approach

as per the framework of ATRS. The entire dataset was split into training set and test set, with 80% training data and 20% test data, to develop a learning model that can effectively predict the atomic activity recommendations for the test data. Some of the major learning models—Random Forest Approach, Naïve Bayes Classifier, Deep Learning Methodology, KNN Classifier, Neural Networks and Decision Trees—were implemented on this dataset to understand and infer that learning model that had the best performance accuracy. These respective systems were developed in RapidMiner. Implementation of one of the approaches—the Random Forest learning approach is shown in Figure 12.3. A comparison study, as discussed in Section 12.6.1, was also done analyzing the performances of each of the different learning models, as discussed earlier, for modeling an "average user."

12.5.1.2 Modeling a "specific user" using the ADL-based task recommendation system

The approach presented in Section 12.5.1.1 involved the discussion of the working of this framework when validated on a dataset where it modeled an "average user." To evaluate the efficacy of this framework for a "specific user," a dataset was developed based on the dataset by Ordóñez *et al.* [37]. This work by Ordóñez *et al.* [37] involved recording different ADLs performed by elderly people in the premises of a smart home. This work consisted of a smart IoT-based environment which comprised multiple sensors that were used to perform activity recognition to sense ADLs in a smart home, over 24 h for a period of 22 days. The different ADLs that were a part of this dataset consisted of the CAs of Sleeping, Showering, Eating Breakfast, Leaving for work, Eating Lunch, Eating Snacks and Watching TV in Spare Time. Multiple occurrences of these CAs from this dataset with respect to time have been shown in Figure 12.4 for illustration. Here, two different activity states corresponding to "0" and "1" were considered in the same manner, as outlined in the previous section. Analysis of one of the activities—the CA of leaving—is discussed in Tables 12.3 and 12.4.

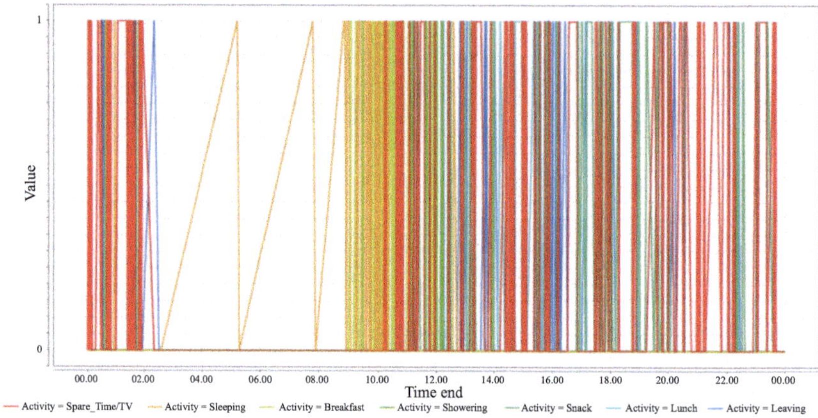

Figure 12.4 Multiple occurrences of the complex activities—Showering, Eating Breakfast, Leaving for Work, Eating Lunch, Eating Snacks and Watching TV in Spare Time from this dataset

Table 12.3 CARALGO study—Leaving (LV)

	Activity name: LV (0.68)
Atomic activities (weights)	At1: getting up (0.12); At2: dressing (0.32); At3: taking a bag (0.30); At4: moving in the direction of door (0.12); At5: going out (0.14)
Context attributes (weights)	Ct1: lights on (0.12); Ct2: availability of dress (0.32); Ct3: availability of bag (0.30); Ct4: door (0.12); Ct5: functionality of door (0.14)
Core γAt and ρCt	At2, At3, At5 and Ct2, Ct3, Ct5
Start AtS and CtS	At1, At2 and Ct1, Ct2
End AtE and CtE	At4, At5 and Ct4, Ct5

Table 12.4 Analysis by ATRS—Leaving (LV)

x_t, total atomic activities, Ati	5
y_t, total context attributes, Cti	5
p_t. total core atomic activities, γAt	3
q_t, total core context attributes, ρCt	3
U_t, total ways the complex activity can be performed	32
V_t, total ways by which the activity end goal would be met	4
W_t, total ways by which the activity end goal would never be met	28

As shown in Table 12.4, there may be four different ways by which this CA may be successfully performed by the user and there may be 28 different ways by which this CA may not be successfully performed by the user. These four different ways refer to those instances when the user would perform all the core atomic

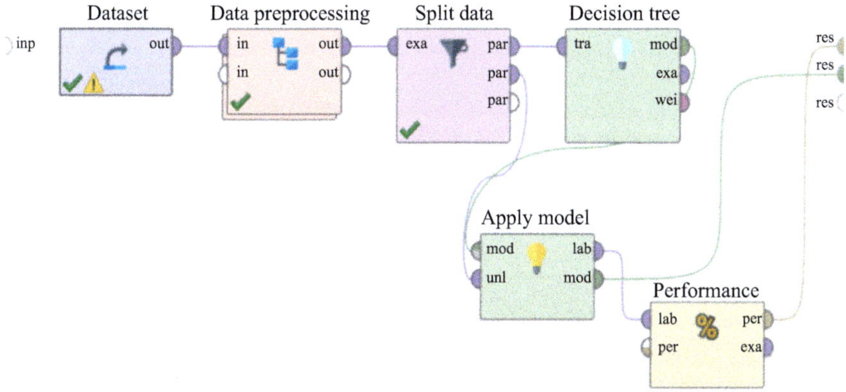

Figure 12.5 Implementation of the ATRS for a "specific user" using the Decision Tree learning approach

activities on the associated core context attributes as well as perform any number of the other atomic activities. The 28 unsuccessful instances refer to those situations when the user does not perform one or more of the core atomic activities on the associated core context attributes and as a result, the CA is not completed in any of those instances. Similarly, different CAs were analyzed, and based on this information, the entire dataset, having a sample size of 450 and consisting of multiple occurrences of these CAs, was split into training set and test set with 80% training data and 20% test data, to develop a learning model that can effectively predict the atomic activity recommendation for the test data. Based on the findings from the implementation of ATRS for modeling a "general user" as discussed in Section 12.6.1, the Decision Tree learning approach was implemented on this dataset as shown in Figure 12.5. The performance characteristics of this approach are discussed in Section 12.6.2.

12.5.2 Fall prediction in the context of ADLs

This section describes the functionality of this framework to predict falls in the context of ADLs. For the implementation of this proposed approach, a dataset based on the UR Fall Detection dataset [40], developed by Michal Kępski, from the Interdisciplinary Centre for Computational Modelling at University of Rzeszow is being used. It is an open-source dataset that contains 30 falls that were performed in the context of 40 ADLs in a smart home environment. All motions carried out by the users, including falls, were recorded by 2 Kinect Cameras. Based on the recordings obtained from these cameras and the data obtained from both wearable and wireless sensors, this work [40] led to the creation of this dataset. The development of this Fall Prediction Framework (FPF) involves the following six steps:

1. Analyze both the macro- and micro-components of the captured motion data.
2. Analyze the best predictors, in terms of attributes of the data.

3. Detect and remove outliers.
4. Split the data into training and test set.
5. Develop a learning model that is able to predict falls in the test set.
6. Implement this approach for different learning models to infer the learning approach that performs best for this data.

To develop this FPF, three distinct states of the user in the context of ADLs were considered—when the user had fallen, when the user was falling down and the user moving around normally. The various attributes in the dataset [40] were studied for the analysis of these different states, and these include bounding box height to width ratio, human height in frame to human height while standing ratio, actual height, distance of person center to the floor and ratio of the number of the point clouds belonging to the cuboid of 40 cm height and placed on the floor to the number of the point clouds belonging to the cuboid of height equal to person's height. Thereafter, RapidMiner was used to analyze each of these attributes to determine the best predictors for fall prediction. Next, the outliers for these predictors were detected and removed. Then, a learning model was developed by following the steps as outlined earlier. Some of the major learning models in terms of Random Forest Approach, Naïve Bayes Classifier, Deep Learning Methodology, KNN Classifier, Neural Networks and Decision Trees were implemented on this dataset to understand and infer which of these learning models had the best performance accuracy. For development of each of these learning models, a sample size of 7,297 was chosen and the data was split into 70% training data and 30% test data. These respective systems were developed in RapidMiner. Implementation of one of the learning approaches—the Decision Tree learning approach—is shown in Figure 12.6. Comparison of these learning approaches in the form of a study, to analyze the best learning approach for performing fall prediction on this dataset, is discussed in Section 12.6.3.

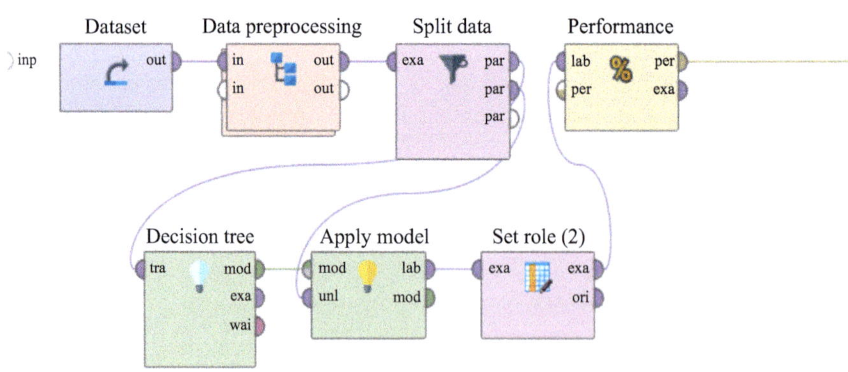

Figure 12.6 Implementation of fall prediction using Decision Tree learning

12.6 Results and discussion

12.6.1 Modeling an "average user" using the ADL-based task recommendation system

A confusion matrix was used for each of the implemented models to understand their performance characteristics in terms of predicting the correct and incorrect activity recommendations. In this context, the correct recommendations refer to those instances when the learning model was able to correctly identify those set of atomic activities (on the corresponding context attributes), for the given CA, that need to be recommended. Incorrect recommendations refer to the ability of the model to correctly identify those sets of atomic activities (on the corresponding context attributes), for the given CA, that do not need to be recommended. The class precisions and overall accuracies of these learning models as obtained from each of these confusion matrices are summarized in Table 12.5.

Table 12.5 Comparison of performance accuracies of different learning models for ATRS for modeling an "average user"

Learning model name	Performance accuracy (recommendations)		
	Class precision—correct (%)	Class precision—incorrect (%)	Overall accuracy (%)
Random Forest	72.09	69.57	70.79
Naïve Bayes	65.91	64.44	65.17
Deep Learning	62.50	63.41	62.92
KNN Approach	68.29	64.58	66.29
Decision Tree	70.00	74.36	71.91
Artificial Neural Networks	62.50	63.41	62.92

After studying these performance characteristics of these respective learning approaches as shown in Table 12.5, it is observed that the Decision Tree learning approach works best on this type of data for making the task recommendations correctly as per the framework of ATRS. It achieves the highest overall performance accuracy of 71.91% with 70% accuracy in predicting the correct atomic activity recommendations and 74.36% accuracy in predicting the incorrect atomic activity recommendations. Therefore, the Decision Tree learning approach is being proposed for implementation of ATRS.

12.6.2 Modeling a "specific user" using the ADL-based task recommendation system

As the Decision Tree approach was found to work best for implementation of the task recommendation as per the framework of ATRS for modeling an "average user," so the same was implemented on the dataset that involved building a task recommendation system based on ADLs for a "specific user." The system, as

Accuracy: 77.78%

	True correct	True incorrect	Class precision
Pred. correct	39	13	75.00%
Pred. incorrect	7	31	81.58%
Class recall	84.78%	70.45%	

Figure 12.7 Performance characteristics expressed as a confusion matrix of ATRS for modeling a "specific user" using Decision Tree approach

implemented in RapidMiner, is shown in Figure 12.5 and its performance characteristics expressed as a confusion matrix are shown in Figure 12.7. As can be observed, the Decision Tree learning method achieves an overall performance accuracy of 77.78% with 75% accuracy in predicting the correct recommendations and 81.58% accuracy in predicting the incorrect recommendations.

On comparing the performance of the ATRS systems for a "general user" and for a "specific user," it is observed that the system achieves a greater performance accuracy for the latter scenario. Thus, the relevance of proposing ATRS for recommending tasks in a personalized manner, for augmenting user performance with a specific focus on elderly people, is justified.

12.6.3 Fall prediction in the context of ADLs

A confusion matrix was used for each of the implemented learning models as described in Section 12.5.2, to analyze their performance characteristics and infer about that learning approach that outperforms the rest. The values of class precisions and overall accuracies of these learning models as obtained from each of these confusion matrices are summarized in Table 12.6. After studying these

Table 12.6 Comparison of performance accuracies of different learning models for developing the fall prediction model

Learning model name	**Performance accuracy (predictions)**			
	Class precision—normal (%)	**Class precision—is falling (%)**	**Class precision—fallen (%)**	**Overall accuracy (%)**
Random Forest	85.89	60.44	96.49	84.00
Naïve Bayes	86.68	47.62	74.38	79.62
Deep Learning	63.05	00.00	99.00	63.89
KNN Approach	87.13	68.90	94.14	86.20
Decision Tree	80.76	64.00	96.01	81.67
Artificial Neural Networks	85.27	68.53	90.36	84.05

characteristics of the respective learning approaches, it is observed that the KNN learning approach works best on this dataset for performing fall prediction as per this proposed framework.

As observed from Table 12.6, the KNN approach achieves the highest overall performance accuracy of 86.20% with 87.13% accuracy in predicting normal motion, 94.14% accuracy in predicting a fall and 68.90% accuracy in predicting that a person will fall. Thus, to conclude, the performance characteristics of different learning models were compared, and the KNN learning approach was found to have the best performance accuracy as per the proposed FPF.

12.7 Conclusion and future work

This work presents a framework that takes a holistic approach toward fostering active aging of elderly people in the future of smart homes and empowering them to perform ADLs in an independent manner. Two specific functionalities presented are (1) a personalized task recommendation system that takes into account the various micro- and macro-actions that a user performs in the context of any given activity and recommends them actions and tasks to complete the same and (2) a fall prediction model that can analyze several dynamics of human motion and predict whether the person is going to fall, even before the fall occurs. These functionalities of the framework have been tested on several datasets and the results are presented and discussed. Comparison studies were also performed for each of these two functionalities where several learning models were implemented. Based on the performance characteristics of these learning models, a learning model that achieves the highest performance accuracy for each of the respective functionalities of this framework was adopted for our study.

To the best knowledge of the authors, no similar approach has been done in this field yet. The results presented uphold the relevance and demonstrate the feasibility for practical implementation of this framework in the future of smart homes. This framework is envisioned to lead toward the development of the future of adaptive behavior intervention technologies in health care and social welfare that will be able to analyze, assess, model, profile and provide efficient personalized behavior interventions to enhance user experiences and foster elderly care while addressing the varying diversities of the elderly population. Future work would involve setting up an IoT-based environment through a host of both wireless and wearable sensors to implement this framework in a real-time context.

References

[1] Bengtsson, T., and Scott, K., "The Ageing Population", 2nd Chapter in Population Ageing – A Threat to the Welfare State, 2010, ISBN 978-3-642-12611-6, eISBN 978-3-64212612-3.

[2] Kon, B., Lam, A., and Chan, J., "Evolution of Smart Homes for the Elderly", Proceedings of the 2017 International World Wide Web Conference Committee (IW3C2), Perth, Australia, Apr. 3–7, 2017.
[3] Manini, T., Development of physical disability in older adults. Curr Aging Sci. 2011;4(3):184–91.
[4] King, M.B., and Tinetti, M.E., Falls in community-dwelling older persons. J Am Geriatr Soc. 1995;43:1146–54.
[5] Tinetti, M.E., Mendes, C.F., Doucette, J.T., *et al.*, Fear of falling and fall-related efficacy in relationship to functioning among community-living elders. J Gerontol. 1994;49:M140–7.
[6] Vellas, B.J., Wayne, S.H., Romero, L.J., *et al.*, Fear of falling and restriction of mobility in elderly fallers. Age Ageing. 1997;26:189–93.
[7] Cumming, R.G., Salked, G., Thomas, M., *et al.*, Prospective study of the impact of fear of falling on activities of daily living, SF-36 scores, and nursing home admission. J Gerontol A Biol Sci Med Sci. 2000;55A:M299–305.
[8] Magaziner, J., Simonsick, E.M., Kashner, T.M., *et al.*, Predictors of functional recovery one year following hospital discharge for hip fracture: a prospective study. J Gerontol. 1990;45:M101–7.
[9] Tinetti, M., Doucette, J., Claus, E., *et al.*, Risk factors for serious injury during falls by older persons in the community. J Am Geriatr Soc. 1995;43:1214–21.
[10] Thakur, N., and Han, C.Y., "An Improved Approach for Complex Activity Recognition in Smart Homes", Reuse in the Big Data Era (pp. 220–31), Lecture Notes in Computer Science, 2019, Vol. 11602. Springer, Cham.
[11] Thakur, N., and Han, C.Y., "A Hierarchical Model for Analyzing User Experiences in Affect Aware Systems", Proceedings of the 9th Annual Information Technology, Electronics and Mobile Communication Conference (IEMCON) 2018, Vancouver, Canada, Nov. 1–3, 2018.
[12] Thakur, N., and Han, C.Y., "Methodology for Forecasting User Experience for Smart and Assisted Living in Affect Aware Systems", Proceedings of the 8th International Conference on the Internet of Things (IoT) 2018, Santa Barbara, CA, Oct. 15–18, 2018.
[13] Thakur, N., and Han, C.Y., "An Activity Analysis Model for Enhancing User Experiences in Affect Aware Systems", Proceedings of the IEEE 5G World Forum Conference (IEEE 5GWF) 2018, Santa Clara, CA, Jul. 09–11, 2018.
[14] Thakur, N., and Han, C.Y., "A Complex Activity Based Emotion Recognition Algorithm for Affect Aware Systems", Proceedings of IEEE 8th Annual Computing and Communication Workshop and Conference (IEEE CCWC) 2018, Las Vegas, NV, Jan. 08–10, 2018.
[15] Thakur, N., and Han, C.Y., "An Approach to Analyze the Social Acceptance of Virtual Assistants by Elderly People", Proceedings of the 8th International Conference on the Internet of Things (IoT) 2018, Santa Barbara, CA, Oct. 15–18, 2018.

[16] Azkune, G., Almeida, A., López-de-Ipiña, D., and Chen, L., Extending knowledge-driven activity models through data-driven learning techniques. Expert Syst Appl. 2015;42(6):3115–28. DOI: 10.1016/j.eswa.2014.11.063, 2015.

[17] Nevatia, R., Hobbs, J., and Bolles, B., "An Ontology for Video Event Representation", CVPRW'04: Proceedings of the 2004 Conference on Computer Vision and Pattern Recognition Workshop Vol 7. IEEE Computer Society, Washington, DC, USA, 2004, p. 119.

[18] Kasteren, T., Noulas, A., Englebienne, G., and Kröse, B., "Accurate Activity Recognition in a Home Setting", UbiComp'08: Proceedings of the 10th International Conference on Ubiquitous Computing. ACM, Seoul, Korea, Sep. 2008, pp. 1–9.

[19] Skocir, P., Krivic, P., Tomeljak, M., Kusek, M., and Jezic, G., "Activity Detection in Smart Home Environment", Proceedings of the 20th International Conference on Knowledge Based and Intelligent Information and Engineering Systems, Sep. 5–7, 2016.

[20] Doryab, A., and Bardram, J.E., "Designing Activity-Aware Recommender Systems for Operating Rooms", Proceedings of the 2011 Workshop on Context-awareness in Retrieval and Recommendation, Feb. 13, 2011.

[21] Abascal, J., Bonail, B., Marco, A., and Sevillano, J.L., "AmbienNet: An Intelligent Environment to Support People With Disabilities and Elderly People", Proceedings of ASSETS'08, Halifax, Nova Scotia, Canada, Oct. 13–15, 2008.

[22] Yared, R., Abdulrazak, B., Tessier, T., and Mabilleau, P., "Cooking Risk Analysis to Enhance Safety of Elderly People in Smart Kitchen", Proceedings of PETRA'15, Corfu, Greece, Jul. 01–03, 2015.

[23] Panhwar, M., Shah, S.M.S., Shah, S.M.Z.S., Shah, S.M.A., and Chowdhry, B.S., Smart phone based fall detection using auto regression modeling in a non-restrictive setting. Indian J Sci Technol. 2017;10:1–6.

[24] Benocci, M., Tacconi, C., Farella, E., Benini, L., Chiari, L., and Vanzago, L., Accelerometer-based fall detection using optimized zigbee data streaming. Microelectron J. 2010;41:703–10.

[25] Wang, L.H., Hsiao, Y.M., Xie, X.Q., and Lee, S.Y., An outdoor intelligent healthcare monitoring device for the elderly. IEEE Trans Consum Electron. 2016;62:128–35.

[26] Cheng, J., Chen, X., and Shen, M., A framework for daily activity monitoring and fall detection based on surface electromyography and accelerometer signals. IEEE J Biomed Health Inform. 2013;17:38–45.

[27] Chen, S.K., Kao, T., Chan, C.T., *et al.*, A reliable transmission protocol for zigbee-based wireless patient monitoring. IEEE Trans Inf Technol Biomed. 2012;16:6–16.

[28] Felisberto, F., Laza, R., Fdez-Riverola, F., and Pereira, A., A distributed multiagent system architecture for body area networks applied to healthcare monitoring. BioMed Res Int. 2015;2015:192454.

[29] Rescio, G., Leone, A., and Siciliano, P., Supervised expert system for wearable MEMS accelerometer-based fall detector. J Sens. 2013;2013:254629.
[30] Lim, D., Park, C., Kim, N.H., Kim, S.-H., and Yu, Y.S., Fall-detection algorithm using 3-axis acceleration: combination with simple threshold and hidden Markov model. J Appl Math. 2014;2014:896030.
[31] Kaptelinin, V., and Nardi, B., "Activity Theory in HCI Fundamentals and Reflections", Morgan and Claypool Publishers (2012) ISBN 9781608457045, eISBN 9781608457052.
[32] Paying for Senior Care Report, "Activities & Instrumental Activities of Daily Living – Definitions, Importance and Assessments", Updated – July 11, 2019, Available: https://www.payingforseniorcare.com/activities-of-daily-living
[33] Chakraborty, S., "A study on Context Driven Human Activity Recognition Framework" – Master's Thesis at University of Cincinnati (2015). Available at: http://rave.ohiolink.edu/etdc/view?acc_num=ucin1439308572
[34] Saguna, Zaslavsky, A., and Chakraborty, D., Complex activity recognition using context-driven activity theory and activity signatures. ACM Trans Comput Hum Interact. 2013;20(6), Article 32.
[35] Biggs, N.L., The roots of combinatorics. Hist Math. 1979;6:109–36, https://doi.org/10.1016/03150860(79)90074-0
[36] Jack, K., and William, K., The UK-DALE dataset, domestic appliance-level electricity demand and whole-house demand from five UK homes. Sci Data; 2014;2:150007.
[37] Ordóñez, F.J., de Toledo, P., and Sanchis, A., Activity recognition using hybrid generative/discriminative models on home environments using binary sensors. Sensors. 2013;13:5460–77.
[38] Yassine, A., Singh, S., and Alamri, A., "Mining Human Activity Patterns From Smart Home Big Data for Health Care Applications", IEEE Special Section on Advances of Multisensory Services and Technologies for Healthcare in Smart Cities, Jun. 2017.
[39] Ritthoff, O., Klinkenberg, R., Fischer, S., Mierswa, I., and Felske, S., "YALE: Yet Another Learning Environment" (2001) DOI: 10.17877/DE290R-15309.
[40] Kwolek, B., and Kepski, M., Human fall detection on embedded platform using depth maps and wireless accelerometer. Comput Methods Programs Biomed. 2014;117(3):489–501, ISSN 0169-2607.

Chapter 13

Design, implementation, and evaluation of a mobile game for blind people: toward making mobile fun accessible to everyone

Muhammad Nazrul Islam[1], Toki Tahmid Inan[1], Nuzhat Tabassum Promi[1], Sadia Zahin Diya[1] and A.K.M. Najmul Islam[2]

Over the last decades, mobile games are increasingly used for entertainment, education, and health. Despite this increased interest, very few games have been developed or adapted to provide new interaction techniques for blind people and those with physical disabilities. Similarly, a limited number of works have been carried out for the entertainment of visually impaired people, especially in developing countries. Thus, a significant number of blind people encounter barriers when playing games. This chapter discusses the design, development, and evaluation of an arcade mobile game (BrickBlaster) for blind gamers in Bangladesh. The usual approach to develop video games for blind people is sensory substitution. Thus, the game was designed to have five levels and was developed using sound cues, vibro-tactile, haptic feedbacks, and voice-over instructions. We conducted an evaluation study with 24 blind people. The findings of evaluation study showed that the game was usable and playable by the blind people. The findings of the evaluation study suggest that completion of each level had reasonable difficulty with respect to completion time and number of attempts. As expected, the difficulty increased as the gamers moved to the next level. Moreover, no significant differences were observed by analyzing the impact of demographics (i.e., gender and high school vs. college) of the players on the playability of the game in terms of *success rate*, *the number of attempts*, *level completion time*, and *player attitude* measures. Furthermore, we found that the game received high evaluation score in terms of its ease of use, ease of learning, enjoyment, level of engagement, and level of excitement. Finally, we found that the participants' willingness to play the game in future as well as to recommend it to others was also high.

[1]Department of Computer Science and Engineering, Military Institute of Science and Technology, Dhaka, Bangladesh
[2]Department of Future Technologies, University of Turku, Turku, Finland

13.1 Introduction

According to the World Health Organization (WHO) [1], disability is an umbrella term, covering impairments, activity limitations, and participation restrictions. Disability is not only a health problem but also a complex phenomenon that reflects the interaction between features of a person's body and features of the society in which they live. The types of disabilities include vision impairment or blindness, deaf or hard of hearing, mental health conditions, intellectual disability, acquired brain injury, autism spectrum disorder, and physical disability [2]. Visual impairment or blindness is one type of disability and it refers to complete or nearly complete vision loss that can be permanent or temporary. A large group of the world population suffers from visual impairment. A statistic shows that over 15% of the world population has some form of disability [3]. Among them, an estimated 253 million people are vision impaired, 36 million people are blind, and 217 million people are moderate-to-severe vision impaired [4]. An estimated statistic from the WHO shows that only the Indian subcontinent, including India, Pakistan, and Bangladesh, has over 23.5% of all blind people in the world [5]. In Bangladesh, 6.6 million people are visually impaired [6].

A significant number of researches have been conducted to design and develop the information and communication technology (ICT)-based assistive applications for visually impaired and blind people [7]. For example, the development of artificial intelligence and IoT-based smart tools to assist blind people to walk more confidently by avoiding any obstacle through voice alert and vibration [8,9], design and development of multimodal (audio and tactical) user interfaces to assist blind people for performing their daily routine tasks [10], the development of learning tool via auditory and haptic feedback [11], and the development of smart home or integrated environment using the audio virtual reality and wearable technologies to guide the smooth movement of visually impaired people [12,13]. However, these development and innovations were for their survivability, not for fun, entertainment, engagement, and the like. Entertainment helps one to get rid of stress and the monotony of daily life. Utilizing the leisure time and having fun are also essential for disable people not only to escape the stress due to their extra barriers but also for improving their self-confidence, happiness, morale, and life satisfaction [14].

The ICT is important for all spheres of a life and plays a key role for the entertainment and fun. ICT in the form of mobile and computer games has become a form of entertainment and a part of everyday life for many people. Mobile games have several benefits to a person's well-being, especially for disable people, as the games not only provide fun but also help one to reduce stress and frustrations [15]. Gaming also has the physical and cognitive therapeutic contribution [16]. Thus, it is essential for visually impaired people to get involved with ICT-based entertainment. However, blind people are deprived of this opportunity as most ICT-based entertainment tools are targeted for normal people. Moreover, earlier studies conducted among the disabled people were focused mainly on the education and everyday task completion rather than leisure.

In recent years, researchers [7] have found that individuals who lack vision are able to develop heightened auditory spatial abilities using senses such as hearing and touch. This suggests that mobile games that require auditory and touch senses can be beneficial for blind and visually impaired people. These games can help them to enhance learning, creativity, and develop problem-solving skills. Also, it will stimulate motivation, therefore, help them to communicate and socialize. However, research and development still lack in designing mobile games, especially targeting blind people. A limited number of research works have been carried out for the entertainment of visually impaired people despite the fact that there are thousands of games developed for blinds.* Furthermore, there has not been any research conducted in Bangladesh for developing a mobile game for blind people.

Therefore, the objective of this chapter is to develop a mobile game for blind people in Bangladesh and to explore its playability. To attain this research objective, a mobile game named "BrickBlaster" is developed and an empirical study with 24 blind people was carried out to investigate (a) usability of the game, whether the levels of BrickBlaster are organized properly (i.e., difficulties increased as the gamers moved to the next level); and (b) players' attitude toward BrickBlaster.

The rest of the chapter is organized as follows. The following section briefly discusses the related works. After that, the conceptual design and the development of the gaming application are presented. Next, participants' profile, study procedure, data collection, and analysis of the evaluation experiment are discussed. The discussion, implications, and idea for future work are presented in the final section.

13.2 Related works

This section briefly discusses the works related to the games, especially designed for visually impaired people.

Braille has conventionally been used for teaching visually impaired children. A braille display is necessary to make the computer and smart devices accessible to the visually impairs. Thus, researchers have constantly been working on minimizing the cost of reading braille on a screen as the conventional brail display costs hundreds to thousands of US dollars. Jayant *et al.* [17] developed a learning app for smartphone using the concept of braille. This application uses vibration and the touchscreen of a smartphone. They called it V-Braille, and it is suitable for both visual and auditory impaired individuals who can only use their tactile sense. Using their V-Braille Milne *et al.* [18] created a braille-based educational word game called VBGhost. This game used audio and haptic feedback to reinforce braille concepts. Later, they developed BraillePlay [19] and included three more games (VBReader, VBWriter, and VBHangman) apart from the VBGhost to promote braille literacy by teaching braille character encodings to the visual impaired people. The game was evaluated with eight participants, and it was found that most

*https://www.theguardian.com/technology/2014/oct/13/video-games-that-let-blind-people-play

participants were able to play the games independently with minimal training. Despite that the game required braille knowledge, reviews and comments from several parents suggested that the game might be helpful for their children to learn braille.

McElligott and van Leeuwen [20] codesigned several toys and computer games by integrating audio with the tactile feedback with the visually impaired children using their disabilities in the design phase rather than compensating for their disabilities. They emphasized the children's experience on existing and conventional hardware devices. Raisamo *et al.* [21] designed a special low-cost vibrotactile gamepad to allow visually impaired and partially sighted children to play a memory game. This game uses a multimodal navigation support with high-contrast visual feedback and audio cues. A motor was used to create force of different magnitude, which allowed the children to differentiate among various kinds of vibrations. To evaluate the usability of the game, they conducted a study with seven blind children, aged from 12 to 13. The outcome of this study showed that the tactile game pad was usable by blind children. The study also received a positive feedback from the feature group. Buzzi *et al.* [22] used vibro-tactile method to develop a game-based Android learning app that allowed children to learn geometry using vibrations and haptic-cues. The purpose of their research was to facilitate the teaching process of geometry, a difficult topic to be taught to the visually impaired children. Their idea was to sense the geometric shapes by dragging a fingertip over the lines. An evaluation study conducted only with one participant found that the vibration could not accurately identify multi-touch interactions. They also proposed some of the improvements that can increase the accuracy.

A Nintendo-Wii-based game was developed by Morelli *et al.* [23] as a fitness game. Exergames use physical activities as an input and considered as a good source of physical exercise. They were able to find out some attributes of exergames that can overcome the limitation of physical activities faced by the visually impaired individuals in physical exercise. Thus, they came up with the motion sensing exergame called *VI-Tennis* and used vibro-tactile and audio cues for giving instructions. It was based on the existing popular exergame, Wii Sports Tennis [24]. They made two versions of this game: audio/tactile and only audio. An evaluation study was conducted on 13 blind children. The result showed that the children received enough physical activity required to maintain a healthy life. They also made a comparison between the two version of the game (audio/tactile and only audio). Although there was no significant difference between the energy spent by the children in these two different versions, children were found to enjoy the audio/tactile more compared to the only audio version. Another Nintendo-Wii-based musical game called AudiOdyssey was designed by Glinert and Wyse [25]. It is a rhythm-based game that used spatial sound cues. It was playable by both sighted and non-sighted gamers as it also had visual interface. This prototype features multiple input control schemes, rhythm-based game play, and fully accessible menus and play levels. It allows everyone to experience a similar gaming experience, regardless of level of vision. There were also other rhythm-based musical games that were developed for various platforms for blind users [26–29].

Apart from musical games, Carvalho *et al.* [30] developed a mobile-based audio puzzle game, where players need to organize audio-clips to complete the musical puzzles instead of images. To play this game, prior knowledge of the musical tracks is required. Other than these, researchers have developed several digital games for recreational purpose available in computers as well as other digital hardware systems that are accessible by the visually impaired individuals [31–33]. Based on the existing and their research works on specific games (both audio-based and haptic), Archambault *et al.* [32] discussed the principal constraints faced while adapting the computer games for the visually impaired children. They developed a python-based gaming platform that allows accessibility in mainstream games. This platform allows customizing the games according to the users' choice and the availability of the devices. Grammenos *et al.* [34] redesigned and developed a universally accessible version of a popular game named *Space Invasion*. They named the new version *Access Invader* that can be played by using both traditional computer inputs (mouse, keyboard, and joystick) and voice control. Researchers from TIM (Tactile Interactive Multimedia) project also developed some computer games for the visually impaired children [31,32]. Their children [31] analysis defined two important things: the soft architecture and the authoring tools and the computer–human interface for children and the resources (audio, tactile). There are some games developed for commercial purpose which are available in Android and IOS platform [35–37]. Developers redesigned and developed some popular games such as *snake*, *ladder game*, and *cricket* so that they are accessible to the blind or partial blind people. Most of the games use the audio and vibration feature of smartphones. Table 13.1 briefly presents a set of mobile games developed for blind people, the features considered to design the game, and the metrics used in evaluating the performance of these games.

In short, the design features considered to design the existing mobile games for blind people were haptic feedback, sound cues, braille dot pattern, voice-over instruction, larger button and interaction space, vibro-tactile, and high-contrast graphical UI. The games were evaluated mainly in respect to its usability/playability (in terms of effectiveness and efficiency) and player experience (fun, enjoyment, engaging, and ease of use/complexity). The previous works mostly focused on either educational or musical game, and only a few were developed for the entertainment purpose, although the arcade and action are the most popular genres for games. Besides, many games were developed for platforms like Nintendo Wii and other physical devices. Only a limited number of gamers are able to avail the games because of unavailability and high cost. Moreover, no mobile game is developed in Bangladesh for blind people. Thus, this research focuses to develop an Android-based arcade game for non-sighted gamers of Bangladesh that will feature sound cues, vibro-tactile, haptic feedbacks, larger button and interaction space, voice-overs instructions in local (Bengali) language.

Table 13.1 Design considerations and evaluation metrics of existing games developed for blind people

Game	Design features	Evaluation metrics (if any)
Blind hero [27]	Haptic feedback and sound cues	Effectiveness (game score, accuracy percentage, and hardware effectiveness), player experience (fun and engagement)
VBGhost [18]	Haptic feedback, sound cues, and braille dot pattern	–
Puzzle games [30]	Sound cues	Effectiveness and player experience (fun, difficulty, challenging, and playable)
Finger dance [26]	Sound cues	Usability test
TapBeats [28]	Haptic feedback, sound cues, and voice-over instruction	Player experience (fun and easy-to-use)
Playing with geometry [22]	Haptic feedback and vibro-tactile	Usability evaluation (effectiveness)
AudiOdyssey [25]	Voice-over instruction	Usability, frustration and challenge levels, overall player experience
Memory game [7]	Sound cues, vibro-tactile, and high-contrast graphical UI	Usability and playability (visual and auditory feedback, localizing, success, and memorizing the vibration)
VI-tennis [23]	Sound cues and vibro-tactile	Effectiveness comparison between the multimodal and unimodal cues (tactile/audio vs. audio)
BraillePlay [19]	Sound cues and larger button/interaction space	Accessibility, ability to engaging, and effectiveness
Digital Clock Carpet MOVA3D [38]	Haptic feedback and sound cues	Usability evaluation
Cardinal direction and modified tower of London [39]	Sound cues and voice-over instruction	Accessibility and usability (game scores, time logs, and survey data)
Game for blind [40]	Binaural technology (voice-over instruction)	Game performance
Rock Vibe	Tactile queues	Accessibility and usability
Access Invaders [34]	Voice-over instruction	Game performance and accessibility

13.3 Game design and development

The visually impaired people are able to develop enhanced auditory and haptic abilities, because the brain areas dedicated to processing vision rewires itself to process touch and sound [41,42]. Different brain-imaging studies have shown that the visual cortex is reassigned to other senses and the ability to reorganize brain is exclusive to blind people. Moreover, the blind people are also better at localizing

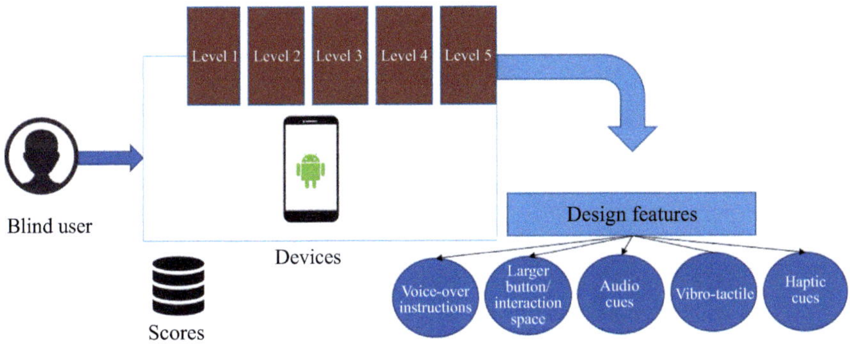

Figure 13.1 A framework to develop the mobile game for blind people

sounds [43]. That is why, to make any game accessible to blind people, auditory and haptic cues should be the major components. For this reason, our goal is to design a game primarily based on touch, vibration, and sound. We also aim to include visual interface for moderate-to-severe visually impaired people.

Our goal was to create an arcade game that uses audio, vibro-tactile, and haptic cues to indicate the events in the game. The main objective of the game was to hit a single object with a ball. The game will give the user a virtual bowling experience. The voice-over instructions are given in Bengali language. When a user touches the ball, there will be a vibration to indicate the location of the ball and also to let the user have a better experience while playing the game. If the user is able to move the ball successfully by swiping the finger on the screen, there will be a ball rolling sound to indicate the movement of the ball. If the ball is able to hit the object, there will be a crashing sound to indicate that the object has been destroyed and a voice-over saying that the level is successfully completed. But if the ball does not hit the object, then the voice-over will instruct that the user did not complete the level and he/she should try again. A framework to develop the game is shown in Figure 13.1.

13.3.1 Game description/features

Unity[†], a cross-platform game engine, has been used to develop the game. The reason behind using Unity is that it has an easily accessible asset store with a wide range of variety. We have chosen Android as the platform because most of the people in Bangladesh now a days use Android mobile phone and blind people are no exception. We have used application program interface (API) level of Android 4.1 Jelly Bean (API level 16) so that blind people with lower version of Android phone can also play the game. The launch/home screen consists of a menu having three buttons. Voice-over instructions will indicate the positions of the buttons on the screen as well as their functionality. We have made the buttons a lot bigger than normal size to make it easier for blind people to tap the desired button accurately by

[†]https://unity.com/

following the instructions. Also, when a button is pressed, a click sound will be heard which will assure the user that the button is correctly pressed. We have divided the screen into three parts for the buttons: the play button covers almost 50% of the screen, and the help and score buttons cover the 20% as shown in Figure 13.2. The "Play" button that is in the upper side of the screen will take the user directly to the game. The lower left "Score" button will take the user to the score option. In this option, there is a voice-over, which will start automatically after clicking the "Score" button and the user will be able to hear the score in the game so far. The lower right button indicates the "Help" button that will enable the user to get the help menu. Help menu also has an automatic voice-over that delineates all the rules and instructions of the game and teaches how to play the game. At any point of the game, the user can come back to the main menu screen by taping the back button of the Android phone (placed at the lower right corner of the most of the phones).

13.3.2 Level description

There are five levels in this game. The game screen looks like an inside of an arena where the player has to roll a ball to hit a certain object(s) placed in different positions. A voice-over will guide the user throughout the game. At each level, the voice-over will instruct whether to tap or swipe across the screen. If the gamer needs to swipe, whether it should be left swipe or right swipe, all detailed instructions would be told. Further, the vibrations and audio sound cues will give

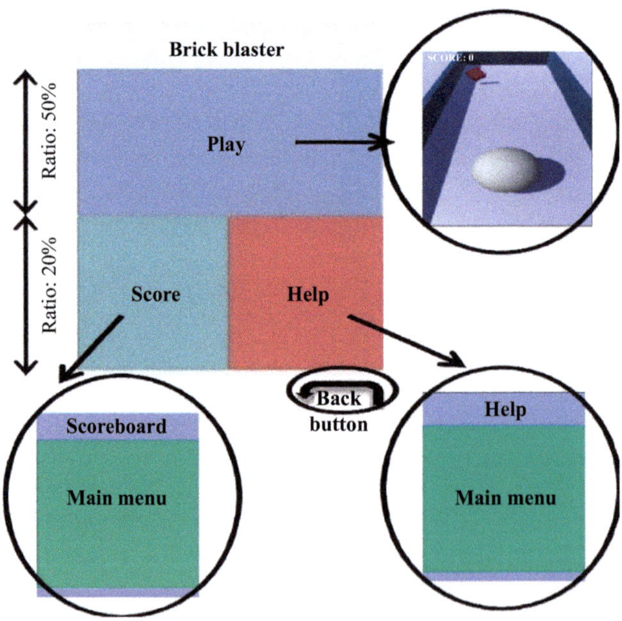

Figure 13.2 Overview of the game

the visually impaired a whole new experience of playing games. At each level, voice-over instructions are given about the object's current positions. The player has to move the ball according to the instructions. If the ball hits the object(s), the player moves onto the next level. And if the ball does not hit the object(s), then the user has the options either retry the same level or go back to the main menu. At any point of the game, if the player wishes to quit the game, he/she can simply press the back button.

During the game development process, we observed that leftmost corner is the easiest to hit. That is why for level 1, we have kept an object at the leftmost corner of the arena. The voice-over instructions will instruct the user to swipe forward from left side of the phone display. When the ball moves forward, a ball-moving sound on the ground will be heard so that the blind player can keep track of the game and do not lose interest. If the ball collides with the object successfully, a crashing sound and voice-over will be played to mark the success and will be taken to the next level. The score will increase automatically upon completing a level successfully. If the user failed to hit the object, the voice-over will immediately inform the blind gamer of the failure and probe him/her to retry the level. The user also has the option to go back to the menu screen. In level 2, we have increased the number of the objects from one to two and kept them at the rightmost corner. This time, the user must hit both of the objects to successfully complete the level. If the user managed to hit only one object, it will count as a level fail and will have the opportunity to try again. This level also has a voice-over guide that will advise on how to use finger to swipe so that the ball hits both of the targets. After hitting each object, a crashing sound will be heard. If the player hears two crashing sound, which means he/she has passed the level and score will increase automatically and will be promoted to the next level. In the case of level 3, only one object is present and it is placed at the center. The reason behind using only one object is that during the development of the game, we have found that the center position is comparatively hard to hit than the leftmost or the rightmost one. Because of this reason, one centered object is used to increase the difficulty a little bit and also to make the game more interesting. Just like previous levels, voice-over guide, audio cues, and vibration, all are present to help the blind player. In the case of levels 4 and 5, we have kept the objects in a leftmost and rightmost diagonal position with respect to the ball, respectively. The diagonal objects can only be hit if the user swipes diagonally, otherwise only one target hit will not be counted as level success. Table 13.2 shows the brief description of each level. The screen shots of the levels are shown in Figure 13.3.

13.3.3 Overview of the gameplay

At the start of the game, the player has three options: Play, Score, and Help. After selecting Play option, level 1 will start. If the player clicks the Score button in the main menu, he/she will get to know his/her current score and if he/she presses the Help button, he/she will get a brief explanation of how to play the game. The data flow of the gameplay is shown in Figure 13.4.

Table 13.2 Description of each level

Level	Description
1	One object is situated at the leftmost corner of the arena
2	Two objects are situated at the rightmost corner of the arena
3	One object is at the center of the arena
4	Two objects are kept in a diagonally left position with respect to the ball
5	Two objects are placed in a diagonally right position with respect to the ball

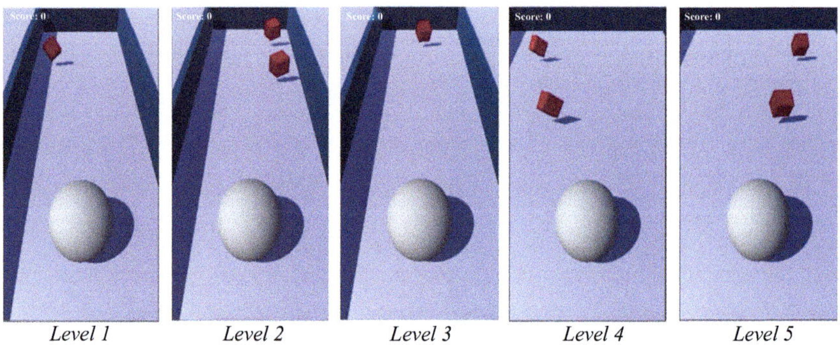

Figure 13.3 Five levels of the game

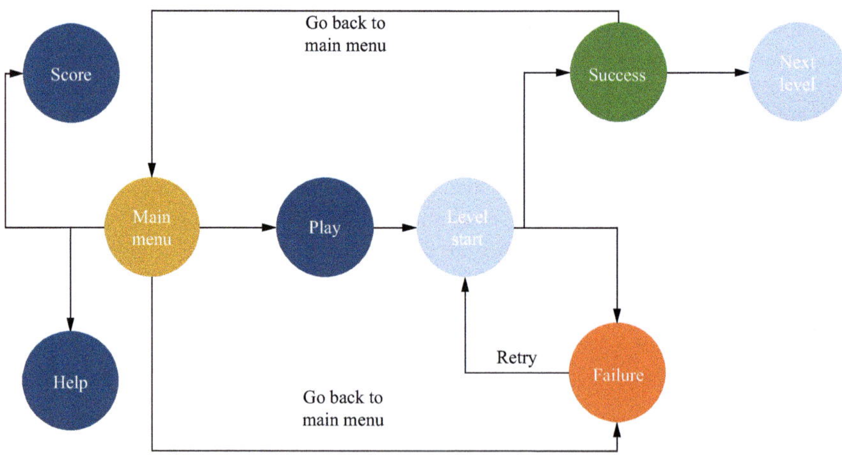

Figure 13.4 Data flow of the gameplay

13.4 Experimental evaluation

13.4.1 Study participants

A total of 24 blind participants were recruited from a formation center in Dhaka, Bangladesh. Among them, 13 were male and 11 female. A total of 12 participants were high-school (class VI–X) students and the rest were college (class XI and XII) students. All of them had experiences with features of phone, whereas only six had experiences with the smartphone mainly for phone call, listening to music, and FM radio. None of them played any mobile game. Their average age was approximately 16.58 ± 3.58 (mean±Std dev) years and ranged between 11 and 21 years. Among them, 13 were blind by birth, 6 were from diseases, and rest of them were by accident.

13.4.2 Procedure and data collection

For each participant, one test-session was carried out following the next steps:

- At first, each participant was briefed about the purpose of the study and his/her roles. The participant was also briefed that the goal of the study is not to assess him/her, rather to evaluate the mobile game, so that he/she behaves normally in the test session and provides honest opinion (if asked) about the game.
- The test participant's demographic information was collected, and a test-consent form was also signed with him/her.
- Then the participant was provided a short training (3–5 min) to demonstrate how to play.
- Next the participant was asked to start playing from level 1 up to level 5.
- At the end, the participant was asked to complete a posttest questionnaire. The questionnaire includes questions, with answer choices, annoying/enjoyable, boring/exciting, not-engaging/engaging, complicated (not-playable)/easy (playable), difficult/easy to learn, confusing/clear, play in future/not play, and recommend to friends/do not recommend. These questions were measured using 5-point Likert-type scale. Apart from these, all participants were asked to share their experiences of playing the BrickBlaster game using an open-ended question.

Each test-session was audio–video recorded for later analysis. A few pictures of data collection are shown in Figure 13.5.

13.4.3 Analysis and result

The study data were analyzed in terms of its playability. According to [44], playability is broad term that captures the guideline of how to implement the necessary elements such as rules to provide the desired sort of gameplay (i.e., interaction). Further, it also captures the evaluation criteria to evaluate the gameplay. Consequently, in order to evaluate the playability of BrickBlaster, we investigate the level organization, and players' overall attitude about their gameplay.

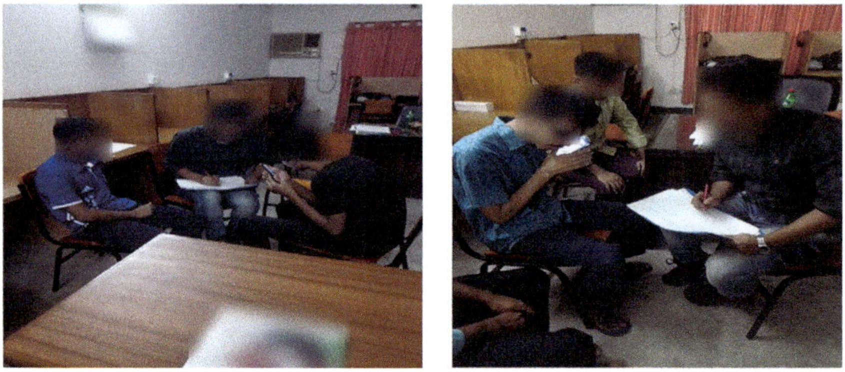

Figure 13.5 Evaluating the game at the formation center in Dhaka

Table 13.3 Success rate in level completion

Level	No. of success	Success rate (%)
1	24	100
2	24	100
3	24	100
4	20	83
5	19	79

13.4.3.1 Game usability

By definition, usability is the ease of use of a system or tool to achieve a particular goal. The International Organization for Standardization (ISO) [45] defines usability as "the extent to which a product can be used by specified users to achieve specified goals with effectiveness, efficiency, and satisfaction in a specified context of use." In this chapter, usability of BrickBlaster was analyzed based on the usability parameters defined by ISO [46], which includes effectiveness, efficiency, and satisfaction.

1. *Effectiveness*—Effectiveness is the accuracy and completeness to achieve goals [47]. Indicator of effectiveness may include the successful completion of tasks with minimum possible attempts [21,48]. In this study, two variables were considered to measure the game's effectiveness: (a) success rate (successful completion of each level of the game) and (b) number of attempts (how many times users tried to complete each level successfully). The level completion status is presented in Table 13.3. The results showed that all the participants successfully completed first three levels. Only four (17%) participants failed to complete the level 4, while five (21%) participants were unsuccessful in completing level 5. In other words, all participants successfully completed the levels 1–3; and only a one-fourth (25%) of the participants

failed at least one level of the game, and among them, 50% failed in both levels (levels 4 and 5). On average, the participants took 3.46, 3.79, 3.04, 4.58, and 5.38 more attempts than the optimal number of attempt (i.e., 1 attempt for each level) to complete the levels 1–5, respectively (see Table 13.4). Players generally take one to three attempts to complete each level of a mobile game when playing the game for the first time. Moreover, due to the complexity of levels 4 and 5, the participants took more attempts on average. The results of success rate and the number of attempts indicate that participants completed their tasks effectively using the application.

2. *Efficiency*—Efficiency is defined as the resources expended to successfully complete a task [49]. In this study, two variables were used to measure the game's efficiency: level completion time and number of times help asked form the researcher [48]. The results of level completion time show that participants took on average 38–131 s to complete any specific level (see Table 13.4), with minimum 17 s and maximum 150 s. Those who completed all the levels took on average 4.80–7.72 min. Ideally, participants are expected to ask no help from the researcher in the case of an efficient UI. Our study results show that on average, the participants asked 1.92, 2.04, 2.0, 2.08, and 2.17 times to complete the levels 1–5, respectively (see Table 13.4). However, the results also show that though the participants took comparatively longer duration and several attempts to complete the levels 4 and 5, the number of times they asked

Table 13.4 Results of evaluation study

Criteria	**Level**	**Mean and SD**	**Min**	**Max**
Number of attempts	1	3.46±0.78	2	4
	2	4.79±1.50	3	7
	3	4.04±0.95	2	5
	4	5.58±0.97	4	7
	5	6.38±1.13	5	8
Level completion time (seconds)	1	38.67±13.96	17	59
	2	47.96±16.24	20	75
	3	44.88±7.67	30	55
	4	113.75±20.0	87	147
	5	131.04±15.62	95	150
Player attitude	Confusing/clear	3.71±0.48	2	5
	Difficult-to-learn/easy-to-learn	3.67±0.64	3	5
	Complicated (not-playable/easy playable)	3.71±0.75	2	5
	Not engaging/engaging	3.96±0.69	3	5
	Boring/exciting	3.92±0.97	2	5
	Annoying/enjoyable	3.92±0.97	2	5
	Play in future/not play	3.67±0.67	3	5
	Recommend to friends/not recommend	3.96±0.69	3	5

for help was a little higher comparing to the case of levels 1–3. This result indicates that participants became confident, enjoyed, and engaged with the BrickBlaster; thus, although they tried several times and took longer time to complete and some of them eventually failed to complete the level, still they did not prefer to ask help to complete the level.

3. *Satisfaction*—Satisfaction is the users' comfort with and positive attitudes toward the use of the game [49]. In this study, nine questions were used to measure satisfaction based on the study of Laugwitz *et al.* [50] and the results are shown in Table 13.4. The players found that the game was enjoyable (mean: 3.92), exciting (mean: 3.92), engaging (mean: 3.96), easy-to-play (mean: 3.71), easy-to-learn (mean: 3.67), and clear (mean: 3.71). Their willingness to play the game in future (mean: 3.67) and recommend it to others (mean: 3.96) was also overwhelming. However, they were less motivated to pay for buying the game (mean: 2.08).

13.4.3.2 Level organization

In this study, two variables were used to measure if the game's levels are organized properly: (a) success rate (successful completion of each level of the game), (b) number of attempts (how many times users tried to complete each level successfully), and (c) level completion time. The level completion rate is presented in Table 13.3. The results showed that the success rates remain the same (100%) for the first three levels and the rate is increased from level 4 (17%) to level 5 (21%). The average attempts taken by the participants to complete each level increase with the level number except the levels 2 and 3 (see Figure 13.6). Like number of attempts, the results showed that the average time taken by the participants to complete each level was increased with the level number except the levels 2 and 3 (see Figure 13.7). Taken together, the results of *success rate, the numbers of*

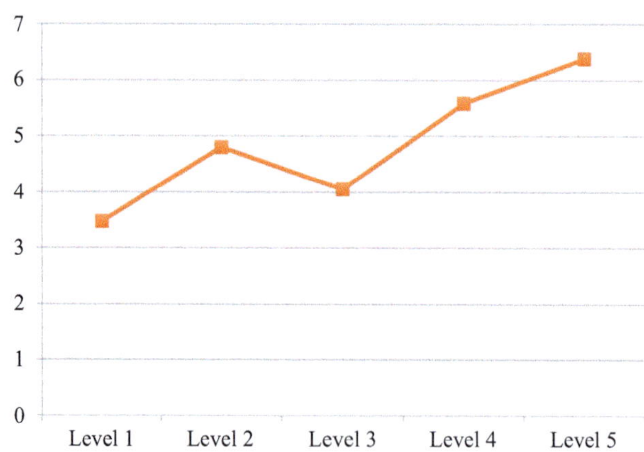

Figure 13.6 Number of attempts

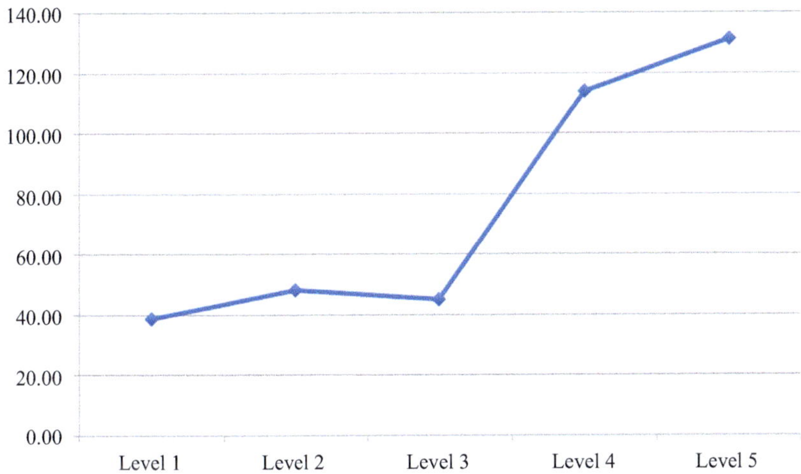

Figure 13.7 Level completion time in seconds

attempts, and *level completion time indicate* that in general the difficulty level increases as the player moves to more advanced level. This indicates that sufficient challenges have been inscribed in designing the level. However, it also suggests that levels 2 and 3 can be swapped for better organization of the game levels.

13.4.3.3 Player's attitude

In this study, nine questions were used to measure players' attitude toward the game and the results are shown in Table 13.4. The results showed that the average score of each measure was high that reflects the positive attitudes toward the use of the game. Some of the sample quotes in this regard from open-ended questions are as follows:

"It's very exciting,"

"I am enjoying the game, and while playing I always wait for the wining sound and vibration,"

"The voice instruction is clear and easy to follow. I learn it within a very short time,"

"Though I have tried severing times, but I failed to complete the level 5. But still I am not feeling boring; I thought I can finish the level if I tried few more times."

Altogether, these results indicated that the players had very positive attitude toward BrickBlaster.

13.4.3.4 Effect of demographics

We investigated the impact of the demographics (i.e., gender and high school vs. college) of the players on the playability of the game. The players were divided into two groups based on gender, and Kruskal–Wallis *H* tests were conducted to investigate differences in terms of success rate, the number of attempts, level completion time, and player's attitude measures. The results did not reveal any significant difference between male and female. Next, the participants were divided on whether they were at high school or college and Kruskal–Wallis *H* tests were conducted. Again, no significant differences were observed in terms of success rate, the number of attempts, level completion time, and player's attitude measures.

13.5 Conclusion

13.5.1 Main outcomes

In this chapter, we have presented design and evaluation of a mobile game (BrickBlaster) for visually impaired people in Bangladesh. The audio, vibro-tactile, haptic cues, large button/interaction space, and voice-over instructions were used to develop the game to make it usable and accessible for the visually impaired people. We measured the effectiveness, efficiency, and satisfaction and showed that the game was playable and usable to the users. The level organization of the (game) was also explored and refined in terms of effectiveness, efficiency, and satisfaction. Moreover, the game proved to be playable for blind people irrespective of their gender and education.

13.5.2 Research contribution

This chapter provides two contributions for designers. First, we empirically show that using sound cues, vibro-tactile, haptic feedbacks, large button/interaction space, and voice-over instructions is important for the design of mobile game for blind people. Several researchers [3,27,28,40] came to similar conclusion in their study. In this sense, we confirm that their findings are valid also in Bangladesh. Second, our game design evaluation shows that the difficulty of playing the game is increased as the gamers move to the next level, which is an indication of proper design of levels. It is worth mentioning here that none of the previous studies have focused on this issue of organizing level to develop a game for blind people. This research has shown how evaluation can be performed to investigate the level organization. If the game levels are not organized based on its complexity measure, players will not get challenge, enjoyment, and more importantly players will not be engaged with the game for a longer time. Altogether, we believe our study provides a template for interactive arcade game design and evaluation on mobile platform that can be used to build any other types of gaming applications for blind people.

13.5.3 Limitations and future work

The research has a few limitations as well. First, a few numbers of levels were designed in this game. Increasing the number of levels may provide different kind of outcomes. Second, though we have recruited participants from different age groups, education levels, and gender the number of participants to each cluster was small. Participants from different groups are required to explore the effects of demographics on gameplay. Our future research will focus on the design and development of more levels of BrickBlaster and then conducting an extensive evaluation study with more participants to verify and generalize the outcomes of this chapter.

References

[1] WHO, "Disabilities," 2018. [Online]. Available: http://www.who.int/topics/disabilities/en/. [Accessed 31 March 20118].

[2] C. f. D. Control, "Disability and Health Overview," 2019. [Online]. Available: https://www.cdc.gov/ncbddd/disabilityandhealth/disability.html. [Accessed 27 June 2019].

[3] WHO, "Disability and Health," 2018. [Online]. Available: http://www.who.int/mediacentre/factsheets/fs352/en/. [Accessed 31 March 2018].

[4] WHO, "Vision Impairment and Blindness," 2018. [Online]. Available: http://www.who.int/mediacentre/factsheets/fs282/en/. [Accessed 31 March 2018].

[5] B. Thylefors, A. Negrel, R. Pararajasegaram and K. Dadzie, "Global data on blindness," *Bulletin of the World Health Organization*, vol. 73, no. 1, p. 115, 1995.

[6] Mazhar, "Blindness Does Not Mean without Sight," 2016. [Online]. Available: http://www.banglanews24.com/economics-business/news/bd/488614.details. [Accessed 31 March 2018].

[7] O. Balan, A. Moldoveanu and F. Moldoveanu, "Navigational audio games: an effective approach toward improving spatial contextual learning for blind people," *International Journal on Disability and Human Development*, vol. 14, no. 2, pp. 109–118, 2015.

[8] M. Helmy Abd Wahab, A. A. Talib, H. A. Kadir, *et al.*, "Smart Cane: Assistive Cane for Visually-impaired People," arXiv preprint arXiv:111051 56. 2011.

[9] M. N. Islam, N. S. Khan, S. Kundu and S. A. Ahsan, "An assistive system of walking for visually impaired," in *4th International Conference on Computer, Communication, Chemical, Material and Electronic Engineering (IC4ME2-2018)*, Bangladesh, 2018.

[10] H. J. Gansmo, "Fun for all = digital competence for all?," *Learning Media and Technology Media and Technology*, vol. 34, no. 4, pp. 351–355, 2009.

[11] O. Lahav, D. Schloerb and M. A. Srinivasan, "Virtual environments for people who are visually impaired integrated into an orientation and mobility program," *Journal of Visual Impairment & Blindness*, vol. 109, no. 1, pp. 5–7, 2009.

[12] L. A. Guerrero, F. Vasquez and S. F. Ochoa, "An indoor navigation system for the visually impaired," *Sensors*, vol. 12, no. 6, pp. 8236–8258, 2012.

[13] M. R. Jahan, F. I. Aziz, M. B. I. Ema, A. B. Islam and M. N. Islam, "A wearable system for path finding to assist elderly people in an indoor

environment," in *The 20th International Conference on Human Computer Interaction (Interacción 2019)*, Basque Country, Spain, 2019.

[14] R. Thorne, "The Importance of Leisure Activities for Disabled Children," 2017. [Online]. Available: http://blog.aidis.org/the-importance-of-leisure-activities-for-disabled-children. [Accessed 27 June 2019].

[15] M. Tollefsen and M. Lunde, "Entertaining software for young persons with disabilities," in *Lecture Notes in Computer Science*, 2004.

[16] Y. Garrido, Á. Marco, J. Segura, T. Blanco and R. Casas, "Accessible gaming through mainstreaming kinetic controller," in *International Conference on Intelligent Technologies for Interactive Entertainment*, 2009.

[17] C. Jayant, C. Acuario, W. A. Johnson, and J. Hollier, "VBraille: haptic braille perception using a touch-screen," in *ASSETS'10 Proceedings of the 12th International ACM SIGACCESS Conference on Computers and Accessibility*, Orlando, FL, 2010.

[18] L. R. Milne, C. L. Bennett, and R. E. Ladner, "VBGhost: a braille-based educational smartphone game for children," in *Proceedings of the 15th International ACM SIGACCESS Conference on Computers and Accessibility*, 2013.

[19] L. R. Milne, C. L. Bennett, R. E. Ladner and S. Azenkot, "BraillePlay: educational smartphone games for blind children," in *Proceedings of the 16th International ACM SIGACCESS Conference on Computers & Accessibility*, 2014.

[20] J. McElligott and L. van Leeuwen, "Designing sound tools and toys for blind and visually impaired children," in *Proceedings of the 2004 Conference on Interaction Design and Children: Building a Community*, 2004.

[21] R. Raisamo, S. Patomäki, M. Hasu and V. Pasto, "Design and evaluation of a tactile memory game for visually impaired children," *Interacting with Computers*, vol. 19, no. 2, pp. 196–205, 2007.

[22] M. C. Buzzi, B. Marina, L. Barbara and S. Caterina, "Playing with geometry: a multimodal android app for blind children," in *Proceedings of the 11th Biannual Conference on Italian SIGCHI Chapter*, 2015.

[23] T. Morelli, J. Foley, L. Columna, L. Lieberman and E. Folmer, "VI-Tennis: a vibrotactile/audio exergame for players who are visually impaired," in *Proceedings of the Fifth International Conference on the Foundations of Digital Games*, 2010.

[24] Nintendo,' "Nintendo Wii Sports Tennis," 2013. [Online]. Available: http://wiisportsclub.nintendo.com/tennis/. [Accessed 7 January 2019].

[25] E. Glinert and L. Wyse, "AudiOdyssey: an accessible video game for both sighted and non-sighted gamers," in *Proceedings of the 2007 Conference on Future Play*, 2007.

[26] M. Daniel, P. Aaron and D. S. A, "Finger dance: a sound game for blind people," in *Proceedings of the 9th International ACM SIGACCESS Conference on Computers and Accessibility*, 2007.

[27] Y. Bei and F. Eelke, "Blind hero: enabling guitar hero for the visually impaired," in *Proceedings of the 10th International ACM SIGACCESS Conference on Computers and Accessibility*, 2008.
[28] K. Joy and R. Jonathan, "TapBeats: accessible and mobile casual gaming," in *The Proceedings of the 13th International ACM SIGACCESS Conference on Computers and Accessibility*, 2011.
[29] T. Allman, R. K. Dhillon, M. A. Landau and S. H. Kurniawan, "Rock Vibe: Rock Band® computer games for people with no or limited vision," in *Assets'09 Proceedings of the 11th International ACM SIGACCESS Conference on Computers and Accessibility*, 2009.
[30] J. Carvalho, T. Guerreiro, L. Duarte and L. Carriço, "Audio-based puzzle gaming for blind people," in *Proceedings of the Mobile Accessibility Workshop at MobileHCI (MOBACC)*, 2012.
[31] A. Buaud, H. Svensson, D. Archambault and D. Burger, "Multimedia games for visually impaired children," in *ICCHP 2002: Computers Helping People with Special Needs*, 2002.
[32] D. Archambault and D. Olivier, "How to make games for visually impaired children," in *ACE'05 Proceedings of the 2005 ACM SIGCHI International Conference on Advances in Computer Entertainment Technology*, 2005.
[33] D. Archambault, T. Gaudy and K. Miesenberger, "Computer Games and Visually Impaired People" *Upgrade*, vol. 8, no. 2, pp. 43–53, 2007.
[34] D. Grammenos, A. Savidis, Y. Georgalis and C. Stephanidis, "Access Invaders: developing a universally accessible action game," in *ICCHP 2006: Computers Helping People With Special Needs*, 2006.
[35] U. A. Studio, "Blind People Game Snake and Ladder," 2018.
[36] U. A. Studio, "Blind Mini Games," 2018.
[37] S. I. Ltd, "Blind Cricket," 2017.
[38] J. Sánchez, M. Saenz and J. M. Garrido, "Usability of a multimodal video game to improve navigation skills for blind children," *ACM Transactions on Accessible Computing (TACCESS)*, vol. 3, no. 2, p. 7, 2010.
[39] D. Song, A. Karimi and P. Kim, "Toward designing mobile games for visually challenged children," in *International Conference on e-Education, Entertainment and e-Management (ICEEE)*, 2011.
[40] K. Drossos, N. Zormpas, G. Giannakopoulos and A. Floros, "Accessible games for blind children, empowered by binaural sound," in *Proceedings of the 8th ACM International Conference on PErvasive Technologies Related to Assistive Environments*, 2015.
[41] M. Bates, "Super Powers for the Blind and Deaf," 2012. [Online]. Available: https://www.scientificamerican.com/article/superpowers-for-the-blind-and-deaf/. [Accessed 15 January 2019].
[42] S. G. Miller and S. Writer, "Why Other Senses May Be Heightened in Blind People," 2017. [Online]. Available: https://www.livescience.com/58373-blindness-heightened-senses.html. [Accessed 15 January 2019].

[43] ScienceDaily, "Blindness" May Rapidly Enhance Other Senses," May 2012. [Online]. Available: https://www.sciencedaily.com/releases/2012/05/120508152002.htm. [Accessed 15 January 2019].

[44] A. Järvinen, A. Heliö, S. Mäyrä and F. Mäyrä, "Communication and Community in Digital Entertainment Services. Prestudy Research Report," 2002.

[45] International Standard, "Ergonomic requirements for Office Work with Visual Display Terminals (VDTs)–Part 11: Guidance on Usability. ISO Standard 9241-11: 1998," International Organization for Standardization, 1998.

[46] International Standard, "Ergonomics of Human-System Interaction – Part 11: Usability: Definitions and Concepts," International Organization for Standardization, 2018.

[47] E. Frøkjær, M. Hertzum and K. Hornbæk, "Measuring Usability: Are Effectiveness, Efficiency, and Satisfaction Really Correlated?," 2000.

[48] M. A. Ahmed, M. N. Islam, F. Jannat and Z. Sultana, "Towards developing a mobile application for illiterate people to reduce digital divide," in *International Conference on Computer Communication and Informatics (ICCCI – 2019)*, India, 2019.

[49] E. Frøkjær, M. Hertzum and K. Hornbæk, "Measuring Usability: Are Effectiveness, Efficiency, and Satisfaction Really Correlated?," in *Proceedings of the SIGCHI Conference on Human Factors in Computing Systems*, 2000.

[50] B. Laugwitz, T. Held and M. Schrepp, "Construction and evaluation of a user experience questionnaire," in *USAB*, 2008.

Chapter 14

Challenges and opportunities of adopting ICTs in the humanitarian sector in Nigeria

Oluwasola Oni[1]

The humanitarian sector is overwhelmed with the constant need to be responsive to situational crises that have arisen in different parts of the world as a result of natural and, sometimes, man-made disasters. The results of which typically include violence, loss of life and a means of livelihood. One of such man-made humanitarian crisis situations is that happened in the north-eastern region of Nigeria. Challenges faced in the region include internal displacement, human casualties and food shortages. Information and communication technologies have played a role in ensuring that productivity is unencumbered by physical restraints. The challenges and opportunities of adopting the necessary technological innovations are explored in this chapter with the aim of providing a framework that might be applicable to similar situations.

14.1 Introduction

According to the United Nations Office for the Coordination of Humanitarian Affairs (OCHA) [1], more than 134 million people across the world need humanitarian assistance and protection. While the intention of humanitarian organisations is to save lives, alleviate suffering and protect human dignity, humanitarian services can be fraught with challenges as well. Along with a key challenge of emergency response and relief work [2], globally, the situation facing humanitarian organisations is critical with incidences of floods, landslides, hurricanes, earthquakes and other natural disasters. Wars and armed conflicts also contribute to humanitarian crises in several countries, and humanitarian assistance can support sustainable peace in a number of ways. These include providing medical treatment, food, water, shelter and sanitation that can reduce the suffering of civilians, which can then increase the ability of the affected population's capacity to overcome further attacks and also begin making efforts towards building peace [3]. However, where aid has been badly planned or executed, such programmes may reduce the

[1]School of Management and Social Sciences, Pan-Atlantic University, Lagos, Nigeria

impact of the humanitarian assistance or exacerbate the suffering [3]. Similarly, some analysts argue that humanitarian assistance sometimes exacerbates conflict rather than contributing to peace [4].

Nevertheless, it is evident that humanitarian assistance has been greatly relied upon reducing the suffering of those caught in conflict stricken areas in Nigeria. In 2019, more than seven million people including more than two million girls, just under two million boys, more than one and a half million women and more than a million men are currently in need of humanitarian assistance in north-east Nigeria as a result of a crisis that has been going on for 10 years [5].

Nigeria is a country of approximately 182,202,000 million people [6] with a wide variety of tribes and cultures with over 500 languages spoken in the country [7]. It is also a country besieged with humanitarian crises in various parts of it. Pockets of violence exist in several states with varying levels of danger attacks on inhabitants. The north eastern Boko Haram crises and the Fulani herdsmen attacks have resulted in the displacement of millions of Nigerians, particularly in the north. A lot of the inhabitants of the areas caught up in the conflict have been subjected to enormous stress, violence, loss of family and friends, loss of homes and communities with very limited access to health or education [8]. Recently, communal conflict between nomadic herdsmen and farmers in the Middle Belt escalated and further worsened the security situation, with at least 1,600 people killed and around 300,000 displaced as a result of the violent clashes [9]. Other disasters in the areas affected by conflict include sexual violence, malnutrition and food insecurity. These have all generated unprecedented levels of chaos and a necessity for a coordinated response to the issues faced by the inhabitants of those areas.

Coordinated action in humanitarian response is a key challenge that has been identified in the domain of ICTs [10]; however, in order to improve coordination and collaboration that will enable humanitarian organisations to successfully and adequately respond to these and similar humanitarian crises, access to accurate, real-time information that can be efficiently managed is a necessity [11]. Tasks included as part of humanitarian response are gathering data about the intensity of a crisis, generating necessary resources, organising the aid effort, supplying goods, staffing operations, ensuring accountability and creating an arrangement with warring factions such that aid can be provided [12]. ICTs have been deployed as a tool to help relief workers to maintain adequate communication and to coordinate their relief efforts. They are instrumental in creating innovative ways for collecting and processing information; ICTs are also creating an avenue for bringing new volunteer networks to the forefront of humanitarian response and are spearheading organisational change within established humanitarian organisations [13].

However, several developments in humanitarian technology seem to be driven by what is achievable as opposed to what is needed [14]. In spite of strides made in the development of technology, humanitarian organisations must note that face-to-face information gathering and sharing still plays a crucial role in humanitarian responses [14] and that technological advancements in information management for humanitarian services are not always guaranteed to generate effective humanitarian relief [15]. Designing an information system (IS) to support humanitarian organisations can be an enormous task [16], but there are countless benefits to be experienced if ICT is adequately deployed to humanitarian relief. However, adoption and adaptation can come with attendant confusion as well. It is therefore worth evaluating the possible challenges within the context of this research area.

This study explores the challenges and opportunities of adopting ICTs in the humanitarian services sector, with the aim of evaluating how the adoption of technological innovations can help to bring relief to areas that have been ravaged by insurgency in north-east Nigeria. This chapter is structured as follows: examples of humanitarian situations that have arisen in northern Nigeria are evaluated in the next section. This is followed by an evaluation of various ICTs and how they have contributed to the aid of the humanitarian sector and relief efforts in different regions. Applicable models will then be discussed in the following section. It is hoped that this work contributes to the literature around supporting humanitarian efforts with the appropriate technology, and towards policy and practice by providing a template that may be applied for adopting and adapting appropriate technology to this ongoing and similar humanitarian crises situations.

14.2 Humanitarian crisis in Nigeria

A humanitarian crisis is described as an event or a series of events that leads to a critical threat to the health, safety, security or well-being of a community or other large group of people, usually over a wide area [17]. A humanitarian relief worker provides relief and protection to victims of humanitarian crisis. Needs are usually different, depending on the crisis in question. Generally, a response evaluation commences by trying to understand the physical and economic conditions of the population before the outbreak of the conflict. Relief workers often risk their lives to reach those affected, under very challenging conditions [18]. Examples of humanitarian crisis include armed conflicts, epidemics, famine, natural disasters and other major emergencies. Such situations often leave survivors in urgent need of life-saving assistance such as shelter, food, water and health care [17].

Armed conflict has contributed greatly to the dire humanitarian situation in northern Nigeria, and a group known as Boko Haram has played a major role in contributing to the state of unrest in the region. This can be described as a complex emergency, because it is a situation that combines internal conflicts with extensive displacements of people and fragile or failing economic, political and social institutions [18]. The grievous harm wrought by Boko Haram on the country is double pronged. First, it espouses an open contempt for Nigeria and its constitution. Second, it underhandedly exposes vulnerable youths to extremist influences. It also recruits and deploys them as an essential part of the group's terror machine [19]. Boko Haram first took up arms against state security forces on 24 December 2003 when it attacked police stations and public buildings in the towns of Geiam and Kanamma in Yobe state [20]. The militant Islamist group is fighting to overthrow the government and create an Islamic state [21]. Although there have been attacks in different parts of the country, the north-east geo-political zone of Nigeria has been worst hit by the Boko Haram insurgency. With the exception of one state in the Geo-political zone, this region has been most susceptible to various magnitudes of terrorist attacks [22].

In its bid to create an Islamic state ruled by sharia and wage war against 'false Muslims' [23], the sect has created a quagmire in north-east Nigeria. Girls have been used as suicide bombers, slavery has been reinstituted and students kidnapped from their dormitories [24]. Boko Haram has also presided over mass killings such as the murder of 59 high-school boys in Buni-Yadi, Yobe state in February 2014 [25]. The group has attacked government storage facilities, fertilizer factories, agricultural input transport systems and farms. The insurgence has reduced agricultural production for Nigerian households, as a result of a decrease through a fall in the cultivation of staple crops such as cassava, soya, sorghum, rice and yam which are key crops in the affected areas [26].

Recent attacks by Fulani herdsmen have also resulted in unprecedented food shortages and displacement of citizens from their homes and farms [27]. Attacks have meant that farmlands have been abandoned, leading to food shortages in areas that are already struggling with security and housing problems. This exposure to armed conflict has worsened food insecurity and malnutrition in the country [28]. In addition, Okoli and Iortyer [22] observe that terrorism, as a whole, in northern Nigeria has resulted in severe humanitarian consequences for the entire country. They include consequences such as human fatalities, internal displacement, human rights abuses, livelihood crisis, as well as public insecurity and a threat to national security. Similarly, Emmanuelar [29] also argues that there is a significant relationship between Boko Haram insurgency and humanitarian crises, leading to human casualties, food insecurity and internally displaced persons (IDPs). The unfortunate condition of these displaced people in their various places of refuge can be described as critically threatening. This is more so considering the fact that they have been dislocated from their family and social capital bases and there are possibilities of further attacks [22].

Additionally, Taylor [30] argues that the effects of the violence in Nigeria are felt not just by the individual victims, but throughout the country and the world as well. The effects are so severe that aside from the Nigerian Civil War, there has not been any other single event of complex emergency in the country that has been as debilitating as the Boko Haram insurgency as far as the humanitarian impact on the country [22]. This is because the drain on economic and social resources is substantial and ends up leaving Nigerian sustainable development at a standstill. More than two million people have been displaced by Boko Haram in Nigeria recently [31] with several hundreds of thousands recorded as newly displaced thus indicating that the situation is dire with no foreseeable end in sight.

Unfortunately, there are no clear laws and policy framework in Nigeria, which defines the roles and responsibilities of humanitarian agencies. This in turn hampers the humanitarian services rendered as there is often a lack of coordination [29]. ICT strategies are required at a national level to provide much needed support for introducing new security, emergency and disaster management frameworks which promote the selective production and use of ICTs. Such strategies can also be helpful in securing new means of financing ICT investment, strengthening scientific and technological research, and helping to improve the capacity to assess the impact of the new international governance regimes for ICT services and technologies [32]. The dire situation previously explored, coupled with the lack of coordination could imply that for a prolonged period, there will be a crises situation that requires a highly concerted effort from the government and humanitarian relief organisations. Considering that the towns and villages under attack are still volatile and can be dangerous for aid workers, there is a need to deploy a coordinated response that could be aided with technology, with the aim of providing some level of protection for volunteers and aid workers. Some efforts have been made in the past to deploy technology to support humanitarian responses with varying success rates.

14.3 ICT and humanitarian response

Information is said to be the lifeblood of any organisation as without information, effective decision-making could be difficult to achieve. Humanitarian organisations also benefit from information as it impacts their ability to make decisions. It is therefore imperative to ensure that the quality of the information used in these sorts of organisations is well managed and not compromised. Some guiding principles have been identified by OCHA-sponsored Symposium on Best Practices in Humanitarian Information Management and Exchange for anyone engaged in any aspect of humanitarian information management. These include interoperability, accountability, verifiability, relevance, objectivity, timeliness, humanity and sustainability [33].

The implications of the terms are explained as follows:

Interoperability: Implies that all data and information that can shared ought to be made available in formats that can be easily retrieved, shared and used by humanitarian organisations.

Accountability: Implies that users are able to evaluate the reliability and credibility of data and information by knowing its source. Information providers should be responsible to their partners and stakeholders for the content they publish and disseminate.

Verifiability: Implies that information should be accurate, consistent and based on sound methodologies, validated by external sources, and analysed within the proper contextual framework.

Relevance: Implies that information is practical, flexible, responsive and driven by operational needs to support decision-making throughout all the phases of a crisis.

Objectivity: Implies that information managers consult a variety of sources when collecting and analysing information such that they provide varied and balanced perspectives for dealing with problems and proffering solutions.

Humanity: Implies that information is never to be used to distort, mislead or cause harm to affected or at-risk populations and should respect the dignity of victims.

Timeliness: Implies that information is collected, analysed and disseminated efficiently and is kept current.

Sustainability: Implies that humanitarian information and data are maintained, sorted and archived, such that it can be retrieved for future use such as for readiness, analysis, feedback and evaluation [33].

Considering the scale of humanitarian crises across the globe, and the number of people affected, it is becoming crucial to find timely and quality information to address the needs of people. The types of crisis and dynamics of situation can unfortunately make data collection a challenging process. As a result, the humanitarian community is addressing the challenges by combining innovations in the use of technology with technical capacity [24]. Effective ICTs are pertinent for humanitarian and emergency response, since consequences are usually life or death [34]. They are critical resources that can facilitate the efforts of humanitarian organisations in assisting local populations and host governments [35]. Humanitarian organisations should regard ICT as a primary service, that is as essential as food, water, shelter and medical care [36]. ICTs have been utilised as tools for coordinating relief efforts to the worst hit areas of a humanitarian crises [37].

Examples of such applications include the Gorkha earthquake in 2015 [38], which emphasises the importance of integrating a variety of ICTs into health systems to make them more resilient in all stages of disaster management and using social media as a mapping platform for natural disasters [39]. They can help to facilitate timely and effective information flows for complex humanitarian activities such as transport, logistics, supply chain management, resource mobilisation, programming, monitoring, evaluation and assessment. ICTs can also facilitate coordination, information creation and sharing within and among humanitarian organisations and their partners [34]. Another application of technology to humanitarian efforts is the use of networks to facilitate collaboration. Meier [13] provides examples of an online platform called 'Digital Humanitarians' that facilitates collaboration between volunteer networks and professional humanitarian organisations and a social-media-based operation dedicated to humanitarian relief called the 'Digital Operation Center' that is run by the American Red Cross, in partnership with Dell.

While the temptation is always there to try the latest offerings from technological innovation and developments, this may not always be beneficial. Technologically advanced humanitarian ISs make claims of being able to access data faster and more accurately.

These claims may however be exaggerated, particularly when considering applications such as visual technology, crisis mapping and big data. In fact, the exuberance about the data exceeds the capacity to substantively analyse it [4]. Where information is a key provision of aid, poor or inadequate data collection processes or loss of data may be potentially harmful [40]. Some instances have been recorded where the availability of information has done more harm than good as mismanagement of data can also be a problem. For example, in Haiti, in a number of instances, potentially valuable information and data were not managed sufficiently to facilitate effective coordination among humanitarian actors. Several times, different actors used different measurements and definitions in data collection and analysis [15]. Disaster relief preparation and response remains a major challenge for organisations, in spite of the progress in humanitarian information management technology development. This is because of the challenge of freely sharing information among stakeholders [41].

Whereas a lack of information diffusion or sharing could hamper efforts being made in the event of a humanitarian disaster, freely sharing could potentially help with efficient coordination. However, information dissemination within a cluster can be accomplished if humanitarian relief agencies are willing to share information, with the cluster lead acting as a filter. Implying that, taking on a central role to simply facilitate information flow improves diffusion. Nonetheless, the willingness of humanitarian agencies to exchange information has a larger impact on information diffusion than that from an information hub [42]. However, as acknowledged by Van de Walle *et al.* [43], this can be a delicate issue. Although information that is processed and analysed by such systems may contribute to improving the livelihoods of people affected by a disaster, it can also endanger their lives if it is inadequate or makes its way into the wrong hands.

While it may be true that the use of innovative ICTs in humanitarian response is imperative and becoming more commonplace, there is a need to consider the practical implementation of such technologies, the ethical implications, policy and standards, including interoperability [11]. Although it is important to adopt technology wisely, it is also expedient to ensure that international standards are established to deal with multi-agency response. Humanitarian organisations also need to test and train with such adopted technologies regularly to ensure that staff and volunteers are able to use it effectively [36]. The development of policy at the organisational, national and international levels to handle ethical requirements for the implementation of such innovative technologies is also expedient. This is so because although these technologies have a potential to aid humanitarian efforts, they may also cause harm if not adequately regulated [11]. Furthermore, ethical consideration of innovative technologies for humanitarian aid ought to be well considered. Vigilance is also necessary to ensure that new applications are developed and implemented in ways that ensure protection for vulnerable individuals and minimise any risk of harm [44]. Minimising the risk of harm is particularly important when handling cases of those caught up in humanitarian crisis situations like that of the north-eastern part of Nigeria. ICTs built to support humanitarian efforts have to take safeguarding principles into account [43].

14.4 Challenges and opportunities for ICT in the Nigerian humanitarian sector

Some examples in the previous section demonstrate how ICT has been useful for responding to humanitarian emergencies and helping to rebuild communities.

There are clearly some opportunities for supporting humanitarian relief work in north-east Nigeria, as new technologies are changing the way assistance is organised and delivered and also changing the relationship between aid givers and aid recipients [45]. Many areas in the country could benefit from ICT as it is crucial for forecasting and can provide early warning signals for conflict prevention. If communities in the affected areas are forewarned about imminent attacks where possible, there is a chance of reducing the impact of such attacks. ICT can also be useful for post-conflict peacebuilding, creating resilient communities that can be better prepared to respond to humanitarian emergencies [46].

ICT can be used to carry out training programs for disaster and emergency preparation. Public websites and mobile applications containing disaster preparedness information in multiple formats, including text, audio and video, could help to circulate information to the target group [47]. A key issue preventing the progress of bringing necessary relief to people caught up in the crisis-affected north-east is the corruption associated with the distribution and management of the humanitarian funding to the Nigerian government. The government's inability to implement the Kampala Convention through the legal, logistical and resources pool framework also accounts for the poor humanitarian condition of the IDPs in the region [48]. However, the introduction of crowd sourcing and social media provides an avenue to evaluate the appropriateness and effectiveness of humanitarian relief projects.

Mobile technologies can also enable self-assessment by people in crisis-affected areas, making them much more active agents in identifying needs and formulating responses [45]. These technologies can also help to address the root cause of violent conflicts by providing access to shared knowledge and mutual understanding which is an indispensable component of conflict prevention and post-conflict reconciliation [49]. Many of the affected areas are remote and rural. However, mobile phones work in those areas. Mobile cash systems can therefore be used where there is a humanitarian crisis, but the cost of transporting or distributing food/non-food items is high or the mode of transportation is insecure [50]. ICTs also offer ways to reveal human right abuses, promote transparent governance and provide access to uncensored information and a means of appealing for help to improve well-being of individuals and communities at risk [49].

It is however worth noting that there are some challenges as well. The infrastructure necessary to support these technologies is not always readily available. For example, as part of an operation to clamp down on Boko Haram, Nigeria's security forces shut down mobile phone networks in Adamawa, Borno and Yobe states in north-east Nigeria between 23 May and 12 July 2013 [51]. This also meant that citizens had no access to their mobile networks and felt cut-off from the rest of the country. Electricity is not always stable, neither is access to internet facilities. The costs of acquiring smartphones and internet access may also be prohibitive. Illiteracy levels also mean that even though there are websites/mobile apps providing necessary content to support humanitarian relief, locals in the region may be unable to understand or apply the knowledge that should have been disseminated on such sites [52].

ICT development in Nigeria is currently experiencing a downturn. Even though at some point the country was regarded as the leader, within the continent [53], internet penetration is at 29% [53]. Considering the fact that the uptake of technology is currently low, it follows that humanitarian ISs uptake and development could also be low. This is because poor uptake of technology is one of the most widely discussed failures where ICT for development is concerned [54]. There is, however, a widespread acknowledgement of the need for more knowledge and skills relating to the application of technology to development programmes at every level, whether this gap is met through training, hiring or partnerships [54]. Unfortunately, and sometimes to their detriment, humanitarian relief organisations generally view ICTs as overhead rather than as infrastructure that must be at the core of their operations. In addition, those donating to relief work tend to view overhead costs as detrimental. Overhead costs are therefore kept to a minimum which in turn has an impact on how much can be spent on ICTs [55].

Technological innovations are not always guaranteed to generate an effective humanitarian response. Other external factors have an impact on the flow of information in humanitarian response. Nevertheless, an effective information and communication infrastructure among humanitarian organisations can offer prospects of ameliorating any constraints that occur as a result of underdeveloped or destroyed physical infrastructure [55]. It is however important to realise that any humanitarian technology will not be flawless and as such, technological failures may translate into humanitarian failures [56]. Therefore, it is expedient for the humanitarian community to consider how and under what circumstances the use of new technologies in a variety of humanitarian contexts may be potentially harmful [57].

As discussed in the previous section, it is evident that humanitarian organisations benefit substantially from using innovative technologies. However, there are challenges as well, especially when considering the nature of humanitarian organisations. It is worth bearing in mind that technological innovation in the context of the humanitarian sector is unique, complex and not driven by the same globally valid objectives and imperatives as other organisations [34]. Furthermore, adequate planning can enhance humanitarian efforts in areas such as coordination, strategic planning, preparedness and accountability. However, these solutions are not typically applicable off the shelf and must be adapted to particular situations and needs [58].

14.5 Adopting ICTs for humanitarian response in north-eastern Nigeria

Technologies that have been adopted for humanitarian purposes include mobile phones, social media platforms, geospatial technologies, crowdsourcing, mobile money and unmanned aerial vehicles (UAVs). Mobile UAVs can be used to supply goods where it could be dangerous to be physically present. However, the use of this means would be dependent on favourable weather conditions. Mobile money can also be used in the place of sending out physical goods to remote or difficult to reach areas. Crowdsourcing can be used to raise funds to support relief work. However, care has to be taken to ensure that the funds are utilised accountably and transparently. Further examples of technologies that would also be beneficial include internet of things, big data and artificial intelligence. Some of these innovations have been applied successfully but some are yet to be adopted, particularly in north-east Nigeria.

14.5.1 Theories explaining the adoption of innovations

There are various theories that can explain how such innovative technologies are adopted and used. Several diffusion studies have been carried out in the field of ISs. Such studies have applied theories ranging from technology acceptance model (TAM), TAM2, theory of planned behaviour (TPB), theory of reasoned action (TRA) to modified versions of the aforementioned. These theories have been used for different purposes. For example, studies considering the adoption of technology for individual have used TAM and TPB for predicting user intentions [59]. TAM is meant to specifically explain user behaviour as far as computing technology is concerned [60], while TPB is used to predict an individual's intention to engage in a particular behaviour at a specific time and place. The intention of the theory is to explain behaviour over which people have the ability to exercise self-restraint [61]. Both models provide good predictions of an individual's intent to use an ICT, although TAM slightly outperformed TPB. TAM is easier to use, while TPB provides more information about the factors users consider when making their choices [59].

The aforementioned theories along with a few others have dominated ICT adoption research but are usually applied to individual use as opposed to the complex network of different groups that are involved in humanitarian relief work. However, one of the most influential works on adoption/diffusion remains that of Rogers [62]. He defines innovation as an idea, practice or object perceived as new by an individual or other unit of adoption, and diffusion is defined as the process by which an innovation is communicated through certain channels over time among members of a social system. Rogers' theory of diffusion of innovations (DoI) seeks to explain how, why and at what rate new ideas and technology spread [62]. Within the theory, the rate of adoption is explained as the relative speed with which the members of a social system adopt an innovation. As a general rule, it is usually measured as the number of individuals who adopt a new idea within a specified period such as a year. The rate of adoption is therefore a numerical indicator of the steepness of an adoption curve for an innovative technology [62].

The DoI theory also seeks to explain how the decision to adopt an innovative technology happens over time. Rather than being an instantaneous action, this decision, according to the theory, happens as a process. This is known as the innovation decision process. Rogers describes this as the process through which an individual, or any other decision-making units, passes from first knowledge of an innovation to forming an attitude towards the innovation, to a decision to adopt or reject, to the implementation of the new idea, and a confirmation of this decision. The process is made up of five stages, which are as follows:

Knowledge: At this phase, there is exposure to an innovative, technology but not enough information has been obtained about it.

Persuasion: At this phase, there is an interest in the innovation, and related information about the technology is actively sought.

Decision: At the decision phase, the advantages and disadvantages of using the innovative technology are actively examined and a decision is made to adopt or reject.

Implementation: At this phase, the innovation is employed and its usefulness is determined. More information about the innovation may also be sought at this stage.

Confirmation: At this phase, a final decision is made to continue using the technological innovation.

Information is sought at various stages in the innovation–decision process in order to reduce any uncertainty about the expected consequences of an innovation. The decision stage leads first to adoption, a decision to make full use of an innovation as the best course of action available, or to rejection, a decision not to adopt an innovation [62]. According to the theory, innovation attributes that explain the rate of diffusion of an innovation are complexity, compatibility, relative advantage, trialability and observability.

Relative Advantage: Relative advantage in the theory of DoI is described as the degree to which an innovation is perceived as being better than the idea it supersedes. The degree of relative advantage is usually expressed in terms of how economically profitable it is to use, social prestige it creates, time saved by using it or any other benefits. The nature of the relative advantage would be dependent on the innovative technology that is being considered for relief work.

Compatibility: This describes how consistent an innovative technology is with the existing values, past experiences and needs of the prospective adopters. Considering the fact that humanitarian organisations sometimes play a highly critical role that involves human lives, it is important to ensure consistency with values as an innovation may not be compatible with the sociocultural values and beliefs of the relief workers or the victims of the humanitarian crisis.

Complexity: This refers to how difficult the innovation is to understand and/or use. For some innovative technologies, it is easy to decipher how they can be applied. Complexity of any innovation introduced to humanitarian relief work has to consider whether that would require new skill sets or training for relief workers. Consideration would also have to be given to how much resources would have to be committed to mitigating the effect of a complex innovation.

Trialability: This describes the degree to which an innovative technology can be tested or experimented with before a commitment is made to adopt.

Observability: This attribute describes the extent to which an innovative technology provides tangible results. It describes the degree to which results of an innovation are visible to others. This implies that the results of some ideas can be easily observed and described, while other innovations can be difficult to observe and describe [62].

The innovation diffusion theory suggests that an innovation is adopted mainly because of its characteristics. This implies that other influences are disregarded. Within the context of innovation diffusion for humanitarian services, considering the interplay of the factors, various organisations, governments, people caught up in the crisis situations, the human and ethical implications of adopting technologies to support humanitarian efforts, it is safe to assume that other influences cannot be disregarded. Innovation is therefore seen as a contradictory and uncertain process. It cannot just be seen as a rational–technical 'problem-solving' process; it also involves 'economic and political' processes in building alliances of interests (among, e.g. supplier firms, technologists, potential users, funding bodies and regulators) with the necessary resources and technical expertise, around certain concepts or visions of as yet unrealised technologies, suggesting that technological innovations are socially shaped [63]. Technological deterministic theories such as DoI presume that particular paths of technological change are inevitable. However, social shaping of technology (SST) studies shows that technology is a social product, which is patterned by its creation and use.

Table 14.1 presents the various theories that can be applied to studies considering the adoption or DoI.

14.5.2 Social shaping of technology

SST studies generally indicate that technologies are created and utilised in specific social contexts and that technological change is intrinsically social, as opposed to being driven simply by a technical logic. Technologies are used in a setting of knowledge, skills, meanings and values, problems and purposed and the objects upon which they act. These studies also indicate that technological change is always part of a sociotechnical transformation and that technology and social arrangements are jointly produced in the same process [64]. Theories that embrace sociotechnical approaches are especially useful when the situation has to consider divergent views and opinions.

It is worth noting that social shaping perspectives do not dismiss technology. There is, however, a strong position that even when there is an emphasis on an innovative technology, people and their beliefs and attitude will play a key role in

Table 14.1 Theories explaining adoption

Theories	**How they have been applied**
TAM	For predicting individual user's intentions
TPB	To predict an individual's intention to engage in a particular behaviour
TRA	To predict how individuals will behave based on pre-existing attitudes
DoI	Explains how diffusion of an innovation happens over time
SST	Complex scenarios presented by a humanitarian context

shaping its adoption, how it is used and the implications within specific social contexts [65]. At the core of SST is the notion that there are choices or options inherent in both the design of innovative technologies and in the direction of innovation programmes. If technology does not emerge from the unfolding of a predetermined logic or a single determinant, then innovation is a path of divergence and convergence, all at the same time. Different routes are available which could lead to different technological outcomes, meaning that these choices could have differing implications for society and for particular social groups [63]. The implications of the technological outcomes are particularly pertinent for humanitarian innovative technologies. This is because such systems are not designed to work in a vacuum. The impact of such systems could indeed mean the difference between life and death. SST research investigates the ways in which social, institutional, economic and cultural factors have shaped the following: the direction as well as the rate of innovation; the form of technology: the content of technological artefacts and practices; and the outcomes of technological change for different groups in society [63].

According to Tusiime and Byrne [34], on the one hand, DoI theory [62] and contemporary innovation frameworks provide a good foundation of well-developed concepts and empirical results applicable to the study of technology adoption and IS innovation. On the other hand, they do not apply well to the complex scenarios and innovation phenomena presented by the humanitarian context. The theory ignores the rationality and history of the origins of the innovation, and as a result, it ignores the rationality and values embedded in that innovation [34]. In contrast, the actor–network theory (ANT) which is based on social theory makes no analytical distinction between the social and the technical. ANT has proved useful for a humanitarian organisation analysis as it helps to explain the mechanisms by which social factors become influential in adoption success or failure [34].

However, critics of ANT believe that it addresses the local and contingent, but that it pays little attention to broader social structures which influence the local [66]. In the humanitarian context, there are several social structures that can have an influence on how an innovation is introduced or used. A different approach has been suggested for IS researchers which is to combine the methodological approach and conceptual ideas of ANT with insights and analyses drawn from theories of social structure [66].

Considering the fact that SST accounts for the differing implications that a particular technology may have for society and for particular social groups, this study suggests a framework combining the innovation attributes, which would consider the nature of the innovation and its characteristics, with elements of SST. This would take into consideration the relevant social groups involved in the humanitarian relief in north-east Nigeria and their views about the nature of innovative technologies that could be used to aid their efforts. The perceived attributes of an innovative technology are clearly factors that can explain whether or not humanitarian organisations choose to adopt the innovation. However, the focus cannot only be on the nature of the technology in question. Not only are the attributes important but the network of relief agencies and other social groups and their views will also have to be considered. This way, the focus is not just on the innovative technology but also on all the groups that are likely to have an influence on its adoption for humanitarian services.

The social groups that could have an influence on the adoption of innovative technologies for humanitarian relief work include the following: United Nations, donor governments, multilateral and bilateral agencies, governments of countries affected by the crises, non-governmental organizations (NGOs), philanthropic foundations, corporations, the giving public and those who have suffered during the crisis [55]. While some of these groups will directly use or be impacted by the adoption of innovative technologies, some will only indirectly impact or be impacted. Humanitarian aid efforts can often be fraught with competition and confusion because the implementation environment of humanitarian emergencies includes a variety of actors, each operating under varying constraints [67]. It is therefore necessary to ensure that there is an adequate representation of the views of all the actors/social groups, in order to ensure that any innovative technology under consideration does not further complicate an already sensitive relief response situation.

14.6 The proposed framework

As previously explained in the preceding sections, this chapter argues that a technologically deterministic view of the adoption of Humanitarian ICTs would not suffice as there are too many influences and interest groups that may be left out. Therefore, the framework proposed in this chapter combines the attributes of an innovation from the theory of DoI, with the views and interpretations of the technology from the network of various relief agencies, victims and other social groups involved in providing support to victims of humanitarian crises. This reflects a SST perspective in the adoption of Humanitarian ICTs, in addition to the features of the innovative technology. The framework ensures that not only are the characteristics of the ICT that facilitate adoption, as identified by Rogers [62] considered, but also the interpretation given to them by the social groups that would use or be impacted by the use of such systems.

The various social groups that have been identified according to existing literature are the United Nations, governments of the countries affected by the crises, donor governments, multilateral and bilateral agencies, NGOs, philanthropic foundations, the giving public and those who have suffered during the crisis. The framework is presented in Figure 14.1.

14.7 Conclusion and further research

This research work provides an exploratory understanding of the use of innovative technologies for humanitarian crises with a particular focus on the Boko Haram situation plaguing the north-eastern region of Nigeria. It examines the challenges and opportunities that may accompany the adoption of technological innovations

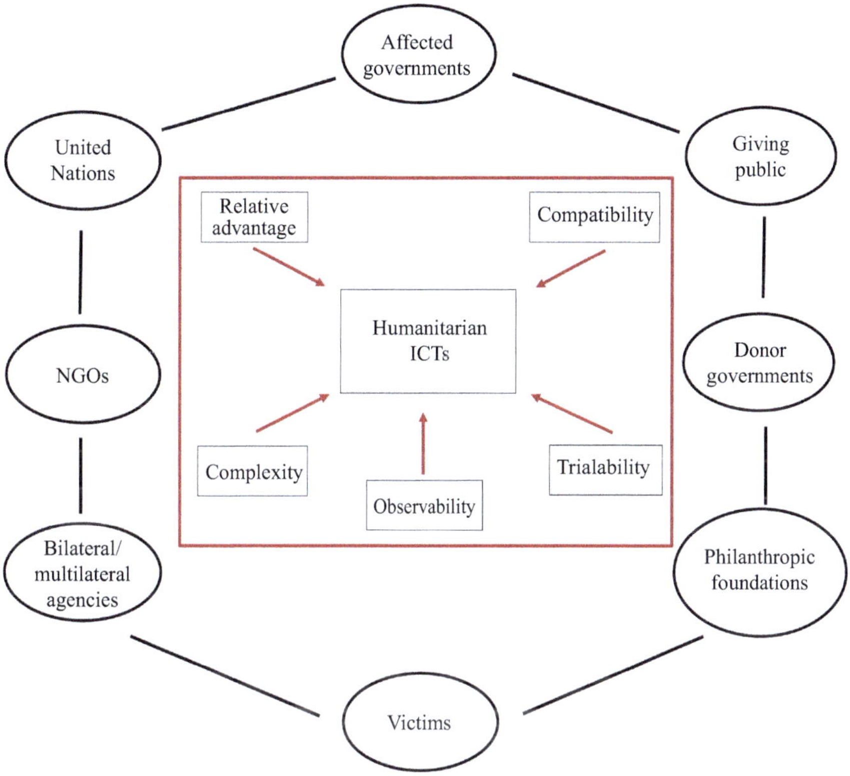

Figure 14.1 Framework guiding humanitarian ICT adoption

that can support relief work in the areas that have been ravaged by Boko Haram attacks. The humanitarian crises constituted by these attacks have greatly caused unrest in the region, resulting in a complex emergency with loss of life, livelihood and mass displacement of inhabitants. Considering the fact that the situation faced by IDPs and surrounding communities is still dire, it is expedient to look into various solutions that can help to abate the crises.

Information is one resource that has been identified as a crucial ingredient for ensuring adequate and cohesive coordination when relief workers respond to a humanitarian crisis. It is also necessary for effective decision-making. A lack of adequate information sharing has also been identified as a way of hampering relief work. Likewise, an abundance of information or sharing with the wrong group can also hamper rather than aid efforts in a humanitarian-crisis-affected area. Privacy and security considerations must also be made. This is particularly important to ensure that the suffering of the victims is not further exacerbated by allowing private information to fall into the wrong hands. For example, the details or locations of victims could inadvertently be shared with would-be attackers or oppressive regimes.

ICT and associated technologies have been identified as crucial for the gathering and processing of this information and the support of humanitarian organisations' work in crisis situations. However, these technologies may not always be effective and can also act as a double-edged sword. This is because such systems are applied in life and death situations and their failure could have dire consequences. In addition, the ubiquity of some of the applications of ICTs that have been identified may have been overstated or exaggerated. It is therefore worthwhile to ensure that clear measurable benefits can be outlined by the relevant social groups before adopting innovative technologies for humanitarian work.

Nevertheless, the use of ICTs and associated technologies has contributed to the aid of the humanitarian sector and relief efforts in different regions. This is because they have the potential to provide solutions that can help to provide succour to victims of humanitarian crisis. Examples of innovative technologies that can further improve humanitarian efforts such as internet of things, mobile money, UAVs, big data, 3D printing were also explored with the aim of examining the benefits experienced from applying such innovative solutions. Some challenges were also identified. The nature of humanitarian work in the north-eastern region of Nigeria is quite unique in the sense that it has been faced with sustained violence in ways that are impossible to anticipate. As such, many technological solutions would have to be tailored to fit the purpose. It was however noted that the design of such technological solutions can be complex and can result in confusion for those involved in humanitarian relief in that region.

The use of innovative technologies in humanitarian response is imperative and must not be seen as a fringe cost but central to relief work. However, considering the spate of violence and displacement, the practical implications of the implementation of such technologies cannot be taken for granted. Interoperability of the technology is important as is the need to ascertain that such adoptions meet with international standards to ensure that such systems can handle the pressure of multi-agency response. The ethical implications of adoption of such technologies also have to be carefully considered. It is therefore important to ensure that at every level, there is an extensive policy in place that clarifies ethical requirements for the implementation of such innovative technologies. Training is another issue that was identified as a concern, as whenever new systems are implemented, new skill sets may be required to ensure that such technologies are adequately optimised to support relief efforts.

Unfortunately, as Nigeria is currently experiencing a downturn in ICT development, this may also serve as a barrier to adopting certain technologies. This is because without the supporting infrastructure in place, many innovations would be unable to serve the purpose for which it was intended. Also, a lack of adequate infrastructure may hamper the application of certain technologies as there could be restricted access to mobile phones and in turn websites. This could in turn mean that people caught up in humanitarian crisis situations might miss out on crucial information. Critical communication may also be curtailed in such a situation.

The north eastern region has experienced so many attacks on educational institutions, which are now mostly abandoned. Illiteracy rates are also high in the region and this would also have to be taken into account when considering the types of innovative technologies to adopt in order to aid relief work in the region. Illiteracy rates could have an impact on communication between local and relief agencies and also on the sharing of vital information.

In order to understand and improve the adoption and effective application of such technologies and based on issues arising from existing literature, some theories were considered, including the TPB, TRA, DoI theory and ANT. As a result of various criticisms and considering the sensitive nature and the impact of relief work, a framework was suggested that combines the innovation attributes from the DoI theory with SST to ensure that not only the characteristics of the innovations under consideration are taken into account, but also how the social groups are able to interpret or shape the innovative technology as well. This framework can be applied by organisations involved in humanitarian relief response as it provides a means of evaluating the contribution of the social groups involved in shaping any technology that is applied to crisis-affected areas.

There are limitations to this study which would lead to further research that aims to apply this framework using primary data obtained from north-east Nigeria and also from the different relevant social groups that have been identified. It is hoped that the data obtained will provide a model for applying technology to the affected areas with minimal risk of harm to the inhabitants of such areas.

References

[1] OCHA. Global Humanitarian Overview 2019 [Internet]. OCHA; 2018 [cited 2019 Sep 15]. Available from: https://www.unocha.org/global-humanitarian-overview-2019.

[2] Müller-Stewens G, Dinh T, Hartmann B, Eppler MJ, and Bünzli F. Humanitarian organizations under pressure. In: Müller-Stewens G, Dinh T, Hartmann B, Eppler MJ, and Bünzli F, editors The Professionalization of Humanitarian Organizations: The Art of Balancing Multiple Stakeholder Interests at the ICRC [Internet]. Cham: Springer International Publishing; 2019 [cited 2019 Sep 15]. pp. 1–7. (Springer Briefs in Business). Available from: https://doi.org/10.1007/978-3-030-03248-7_1.

[3] Lange M, and Quinn M. Conflict, Humanitarian Assistance and Peacebuilding: Meeting the Challenges. London: International Alert; 2003.

[4] Smock DR. Humanitarian Assistance and Conflict in Africa. Washington, DC: US Institute of Peace; 1996.

[5] ReliefWeb. Nigeria: 2019-2021 Humanitarian Response Strategy (January 2019–December 2021) – Nigeria [Internet]. ReliefWeb; 2019 [cited 2019 Sep 11]. Available from: https://reliefweb.int/report/nigeria/nigeria-2019-2021-humanitarian-response-strategy-january-2019-december-2021.

[6] DataBank. World Development Indicators [Internet]. DataBank; 2015 [cited 2019 Sep 15]. Available from: https://databank.worldbank.org/source/world-development-indicators.

[7] Gordon, RG., Jr. editor, Ethnologue: Languages of the World, Fifteenth edition. Dallas, TX: SIL International; 2005. Online version: http://www.ethnologue.com/.
[8] Emergencies [Internet]. Unicef.org. 2019 [cited 2019 Sep 20]. Available from: https://www.unicef.org/nigeria/emergencies.
[9] Human Rights Watch. World Report 2019: Rights Trends in Nigeria [Internet]. Human Rights Watch; 2018 [cited 2019 Sep 20]. Available from: https://www.hrw.org/world-report/2019/country-chapters/nigeria.
[10] Saab DJ, Tapia A, Maitland C, Maldonado E, and Tchouakeu LM. Inter-organizational coordination in the wild: Trust building and collaboration among field-level ICT workers in humanitarian relief organizations. Voluntas: International Journal of Voluntary and Nonprofit Organizations. 2013;24(1):194–213.
[11] Greenough PG, Chan JL, Meier P, Bateman L, and Dutta S. Applied technologies in humanitarian assistance: Report of the 2009 Applied Technology Working Group. Prehospital and Disaster Medicine. 2009;24(S2):s206–9.
[12] Adan, A. A., and Kising'u T. Factors influencing humanitarian aid delivery in Mogadishu, Somalia case study SACID Non-Governmental Organization. Strategic Journal Business Change Management. 2018; 5(2):106–25.
[13] Meier P. New information technologies and their impact on the humanitarian sector. International Review of the Red Cross. 2011;93(884):1239–63.
[14] Read R, Taithe B, and Mac Ginty R. Data hubris? Humanitarian information systems and the mirage of technology. Third World Quarterly. 2016; 37(8):1314–31.
[15] Altay N, and Labonte M. Challenges in humanitarian information management and exchange: Evidence from Haiti. Disasters. 2014;38(S1):S50–72.
[16] Bui T, Cho S, Sankaran S, and Sovereign M. A framework for designing a global information network for multinational humanitarian assistance/disaster relief. Information Systems Frontiers. 2000;1(4):427–42.
[17] Humanitarian Coalition. What Is a Humanitarian Emergency? [Internet]. 2015 [cited 2019 Sep 22]. Available from: https://www.humanitariancoalition.ca/what-is-a-humanitarian-emergency.
[18] Weiss TG. Humanitarian Challenges and Intervention. New York: Routledge; 2018.
[19] Omilusi M. The multi-dimensional impacts of insurgency and armed conflicts on Nigeria. Asian Journal of Social Science, Arts and Humanities. 2016;4(2): 29–39.
[20] Onuoha FC. The Islamist challenge: Nigeria's Boko Haram crisis explained. African Security Review. 2010;19(2):54–67.
[21] Chothia F. Who are Nigeria's Boko Haram Islamists. BBC News. 11 Jan 2012.
[22] Okoli AC, and Iortyer P. Terrorism and humanitarian crisis in Nigeria: Insights from Boko Haram insurgency. Global Journal Human Social Science-GJHSS-F. 2014;14(1):39–49.
[23] Walker A. What is Boko Haram? Vol. 17. Washington, DC: US Institute of Peace; 2012.

[24] Eme OI, Azuakor PO, and Mba CC. Boko Haram and population displacement In Nigeria. Practicum Psychologia. 2018;8(1):75–96.
[25] Oriola TB, and Akinola O. Ideational dimensions of the Boko Haram phenomenon. Studies in Conflict and Terrorism. 2018;41(8):595–618.
[26] Adelaja A, and George J. Effects of conflict on agriculture: Evidence from the Boko Haram insurgency. World Development. 2019;117:184–95.
[27] Kah HK. 'Boko Haram is losing, but so is food production': Conflict and food insecurity in Nigeria and Cameroon. Africa Development. 2017;42(3): 177–96.
[28] Fadare O, Akerele D, Mavrotas G, and Ogunniyi A. (2019). Effect of conflict and food price shocks on calorie intake and acute malnutrition in Nigeria: A micro-panel data analysis. Paper presented at the 93rd Annual Conference of the Agricultural Economics Society, University of Warwick, England 15–17 April 2019. Retrieved from https://EconPapers.repec.org/RePEc:ags:aesc19:289676
[29] Emmanuelar I. Insurgency and humanitarian crises in Northern Nigeria: The case of Boko Haram. Academic Journals: African Journal of Political Science and International Relations. 2015;9(7):284–96.
[30] Taylor LL. Boko Haram terrorism: Reaching across international boundaries to aid Nigeria in the humanitarian crisis. ILSA Journal of International and Comparative Law. 2014;21:1.
[31] IDMC. GRID 2019, Global Report on Internal Displacement 2019; 2019 [Internet]. [cited 2019 Sep 15]. Available from: http://www.internal-displacement.org/global-report/grid2019/
[32] Garba EJ, and Ahmadu AS. ICT implementation framework for integrated national security, emergency and disaster management for Nigeria. West African Journal of Industrial and Academic Research. 2015;15(1):27–37.
[33] Maitland C, and Tapia A. Outcomes from the UN OCHA 2002 symposium & HIN workshops on best practices in humanitarian information management and exchange. Prepared for United Nations Office for the Coordination of Humanitarian Affairs. 2007 Oct 13.
[34] Tusiime E, and Byrne E. Information systems innovation in the humanitarian sector. Information Technologies & International Development. 2011;7(4):35.
[35] Rameshwar D, Zongwei L, Angappa G, Shahriar A, Hazen BT, and Douglas MA. Big data and predictive analytics in humanitarian supply chains. Internal Journal of Logistics Management. 2018;29(2):485–512.
[36] Nelson CB, Steckler BD, and Stamberger JA. The evolution of hastily formed networks for disaster response: Technologies, case studies, and future trends. In: 2011 IEEE Global Humanitarian Technology Conference; 2011. pp. 467–75.
[37] Wentz L. An ICT Primer: Information and Communication Technologies for Civil-Military Coordination in Disaster Relief and Stabilization and Reconstruction. Washington, DC: Center for Technology and National Security Policy at the National Defense University; 2006.

[38] Crane O, Balen J, Devkota B, Ghimire S, and Rushton S. Use of information and communication technologies in the formal and informal health system responses to the 2015 Nepal earthquakes. Health Policy and Planning. 2017;32(Suppl_3):iii48–58.

[39] Middleton SE, Middleton L, and Modafferi S. Real-time crisis mapping of natural disasters using social media. IEEE Intelligence Systems. 2014;29(2): 9–17.

[40] Sandvik KB, Jumbert MG, Karlsrud J, and Kaufmann M. Humanitarian technology: A critical research agenda. International Review of the Red Cross. 2014;96(893):219–42.

[41] Saab DJ, Maldonado E, Orendovici R, *et al.* Building global bridges: Coordination bodies for improved information sharing among humanitarian relief agencies. In: Proceedings of ISCRAM 2008 – 5th International Conference on Information Systems for Crisis Response and Management. Information Systems for Crisis Response and Management, ISCRAM; 2008. pp. 471–83.

[42] Altay N, and Pal R. Information diffusion among agents: Implications for humanitarian operations. Production and Operations Management. 2014; 23(6):1015–27.

[43] Van de Walle B, Van Den Eede G, and Muhren W. Humanitarian information management and systems. In: International Workshop on Mobile Information Technology for Emergency Response. Berlin, Heidelberg: Springer; 2008. pp. 12–21.

[44] Hunt M, Pringle J, Christen M, Eckenwiler L, Schwartz L, and Davé A. Ethics of emergent information and communication technology applications in humanitarian medical assistance. International Health. 2016;8(4):239–45.

[45] Bennett C, Foley M, and Pantuliano S. Time to Let Go: Remaking Humanitarian Action for the Modern Era. London: ODI; 2016.

[46] Momoh AM, and Majaro-Majesty HO. ICT, Government Bureaucracy, and Peace Building in Nigeria [Internet]. 2015. [cited 2019 Sep 03]. Available from: http://citeseerx.ist.psu.edu/viewdoc/download?doi=10.1.1.565.8667&rep=rep1&type=pdf.

[47] Firdhous MFM, and Karuratane PM. A model for enhancing the role of information and communication technologies for improving the resilience of rural communities to disasters. Procedia Engineering. 2018;212:707–14.

[48] Lanshie EN, and Henry YB. Boko haram insurgency, internally displaced persons and humanitarian response in northeast Nigeria. International Journal of Humanities and Social Science. 2016;4(8):141–50.

[49] Mubaraka CM, Kalulu RA, and Salisu MJ. Information Technology and Humanitarian Emergency Response Management in WFP Uganda: a Behavioral Perspective; 2013.

[50] Abushaikha I, and Schumann-Bölsche D. Mobile phones: Established technologies for innovative humanitarian logistics concepts. Procedia Engineering 2016;159:191–8.

[51] Jacob J, and Akpan I. Silencing Boko Haram: Mobile phone blackout and counterinsurgency in Nigeria's Northeast region. Stability: International Journal of Security and Development. 2015;4(1):1–17.
[52] Yoro RE. Application of information and communication technology (ICT) in Nigerian education: Issues and challenges. Multidisciplinary Journal of Research Development. 2010;15(1):47–53.
[53] Gillwald A, Mothobi O, and Rademan B. The State of ICT in Nigeria 2018; 2018.
[54] Haikin M, and Flatters G. Digital Development: What is the Role of International NGOs? ICT for Development Programmes and Opportunities in the Horn, East and Central Africa; 2017.
[55] Maiers C, Reynolds M, and Haselkorn M. Challenges to effective information and communication systems in humanitarian relief organizations. In: IPCC 2005 Proceedings International Professional Communication Conference, 2005. IEEE; 2005. pp. 82–91.
[56] Jacobsen KL. The Politics of Humanitarian Technology: Good Intentions, Unintended Consequences and Insecurity. Abingdon: Routledge; 2015.
[57] Jacobsen KL. Humanitarian Technology: Revisiting the 'Do No Harm' Debate; 2015.
[58] Maiers C, Reynolds M, and Haselkorn M. Challenges to Effective Information and Communication Systems in Humanitarian Relief Organizations; 2005. pp. 82–91.
[59] Mathieson K. Predicting user intentions: Comparing the technology acceptance model with the theory of planned behavior. Information Systems Research. 1991;2(3):173–91.
[60] Davis FD. User acceptance of information technology: System characteristics, user perceptions and behavioral impacts. International Journal of Man-Machine Studies. 1993;38(3):475–87.
[61] Ajzen I. The theory of planned behavior. Organizational Behavior and Human Decision Processes. 1991;50(2):179–211.
[62] Rogers EM. Diffusion of Innovations. Simon and Schuster; New York: Free Press. 2010.
[63] Williams R, and Edge D. The social shaping of technology. Research Policy. 1996;25(6):865–99.
[64] Russell S, and Williams R. Social shaping of technology: Frameworks, findings and implications for policy with glossary of social shaping concepts. Shaping Technology Guiding Policy Concepts Spaces Tools. 2002:37–132.
[65] Dutton WH. The social shaping of digital research. International Journal Sociology Research Methodology. 2013;16(3):177–95.
[66] Walsham G. Actor-network theory and IS research: Current status and future prospects. In: Information Systems and Qualitative Research. Boston, MA: Springer; 1997. pp. 466–80.
[67] Stephenson Jr M, and Schnitzer MH. Interorganizational trust, boundary spanning, and humanitarian relief coordination. Nonprofit Management Leadership. 2006;17(2):211–33.

Chapter 15

Deciphering adoption and adaption of ICTs in humanitarian services: the southern African context

Paul Sambo[1] *and David Chikodzi*[2]

Globally, the rate and frequency at which both natural and man-made disasters have been occurring are on the increase. The southern African region has its own fair share of disasters such as floods, droughts, wars and health issues, hence a need to establish structures for emergency preparedness. With the adoption of information communication technologies (ICTs), humanitarian services, for interventions during emergencies such as disaster preparedness, responses and awareness would be enhanced. This chapter discusses the nature of humanitarian issues affecting the region and how best adoption and adaption of ICTs have and/or can close the gaps of these issues, thereby maximising the efficiency of the humanitarian sector in mitigation, management and recovery efforts. The actor–actant–network theory was used as a theoretical framework for the study. The framework managed to outline how actors, their institutions and actants (technology) interact through their networks in disaster mitigation. The study revealed the ICTs that are used in the southern African region to mitigate, manage and recover from floods, droughts, health and conflict humanitarian disasters. The chapter further explored the challenges faced by the southern African regional countries such as funding, infrastructure, emerging technologies and computer literacy, among others that are affecting the adoption and adaption of ICTs in the humanitarian sector. Recommendations for adopting and adapting ICTs for the region by the policy makers are also discussed.

15.1 Introduction

Globally, there has been an increase in the occurrence and magnitude of both natural and human-induced disasters. ICTs have become an increasingly critical

[1]Department of Mathematics and Computer Science, Great Zimbabwe University, Masvingo, Zimbabwe
[2]Department of Physics, Geography and Environmental Sciences, Great Zimbabwe University, Masvingo, Zimbabwe

component of modern day humanitarian response to disasters [1]. ICTs are being adopted in a diversity of ways by a battery of different actors during humanitarian situations. These actors include governmental, non-governmental and the affected communities. Some of the most widespread and ongoing adoption of ICTs in humanitarian contexts include issuing of early warning to both the at-risk populations and the responders, collecting and communicating information in order to dispel rumours and improve situational awareness, aiding improved communication between the affected communities and response activities and also in the general coordination in disaster management.

Due to the increasing speed at which adoption of ICTs by both responders and crisis-affected communities takes place, the United Nations Office for the Coordination of Humanitarian Affairs [2] has noted them to hold much promise for the improved delivery of humanitarian programmes. This led [2] to officially proposing the recognition of information management during disasters and the corresponding ability to communicate it as a basic humanitarian need [2]. Humanitarian communication can be defined as the technical capacity building, information collection and dissemination, preparedness activities and data analysis for the purposes of saving lives, alleviating suffering and protecting the dignity of crisis-affected populations when performed in accordance with international standards of humanity, impartiality, neutrality and independence. ICTs are therefore increasingly becoming a defining component of humanitarian communication, and aid organisations are increasingly recognising and prioritising communication as a form of assistance as important as water, food and shelter [1]. With limited access to information, extreme event survivors will neither access the assistance that they require nor make knowledgeable choices about their recovery.

For use of ICTs to be considered humanitarian, three ‘P’s’ must be observed. These are populations, principles and purposes. In terms of population, communication using ICTs during humanitarian crises should be intended for the benefit of the affected populations. In terms of principles, the actors and their activities must uphold all the four core humanitarian principles of humanity, impartiality, neutrality and independence. The primary objective of using the ICTs in communication should be to save lives, alleviate suffering and to protect human dignity during and after the humanitarian crisis [3].

Other initiatives at the global level such as United Nations Sendai Framework for Disaster Risk Reduction 2015–2030 have noted the importance of adopting ICTs in disaster risk reduction. The framework advocates for a substantial increase in the availability of and access to multi-hazard early warning systems and disaster risk information and assessments to people by 2030. It also advocates for the promotion of near real-time access to dependable information, adoption of space based and in situ data, including geographic information systems (GIS) and communications technology innovations such as social media, mobile telephones and radios to enhance the collection, analysis and dissemination of data related to disasters [4]. The southern African region has encountered both natural and man-made humanitarian disasters. These humanitarian situations have slowed down the economic development rate in the region as the countries affected by the disasters concentrate on recovering and rebuilding against the disasters. These disasters include floods, droughts, wildfires, disease outbreaks, wars and land degradation. As humanitarian actors in the southern African region are becoming ever more dependent on ICTs as part of their work plans, it becomes necessary to decipher the extent to which there has been adoption and adaption of ICTs in humanitarian services in the region. With this aim, the paper will specifically unravel the adoption and adaption of ICTs during flood, drought, war and epidemiologically induced humanitarian situations in the region with the view of analysing the gaps and blind spots in their adaptation as well as highlighting actionable best practices.

15.2 Methodology

15.2.1 Study area

The study will focus on case studies from the southern African region. The region has a total of 15 countries as shown in Figure 15.1. The region extends from 0° to almost 35° south of the equator and consequently experiences a wide range of tropical to temperate climatic conditions. While southern African countries largely have similar climatic conditions, there are also some subregional variations. Rainfall in the region is mainly driven by the Inter Tropical Convergence Zone whose influence diminishes southwards towards South Africa, where rainfall activity is influenced mainly by southern hemisphere fronts. Temperatures in the

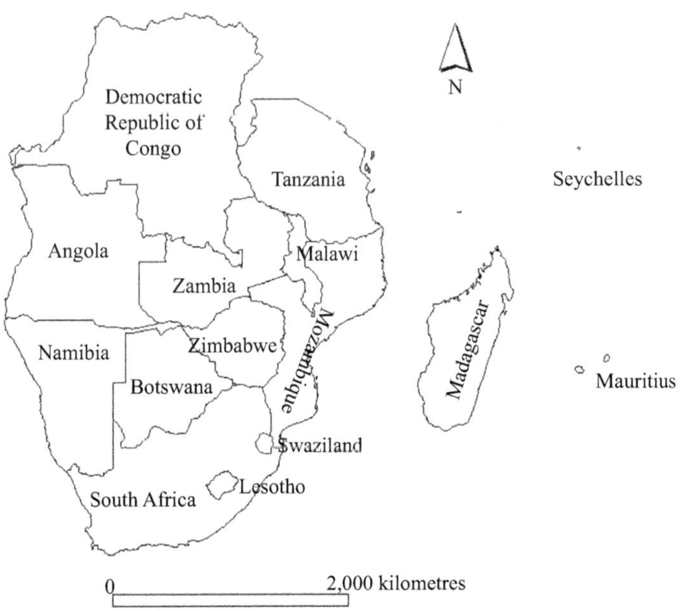

Figure 15.1 Study area – southern Africa

region vary according to season and also altitude but range from −18°C in winter in the Maloti Highlands of Lesotho to over 40°C in the summer of the Kalahari in Botswana.

Southern Africa's major economic activities are led by services at 51% of the gross domestic product (GDP) followed by industry at 32% then agriculture at 17%. The region has managed to post a positive economic growth since 2000 with the highest being 6% in the years 2006 and 2007. While there are some similarities, the countries to some extent differ in terms of history and socio-economic achievements. Some countries like Mauritius rank high on the global human development index, at rank 80 on the global index, Mauritius is also ranked highest in Africa. Others like Mozambique on the other hand are ranked very low at 185 out of 187 countries worldwide. The southern African region includes four countries officially classified as middle-income countries [5]. These countries are Botswana, Mauritius, Namibia and South Africa.

Although the region has made some key progress on human development, this progress has been constrained by a number of challenges that have important implications for its sustainability of development. These challenges such as land degradation are caused by overcrowding, overstocking, poor land management and soil erosion and have a direct impact on livelihoods and food security. On the economic front, key challenges include extreme poverty, income inequality and limited beneficiation of key outputs like minerals and agriculture, high unemployment and restricted economic diversification [5]. For example, some of the southern African countries like Botswana and Zambia are nearly mono mineral economies dominated by diamonds and copper, respectively. This can pose a danger to their economies should the prices of these minerals tumble on the world market. On the environment, there has been an extensive loss of forests cover and related environmental impacts, deteriorating biodiversity and quality of the ecosystems, and limited environmental conservation. On governance, the region has handicapped institutional capacity in service provision especially those involved in environmental management and urban planning, poor public sector and rural management, overlap, duplication and competition between institutions.

15.2.2 Data collection

In this chapter, document analysis informed the broader data collection and analysis research methodology. Document analysis is a form of qualitative research in which documents are interpreted by the researcher to give voice and meaning around the topic being assessed [6]. Documents analysing incorporates coding content into merging themes akin to focus group discussions and in-depth interviews. O'Leary [7] identifies three primary types of documents that are: 1. public records like official, ongoing records of a public organisation's activities; 2. personal documents like first-person accounts of an individual's actions, experiences and beliefs from incident reports, reflections, journals and newspapers; and 3. physical evidence which include but not limited to posters, agendas, handbooks and training materials.

The actor–actant–network theory also offers a framework of enquiry in the study. The actor–actant–network theory has been used successfully to trace how actors like humans and their institutions together with their actants (non-human phenomena) interact in space and time through their networks on matters of disaster mitigation [8] or exercise their power in politics [9]. As such this method was reckoned to be suitable for establishing how different actors adopted and diffused ICTs in humanitarian services in southern Africa. This method mainly complemented document analysis in the study. In order to seek convergence and corroboration, researchers usually adopt at least two resources using different data sources and methods. The purpose of triangulating is to provide a confluence of evidence that breeds credibility [6]. Corroborating findings across data sets can reduce the impact of potential bias by examining information collected through different methods.

Case studies of specific humanitarian situations were used in the chapter to illustrate the adoption and adaption of ICTs in humanitarian services in the region. Four key humanitarian situations resulting from floods, droughts, wars and diseases were analysed. Case studies are important when a researcher wants to gauge the effectiveness and also weaknesses of an intervention, hence appropriate in this study, which seeks to investigate the extent to which ICTs have been adopted and adapted in humanitarian services specifically looking at the southern African context.

15.3 Findings and discussion

15.3.1 ICTs and flood humanitarian issues

Flooding has become an increasingly severe and more frequent problem in most southern African countries, coming with negative repercussions for both rural and urban development [10]. The problem of flooding is compounded by climate change, low adaptive capacities of victims, lack of preparedness, increasing population densities, poverty and conflicts [11]. The African continent ranks the second hardest hit by flooding after Asia in terms of number of events, damages and deaths. However, in the last decade, the number of flood events recorded in Africa is higher than the rest of the world [12]. Flooding impacts are wide ranging, potentially interrupting food and water supplies, damaging houses, roads, electricity and other essential infrastructures, thus affecting economic development of a nation. Floods also affect health provisions which often leads to outbreaks of diseases in the affected areas.

In southern Africa, recent flood events that have created humanitarian crises include flooding in the 2019 KwaZulu-Natal flood in South Africa, which left almost 60 people dead; the 2019 cyclone Idai in Malawi, Mozambique and Zimbabwe, which recorded casualties of over 1,000 people dead and thousands missing; floods in the Shire basin of Malawi in 2019, which left over 30 people killed and over 230,000 people homeless. Floods also occur frequently in the Katima Mulilo area of Namibia and in 2007 this led to the displacement of more than 1,000 people and death of 6 people. In all these humanitarian situations, ICTs have played an increasingly important role in improving the effectiveness of humanitarian actors before, during and after the disaster. Most of the countries in this region have noted the use of ICTs in their policies as being the key in reducing disaster vulnerability.

Remote sensing and GIS are amongst the most adopted ICTs in managing flood disasters. The utility of these technologies that comes in their ability gives near real-time information to support decision-making in things such as showing flood risk and flood safe areas to assist those evacuating the victims. These technologies also give a repetitive view of the affected areas in order to give the situation before, during and after the flood events. This will aid both rescue operations and post-flood recovery. Remote sensing has also been used in the region to monitor the development, progress and paths of weather systems likely to cause flooding and in the process develop flood early warning systems under different likely scenarios for the benefit of the communities likely to be affected as well as those likely to respond to the humanitarian situation which normally follows widespread flooding.

The southern African countries have noted the importance of these Earth observation technologies in managing floods hence embarked on a project called Monitoring of the Environment for Security in Africa (MESA) for the southern African region [13]. The project aimed at improving and coordinating Earth (land and water) observation systems, to provide easier and more open satellite data access, to foster use of Earth observation data and to build capacity on scientific applications of satellite data to answer society's need for informed decision-making. Flooding was one of the thematic areas that the MESA project covered. The project was aimed at improving access to the Earth observation data and consolidation of geo-information service during the period of September 2013 to December 2017. The project to a large extent improved the southern African's region capacity to adopt remote sensing in GIS technologies for managing flood disasters. Figure 15.2 shows satellite imagery of the situation before and after the floods which occurred in the Katima Mulilo area of Namibia in the 2006/2007 rainfall season. The images in Figure 15.2 can therefore help one to determine both the extent of flooding and the damage due to the floods. This can assist in future settlement planning by avoiding areas vulnerable to flooding and also to build flood-resilient infrastructure in such areas.

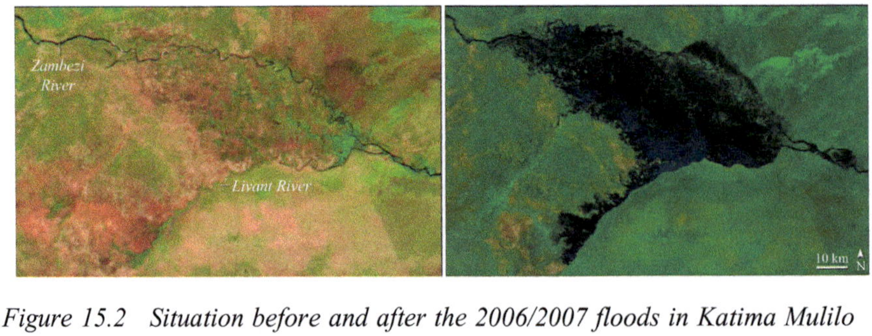

Figure 15.2 Situation before and after the 2006/2007 floods in Katima Mulilo [Source: NASA (2007)]

Southern African countries have also been benefitting from satellite imagery and maps produced and delivered to them during flood humanitarian situation through the international space and major disasters charter. The charter started its operations in November 2000 and is composed of members from the European Space Agency, French Space Agency and Canadian Space Agency, who are mainly the owners of space assets such as European Remote Sensing satellite. The charter members have pledged to support, by means of space assets and the associated information and services, the provision of a unified system of data acquisition and delivery to those affected by both natural and man-made disasters. They endeavour to make their satellite resources available to supply, during the period of crisis, emergency organisations, essentially the national civil protection agencies, with a coordinated and free of charge access to space systems [14]. Figure 15.3 shows the impact map of cyclone Idai in the Chimanimani area of Zimbabwe in 2019. The map clearly shows where floods and landslides occurred and the affected human settlements.

Figure 15.3 was therefore useful for both rescue efforts and coordination in guiding decision-making on safe areas to evacuate survivors, recovery of bodies, assessing the extent of damage to infrastructure and houses.

Throughout southern Africa, ICTs have also been used for humanitarian communication. This communication is meant to benefit the communities in danger of disaster. In most cases across the region, there is a notable ICT divide between rural and urban settlements. A study, which analysed the gaps in the flood early

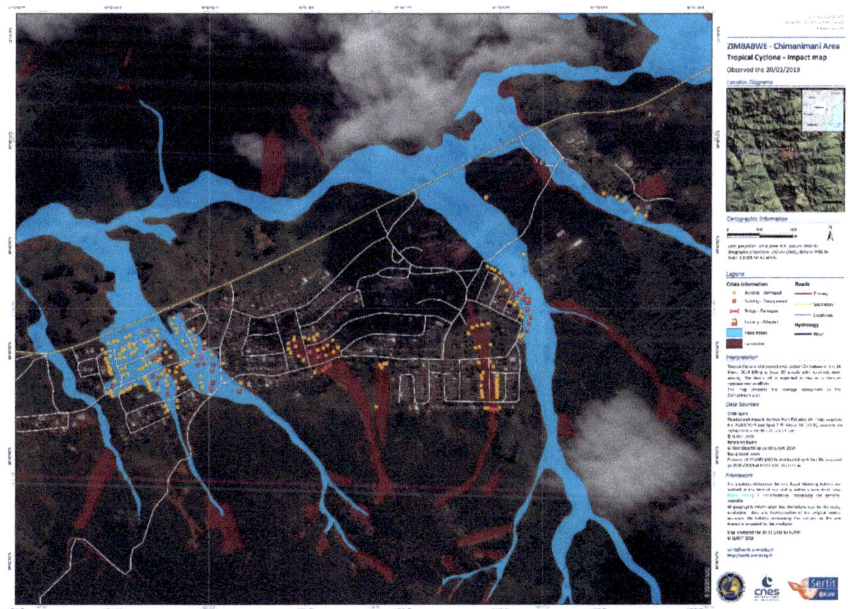

Figure 15.3 Tropical cyclone impact map in Chimanimani, Zimbabwe [Source: Space and major disasters charter (2019)]

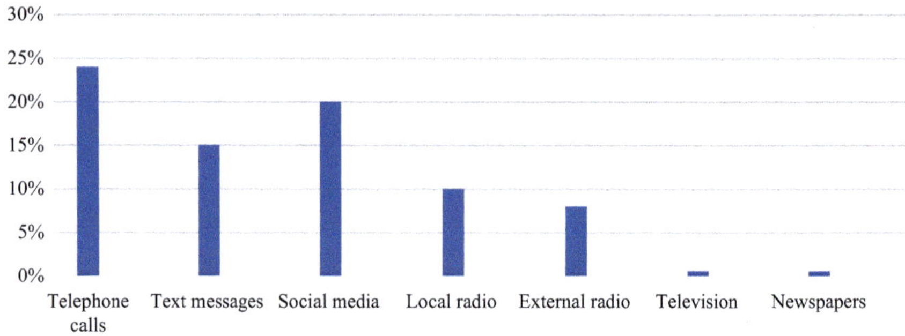

Figure 15.4 ICTs used in flood early warning in Tsholotsho, Zimbabwe [Source: Sibanda (2018)]

warning system used in the Gwai river valley of rural Tsholotsho district in Zimbabwe, observed that the adoption of ICTs in the dissemination of flood information was important if the communication was to reach the possible victims with enough lead time to move flood safe areas. Figure 15.4 shows the ICTs used in the dissemination of information on early flood warning in Tsholotsho, Zimbabwe. Twenty-four per cent of the respondents mainly relied on individual telephone calls from relatives in other places on the pending floods. Another 15% relied on bulk text messages sent by the government civil protection unit. The effectiveness of such texts was, however, observed to be handicapped by the language in which they are sent, which is mainly English. English is not the first language for most of the vulnerable groups in the areas. Of major concern was the fact that only 10% of the people rely on the local radio services for their flood information. This was because of poor local radio station reception, no access to the national electricity grid and a huge number of families not owning radio receiver gadgets.

In recent years, there has been an increase in the use and reliance on social media as it is a very cheap means of communication, in particular WhatsApp and Twitter. Twenty per cent of the participants used social media as a source of information on floods in Tsholotsho. However, its use comes with adeptness to navigate applications on mobile Smartphones. This may be a challenge for people with a very limited education and the fact that mobile phones are also expensive. Television, radio and mobile phones are therefore the most useful conduit for flood alerts in most rural areas with poverty and literacy problem. Authorities and other stakeholders like service providers therefore need to install more mobile base stations and rural electrification to support the ICTs in the remote areas so that communities are able to harness technology and utilise its advantages for timely access to vital emergency information to reduce their vulnerability.

15.3.2 ICTs and drought humanitarian issues

Drought is one of the main natural hazards that the southern African region experiences frequently [15]. Drought is any period when an area experiences moisture and/or water deficiency, which deviates from the long-term normal climatic mean [16]. Drought periods range from weeks to decades and create socio-economic impacts that vary from famines and food insecurity to decreases in GDP and loss of ecosystem services [17]. In the southern African region, intensity and frequency of droughts have increased to such a level that at least one part of the region has drought, although other regional areas have generally good rains.

ICTs have been adopted in the region to manage drought-induced humanitarian situations and have utility in predicting and monitoring the severity of drought episodes and provide baseline information to facilitate planning of drought-assessment projects. ICT products have been used to assist in easily identifying drought hotspots and to support decision-making in prioritising drought relief by providing complete drought information coverage of a country or region, which in turn can be compared with socio-economic statistics and historical data to give a clearer picture of areas and groups affected by drought [13].

In southern Africa, the Famine Early Warning Systems Network (FEWS NET) is one of the leading providers of information and analysis on the level of food security for each country. FEWS NET has a vast community of users in the region from governments, NGOs and researchers responding to humanitarian situations. The network publishes on their website up-to-date information on current and projected food insecurity for the region and produces maps of the affected areas. Their strength comes from the fact that they use multiple parameters to establish food insecurity. The factors include weather, climate, agricultural production, socio-political issues, markets, livelihoods and humanitarian response. In 2019, FEWS NET noted that some parts of the region will be affected by food shortages due to erratic rains leading to below average harvest, the tropical cyclones Idai and Kenneth, which affected some parts of the regions, and the general increase in the cost of living especially in Zimbabwe. They produced interactive maps for the regions of food insecurity phases ranging from minimal, stressed and crises to famine and emergency [18].

To improve water resources management within the Zambezi river basin, riparian countries created the Zambezi Watercourse Commission (ZAMCOM). The mandate of this commission is to collect, evaluate and share all data and information on the Zambezi Watercourse [19]. To fulfil this mandate, the countries developed the Zambezi Water resources Information System (ZAMWIS) as a database, a platform that can be used in supporting informed decision-making and planning processes in water-related challenges. ZAMWIS is therefore an interactive, web-based data and information portal for the ZAMCOM based on near real time, contemporary and historical spatial data, hydrological time series, Earth observation information and other related information. The system and data sets are open to any individuals, communities and organisations with an interest in water-related issue. The system provides early warm for future water shortages (hydrological droughts) well in time to evoke measures to manage the shortages before they occur, hence an important ICT with the management of hydrological drought in the Zambezi basin.

During the 2017–18 season, Cape Town, South Africa, was facing critical water shortages due to consecutive years of drought in the Western Cape province. According to PwC [20], Cape Town was experiencing one of the most severe droughts in living memory, with dam water levels at their lowest in nearly 20 years. The possibility of the city water supply dams failing and taps running dry became synonymous with the term Day Zero. Nhamo and Agyepong [21] observed that Day Zero was not a static date but influenced by dynamic factors such as consumption, precipitation, evaporation rates and the availability of new supply volumes and their timing. Based on these factors, 12 April 2018 was to be the date of the largest drought-induced municipal water failure in modern history. Faced with the possibility of water shortages due to drought, Cape Town developed the Critical Water Shortages Disaster Plan which included phases of water rationing, disaster restrictions and full-scale disaster implementation [22].

In 2017, the City of Cape Town began intensifying its drought-awareness campaign, publishing weekly updates on regional dam levels and water consumption derived from smart water meters and using electronic boards around the city, including on freeways to notify drivers of how many days of water supply Cape Town had left. By January 2018 and with Day Zero looming, the city heightened its campaign for residents to save water. In addition to its countdown to Day Zero, the city launched a city wide electronic water map to show water consumption on a household level, allowing people to compare their consumption to their neighbours' and the rest of the city [21]. The city's weekly water report became a regular topic at both formal and social institutions and on various electronic media, including radio stations. Governmental and civic organisations also joined the water-saving campaign and circulated water-saving techniques and people exchanged these tips using social media. In the end, the city managed to significantly reduce its water consumption to avoid Day Zero and was also saved by the sudden down pour of rains at the last minute. Cape Town, therefore, became a good example of how the use of different ICTs such as smart water meters, electronic maps, social media, radio, television and smartphones together with the availability of the internet can be used for community environmental education to avoid disaster and ensuing humanitarian crisis.

15.3.3 ICTs and epidemiological humanitarian issues

ICT is playing a significant role in the management of health issues in the southern African region. ICTs are used for preparing for, mitigating, reacting to and recovery from health disasters. The southern African region has not been spared from the spread of communicable diseases and effects of non-communicable diseases. Communicable diseases such as tuberculosis (TB), malaria, human immunodeficiency virus (HIV) and acquired immune deficiency syndrome (AIDS) have had a devastating effect in the region [22]. The southern African region has the highest prevalence rate of HIV in the world, with one-third (12.8 million) of the people living with HIV in the world [23]. Cholera is another prevalent disease in the region with most reported cases in 2019 affecting Zimbabwe, Zambia, Malawi and

Mozambique with more than 15,042 reported cases and 237 deaths confirmed [23]. Other present communicable diseases are hepatitis E in Angola and Namibia, Guinea worm in Angola (first case reported in June 2018), measles in Zambia, typhoid in Zimbabwe and listeriosis in South Africa [23]. The three most prevalent non-communicable diseases with a far-reaching effect in the southern African region are cervical cancer, hypertension and diabetes [24].

The SADC protocol on health issues [25] recommended the following to be adopted by the respective countries: health information systems policy framework, development of common definition and common data dictionary, establishment for mechanisms of information exchange, establishment of SADC Regional Data of Health and Social Services Indicators and development of telehealth applications. This had led to the development, implementation and adoption of ICTs in the SADC member countries' health sector to prevent, manage and recover from communicable and non-communicable diseases.

15.3.3.1 Communicable diseases

Tuberculosis

TB is one of the leading causes of morbidity and mortality in southern Africa. Member states are home to 25% of sub-Saharan Africa's population but account for 50% of TB cases reported in the region. Five member states are among the 22 countries globally with the highest burden of TB. Ten of the 15 member states also have very high TB prevalence with prevalence rates ranging from 100 to 300 cases per 100,000 populations. TB is caused by a type of bacterium called *Mycobacterium tuberculosis* [26].

In South Africa through innovation, a technology to test TB meningitis was developed by Antrum Biotech in association with the University of Cape Town. The newly developed technology is used for rapid tests of TB meningitis and the results are produced within 2 hours. In order to share and implement the new innovative technology within the SADC region, South Africa is collaborating with SANBio for the adoption of the newly developed health systems [27].

Surveillance and monitoring technology, laboratory information systems and e-Learning platforms are used to detect, prevent, control and recover from TB disease. Surveillance ICTs are used to detect the spread of TB in order to apply and assess preventive and control measures. The surveillance systems such as event-based surveillance systems are used by clinicians outside national TB programmes to notify TB cases and access to automatically produced standard charts and maps showing TB trends and treatment by SADC countries. The most commonly used event-based surveillance system in the SADC region is the Global Public Health Intelligence Network which scans a multitude of informal sources of information such as websites, electronic discussion forums, news reports and online newspapers for infectious diseases and produces maps showing the affected areas [28]. Laboratory information systems are used to store data from diagnostic devices and data entry applications. The information systems are also capable of sending short message service (SMS) notifications to clinicians and patients and are used for stock management applications. e-Learning is used for training purposes by healthcare professionals such as preventative measures and symptoms and treatment of TB. The e-Learning can be in the form of podcast, web application, mobile application or even broadcasting services [29].

Ebola

Ebola is a viral disease which can cause immediate death when one is infected. Once the Ebola virus (EBV) enters the body it destroys cells, making some of them to explode. The virus destroys the immune system, causing internal bleeding, and damages almost every organ. The EBV was first detected in 1976 in The Democratic Republic of Congo (DRC).

ICTs have been developed to detect, prevent, control and recover from such a disease. One of the technologies that is being used to detect Ebola in the DRC is GeneXpert, which was primarily developed to detect cases of TB, and has been adopted for the testing of pathogens for diseases such as HIV, malaria and sexually transmitted infections (STIs) [30]. The GeneXpert system is used to diagnose and detect the EBV within an hour. Other devices such as walk-through thermal scanners and handheld scanners are used to detect, manage and control the spread of EBV at national ports of entry, including airports.

HIV and AIDS

One of the SADC countries, Eswatini, with a population of 1.1 million has about 200,000 people living with HIV [31]. Figure 15.5 highlights the prevalence of HIV in the SADC member states with Eswatini recording the highest prevalence percentage, and Madagascar has the least prevalence percentage in the region. In March 2016, the government of Malawi and United Nations International Children's Fund jointly tested the use of unmanned air vehicles (UAVs) to transport samples of blood for infants for HIV testing. Malawi is one of the least developed countries, ranking 170 out of 188 on the Global Index. Over 70% of the population live below the poverty datum line, and approximately 63% of children

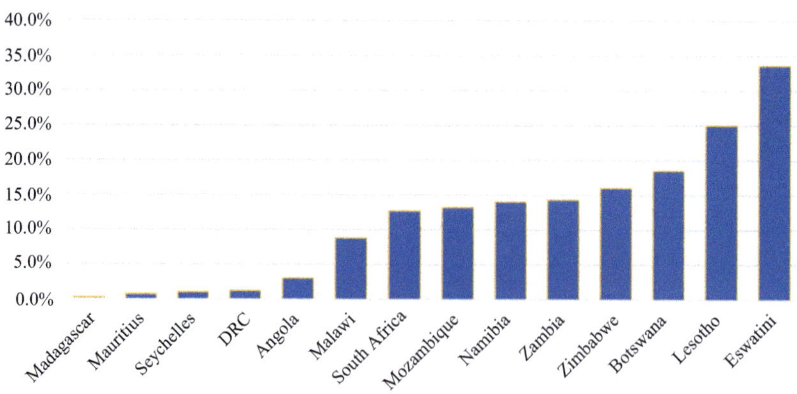

Figure 15.5 Prevalence of HIV (%) in the SADC member states [Source: SADC Member States (2018)]

live in poverty [32]. In 2015, an estimated 110,000 children (age from 0 to 14 years) were living with HIV of whom 49% (53,400) were receiving antiretroviral treatment [33]. Before the introduction of UAV technology, it was taking an average of 11 days to get samples from the health centre to the testing lab and about 8 weeks to get feedback [34]; thus the use of this technology was meant to reduce the processing time and results of HIV. UAVs are used to transport medicines in disasters areas. This technology has a fast response in areas which are not accessible during disasters. Malawi also implemented some health situation room software which enables the government to make informed decision about policies and health issues, including HIV. The Malawian national information system processes information from multiple data sets through real-time visualisation, and this makes the decision-makers to make informed decisions to prevent, control and recover from an epidemiological disaster [35].

Cholera

Cholera is an infectious disease that causes severe watery diarrhoea, which can lead to dehydration and can even cause death if left untreated. Cholera is caused by eating food or drinking water contaminated with a bacterium called *Vibrio cholerae* [36]. In 2019, the southern African region reported the highest number of cholera cases in the sub-Saharan region with countries such as Angola, Malawi, Mozambique, Tanzania, Zambia and Zimbabwe affected most. As on 31 July 2019, Mozambique had 6,931 reported cases and 8 deaths since the declaration of the cholera epidemic on 27 March 2019. According to the World Health Organisation (WHO) cholera analysis of the southern African region 2019, most reported cases are from urban areas as compared to rural areas.

Laboratory information systems and the social media are used as ICT platforms for the prevention, management and recovery from cholera endemic. The southern African region uses laboratories to detect Deoxyribonucleic Acid (DNA) sequences of specific for *V. cholerae*. Bacterial subtyping provides a measure of the genetic relatedness of strains and is vital for accurate epidemiological investigations of bacterial infections. Subtyping methods provide data suitable for laboratory-based surveillance and outbreak detection and investigation. In the southern Africa region, the laboratories use pulsed-field gel electrophoresis as the primary method to subtype toxigenic *V. cholerae* and strictly follow PulseNet protocols and methods. PulseNet Africa was established in August 2010 and is coordinated by the Centre for Enteric Diseases, National Institute for Communicable diseases, South Africa.

Malaria

Malaria remains a disease of public health significance in the southern African region. Malaria is a human disease that is caused by sporozoan parasites (genus Plasmodium) in the red blood cells. The disease is transmitted by the bite of anopheline mosquitoes and is characterised by periodic attacks of chills and fever. Malaria is responsible for 20% of childhood death and in excess of 30% and 40% of outpatient visits and hospitalisations, respectively. WHO has estimated that three-quarters of the population residing in this southern African region is at a risk of contracting malaria, including 25 million children younger than 5 years of age and approximately 8.5 million pregnant women. As national borders become increasingly porous, a harmonised and coordinated effort within the region is essential for malaria control.

Cutting-edge nanotechnology is being used to develop an affordable, easy-to-use smartphone-like device able to analyse malaria strain DNA from a finger-prick of blood in 15 min. This enables a personalised prescription of drug combinations to be given to patients straight away. The device was developed by Nanomal consortium, led by St George's University of London and biotech company QuantuMDx, in response to WHO warnings of the growth of drug resistance due to inaccurate drug prescription.

The sterile insect technique (SIT) is based on the principle of suppressing a target insect population by mass males, sterilising them and then releasing them into target pest/vector population. If successful, mating between sterilised males and wild females will not result in the production of offspring, leading to the population suppression. STI programmes have been considered in the elimination and control of malaria. The main objective of the SIT project coordinated by the Vector Control Reference Laboratory is to ascertain whether it is feasible to use mosquito colonies held in South African facilities to field SIT programmes for malaria vector control. Baseline data collection in northern KwaZulu-Natal has produced valuable biological information on the target species, *Anopheles arabiensis*. The project was funded by the International Atomic Agency, the Industrial Development Corporation and the South African Nuclear Energy Corporation through its Nuclear Technologies in Medicine and the Biosciences Initiative – a national platform funded by the Department of Science and Technology [37].

LAMP (loop-mediated isothermal amplification) technology is being tested for use in South Africa to eliminate malaria carriers (individuals infected with malaria but who do not display signs of the disease and patients with parasite loads below the detection limits of microscopy and rapid diagnostic tests). LAMP is being tested as a point-of-care device capable of rapidly detecting extremely low parasitaemias in rural settings, enabling prompt diagnosis and treatment. Data obtained will be used to inform LAMP diagnostics policy for the malaria-endemic provinces [37].

15.3.3.2 Non-communicable diseases

In order to mitigate the rising suffering of non-communicable disease in the SADC region, member states agreed to address the challenge through monitoring and risks in Article 13 of its Protocol on Health. This article encourages member states to 'adopt appropriate strategies for the prevention and control of non-communicable diseases' [25].

Cervical cancer

Malawi and Mozambique have a high prevalence rate of non-communicable disease such as cervical cancer in the world. Cancer is caused by changes (mutations) to the DNA within cells. To reduce cervical cancer in Malawi and Mozambique, USAID funded the following activities: accelerating the introduction of a screen–and–treat strategy for human papillomavirus (HPV) and cervical precancerous lesions in the Republic of Malawi. In Mozambique, the USAID is evaluating the innovative technologies and approaches to addressing cervical cancer. A consortium between the University of Texas MD Cancer Centre and the University of Eduardo Mondlane in Maputo is integrating existing screening and treatment technologies with voluntary family planning programmes, while testing the feasibility and efficacy of innovative tests for HPV [38]. As prevention of cervical cancer expands, there is still immense opportunity to identify and scale promising new diagnostic technologies.

Hypertension
Hypertension is a long-term medical condition in which the blood pressure (BP) in the arteries is persistently elevated, increases the risk of heart attacks and strokes and contributes to heart and kidney diseases. Hypertension is a leading cause of the global epidemic of cardiovascular disease. Hypertension patients use home BP devices for monitoring the BP. Telehealth care and mobile health (m-health) are used as strategies for improving outcomes of people living with hypertension. Clinicians use telephones and mobile phones to communicate and monitor the health condition of hypertension patients. m-Health services are used to augment clinician with automated services delivered through Smartphones, smart watches, short message and automated calls [39].

Diabetes
There has been an increase in the number of people suffering from diabetes in the world, 15% of the age 24–64 years are in the African region. Diabetes occurs when the pancreas does not produce enough insulin or when the body cannot effectively use the insulin it produces. The main symptoms of diabetes include increased thirst, needing to urinate more often than usual and increased hunger [40]. Patients with diabetes can experience episodes of low blood sugar which are manifested by weakness, headaches and in severe cases loss of consciousness.

ICTs through medical radio-shows, mass messaging take part in educating the public on how to identify such symptoms and perform first aid that is giving the patient carbohydrates. ICTs have also been useful in the detection of blood sugar levels instantly with easy to use' portable glucometers. Laboratory services aid in the detection whether the patient was compliant with medication over a period of three months by measuring HbA1c, glycosylated haemoglobin [41].

15.3.4 ICTs and conflict humanitarian issues

Man-made disasters such as armed conflicts pose serious humanitarian disasters on the southern African region. Conflicts in the southern African region emerged from large-scale wars, some from the Cold War era and others from anti-colonial rule. However, small insurgencies have emerged in some of the countries in the region. Wars that erupted during the Cold War and anti-colonial wars had an international flare, while the current insurgencies are national in outlook. Small uprisings have emerged in the DRC, Mozambique and Angola [42]. The root cause of the small uprisings is associated with governments failing to accommodate opposition parties in power-sharing agreements, forcing the opposition parties to seek regional autonomy in their stronghold areas. Tanzania has experienced sporadic incursions from Somalia. The Al Shabab militants from Somalia often enter Tanzanian territory to carry out small insurgencies [43]. In the eastern region of DRC, there have been insurgencies that have been going on in the Goma area. In 2012, the Mozambique opposition leader Afonso Dhlakama led an armed mutiny for territorial autonomy, political and economic independence against the government [44]. Although Afonso Dhlakama died in 2018, the rebellion is still on-going in Mozambique. In 2016, there was an armed insurrection against government troops in Angola's Cabinda area; this disrupted activities in the oil-rich area [45]. However, all the conflicts in the southern African region have not resulted in full-scale national wars [42].

Conflicts are also associated with political and socio-economic problems. In the southern African region, Zimbabwe has had socio-economic problems for the past two decades, with its citizens migrating to other neighbouring countries as economic refugees. This has resulted in countries such as South Africa, Mozambique and Botswana having an influx of immigrants. This has exerted a lot of pressure on the neighbouring countries' economies, forcing the citizens of these SADC countries to fight against immigrants. The South African citizens have failed to accommodate foreigners who are coming from conflict zones; especially nationals from neighbouring countries such as Zimbabwe, DRC and Mozambique have often been attacked necessitating humanitarian need. Between 2000 and March 2008, at least 67 deaths associated with xenophobic attacks were reported in South Africa. In May 2008, 62 deaths also related to xenophobic attacks were reported; among the dead were 21 South Africans. In 2015, South Africa experienced a nationwide spike in xenophobic attacks against immigrants, which prompted southern African governments such as Malawi, Zimbabwe and Mozambique to repatriate their citizens. In a report by the Human Sciences Research Council on the causes of South African xenophobic attacks, the identified reasons causing the violence were that the citizens felt relatively deprived specifically from intense competition for jobs, commodities and housing; group processes, including psychological categorisation processes that are nationalistic rather than superordinate, and South African exceptionalism and exclusive citizenship.

The most commonly used ICTs in regions of conflict zones are radio and television broadcasting services, mobile phones services and social media services [46]. Radio broadcasting services are used to warn people travelling to and from conflict zones of the pending dangers. Voice calls can be made using mobile devices; this enables communication with families, friends and relief workers during the times of crisis, with immeasurable life-saving impacts. Mobile voice calls are a good source for tracking the movement of people during a disaster. The challenge is that most voice calls are overwhelmed by high traffic during disasters, thus limiting the communication between the interested parties. In this case, SMS is more effective as it can be temporarily stored and continues to retransmit the message for some time before it sends back a message advising the sender about the status of the message. Specific SMS codes are created to send distress messages very fast during disasters. The advantage of using mobile phones during disasters is that the phone is able to generate data such as Global Positioning System (GPS), accelerometers and gyroscopes which are vital for rescue efforts. GPS enables rescue workers to identify the exact area where the person is located. By using the GPS coordinate system, rescue efforts can easily be pursued. The mobile devices may also be used to record events and send images about the real situation as being obtained on the ground [47].

During a workshop held in Johannesburg, South Africa, from 18 July to 19 July 2016, which discussed the adoption of a coordinated regional transportation plan for humanitarian relief cargo in the SADC region, it was agreed that a GIS mapping system should be installed at regional ports, railway and road networks. The vehicles or trains transporting humanitarian relief cargo were to be equipped with vehicle tracking and monitoring system [48]. In the case of breakdowns, accidents and conflict disturbances, the relief workers will be able to track where the vehicle was last spotted and carry out rescue efforts.

Since 2015, the United Nations has been working on developing machine-learning capabilities that can identify and analyse xenophobic tweets during major events of terrorist incidents or refugee crises to gauge attitudes towards the people involved across the globe, including southern African region. The data is collected from the social media applications such as Twitter through the use of machine learning. Big Data analysis is then used to analyse the attitude of the people and identify areas where the xenophobic attacks will be taking place. The results from the Big Data analysis will assist humanitarian efforts to prevent, respond and recover from xenophobia attacks.

Governments in the southern African region are also using Big Data analytics to prevent, manage and carry out recovery efforts for disasters caused by other conflicts such as social unrest and wars. Big Data tools are used to process large amounts of crisis-related data to support more effective disaster response. Data sources identify the various generators of online activities especially on social media, sensors and data exhaust. High-performance computing system is used to capture and analyse the data. It has been observed that under extreme stress conditions, human beings tend to exaggerate information [49]. The governments make various efforts to validate the authenticity of the data before taking action on the information. In Zimbabwe during the January 2019 fuel hike protests, the government shut down the Internet to restrict social activities on Twitter, WhatsApp and other online activities [50].

15.3.5 Challenges of adopting ICTs in humanitarian services in the region

Across the southern African region, the challenges to realising the potential of ICTs in humanitarian are mainly financial, technical and institutional in nature. In most of the countries in the region, the first responders to emergencies are the local governmental authorities. Most of the ICTs that are used in the southern African region are imported, prompting the need to train the local manpower on how to use the equipment. Generally, this has been met with difficulties because of digital literacy. Most of the countries in the region do not manufacture any ICTs with the exception of South Africa; hence, funding for the development of these systems becomes a challenge as there are so many areas that will be competing for the same financial resources like electricity, education and housing. There are also limited facilities such as laboratories for testing and treatment of communicable diseases like HIV, TB and malaria, which result in delays of tests, treatment and mitigations against these diseases, hence the need for more laboratories.

Another challenge faced by the southern African region is that it is constituted by some of the poorest countries in the world such as Malawi and Mozambique, with limited infrastructure; especially in the rural areas, there is a need to fund these nations to improve accessibility to the basic needs such as radio broadcasting services, Internet and laboratory facilities. Also equipping the local authorities with ICT capacity is therefore recommended if communities are to benefit from their use during disaster periods.

The digital divide is still very high between the rural and urban areas with high-speed broadband coverage still patchy, with most rural and remote areas remaining uncovered. Coverage is mainly concentrated in affluent urban areas. Computer literacy is still a challenge especially among the elderly as compared to the young generation who are techno savvy; the youth are growing up in an environment where there is a high proliferation of ICTs. This is also exacerbated by inadequate electricity as the national power grids do not cover all areas in different countries hence leaves out a significant population dependent on alternative power sources which in most cases are more expensive. Even those on the national grid time and again experience erratic power supplies; this has a negative effect on the development and adoption of ICTs. Some of the factors limiting the adoption of ICTs for humanitarian services include weak enabling policies, limited capacity of social intermediaries to scale up and secure sustainability and weak partnerships among local stakeholders [1]. Policies that promote e-Health services should be developed and promoted through the media and Internet.

Growing access to ICTs and the increasing applications of sensors are generating massive volumes of data. The amount of growing data poses challenges for data management, analysis and verification. The other challenge of adopting humanitarian ICTs is about dealing with false information for Big Data analytics which is created by the rapidity of the data flows on the Internet; this makes it difficult to verify information leading to disastrous consequences. This is caused by the fact that under immense stress the public often exaggerate information and also on over-reliance on Big Data [47]. Before the dissemination of information to the public, the government or other public entities, especially the media, should make an effort to verify information before it releases it to the public to avoid causing chaos in the nation.

The use of drones and UAV is a security concern in some SADC countries, and the legislation requirements limit the range and usage of such technology. There is a need to relax on the laws for using UAV; especially during disasters, UAVs are useful tools to monitor and transport food and medicine in needy areas.

15.4 Conclusions

A case study of the southern African region revealed that ICTs are being used to mitigate against floods, droughts, diseases and in conflicts for humanitarian needs. The chapter has also shown that the access and use of ICTs across the region is not uniform and differs from country to country and from region to region.

The challenges that are impeding the adoption and adaption of ICTs such as lack of infrastructure, computer literacy, erratic power supplies, and legal and security issues were identified, and policymakers should develop polices to address the issues. The research concludes that ICTs are an indispensable component of humanitarian services, and the adoption by actors in the sector must be encouraged.

References

[1] World Bank. *Municipal ICT Capacity and its Impact on the Climate-Change Affected Urban Poor: The Case of Mozambique*. Washington, DC; 2012.

[2] OCHA. *OCHA on Message: Humanitarian Principles*. UN Office for the Coordination of Humanitarian Affairs, June 2012. 2013.

[3] Tusiime E., and Byrne E. 'Information systems innovation in the humanitarian sector'. *Inf Technol Int Dev*. 2011;7(4):37.

[4] United Nations. *Sendai Framework for Disaster Risk Reduction 2015–2030*. UNISDR. Geneva; 2015.

[5] United Nations Commission for Africa. *Sustainable Development Goals for the Southern Africa Subregion, Summary Report*. Economic Commission for Africa. Addis Ababa; 2015.

[6] Bowen G.A. 'Document analysis as a qualitative research method'. *Qual Res J*. 2009;9(2):27–40.

[7] O'Leary Z. *The Essential Guide to Doing Your Research Project*. 2nd ed. SAGE Publications, Inc. Thousand Oaks, CA; 2014.

[8] Nhamo G. 'Copenhagen 18/12: Africa must continue calling for a fair, ambitious and binding climate deal'. *Afr Asian Stud*. 2011;10(2–3):100–20.

[9] Davies A.R. 'Power, politics and networks: Shaping partnerships for sustainable communities'. *Area*. 2002;34:190–203.

[10] Douglas I., Alam K., Maghenda M., McDonnell Y., Mclean L., and Campbell J. 'Unjust waters: Climate change, flooding and the urban poor in Africa'. *Environ Urban*. 2008;20:187–205.

[11] Adelekan I.O. 'Vulnerability of poor urban coastal communities to flooding in Lagos, Nigeria'. *Environ Urban*. 2010;22(2).

[12] Subbarao I., Bostic N., and James J. 'Applying yesterday's lessons to today's crisis: Improving the utilization of recovery services following catastrophic flooding'. *Disaster Med Public Prep*. 2008;2(3):132–3.

[13] MESA-SADC. *Monitoring of the Environment for Security in Africa (MESA) Project*; 2012.

[14] Bessis J.L., Béquignon J., and Mahmood A. 'Three typical examples of activation of the international Charter "Space and Major Disasters."'. *Adv Space Res*. 2004;33:244–8.
[15] CEDRISA. *Droughts and Floods in Southern Africa: Environmental Change and Human Vulnerability*. UNEP and SARDC. Nairobi; 2009.
[16] Warren A., and Khogali M. *Assessment of Desertification and Drought in the Sudan–Sahelian Region 1985–1991*. UNSO. New York; 1992.
[17] Mutowo G. and Chikodzi D. 'Remote sensing based drought monitoring in Zimbabwe'. *Disaster Prev Manage.* 2014;23(5):649–59.
[18] FEWS NET. *Southern Africa Food Security Outlook*; 2019.
[19] ZAMCOM. *ZAMWIS Brochure*; 2019.
[20] PwC. *Western Cape Water Crisis: How Resilient Is Your Organisation in the Face of the Current Water Crisis?*. PricewaterhouseCoopers. Cape Town; 2017.
[21] Nhamo G., and Agyepong A.O. 'Climate change adaptation and local government: Institutional complexities surrounding Cape Town's Day Zero'. *Jamba J Disaster Risk Stud*. 2019;11(3).
[22] City of Cape Town. *Critical Water Shortages Disaster Plan: Public Summary, Cape Town*. Cape Town; 2017.
[23] FAO. 'Southern Africa resilience strategy: 2018-2021'. Health Policy Plan [Internet]. 2017;27:1–23. Available from: http://documents.worldbank.org/curated/en/844381468209068874/pdf/792930WP0WB0NC0l0Box07929300PUBLIC0.pdf.
[24] IFRC. *Southern Africa IFRC Country Cluster Support Team 2019 Programme Overview*; 2019.
[25] O'Kane C., and White H. *The Three Non-Communicable Disease*s; 2017.
[26] SADC. *SADC Protocol on Health*; 1999.
[27] Southern African Development Community. *Assessment Report on the Status of HIV and AIDS, Tuberculosis, and Malaria Surveillance Systems in the SADC Region* [Internet]; 2009: 2–33 p. Available from: https://www.sadc.int/files/4314/1172/0046/Assessment_Report_on_the_Status_ofHIV_Testing_and_Counselling_Policies_inthe_SADC_Region.pdf.
[28] UCT. *New TB Meningitis Test Beats the Rest*; 2018.
[29] Carrion M, and Madoff L.C. 'ProMED-mail: 22 years of digital surveillance of emerging infectious diseases'. *Int Health*. 2017;9(3):177–83.
[30] WHO. *The Role of e/mHealth in Tuberculosis and Tobacco Contro*l; 2015:25–7.
[31] WHO. *New Technology Allows for Rapid Diagnosis of Ebola in the Democratic Republic of the Congo* [Internet]. WHO; 2017. Available from: http://www.who.int/emergencies/ebola-DRC-2017/articles/rapid-diagnosis/en/.
[32] UNICEF. The Situation of Children and Women in Malawi: Facts and Figures [Internet]. UNICEF; 2018. Available from: www.unicef.org/malawi/situation-children-and-women-malawi.
[33] AVERT. *HIV and AIDS in Malawi AVERT*. AVERT; 2016.

[34] UNICEF. *Malawi Tests First Unmanned Aerial Vehicle Flights for HIV Early Infant Diagnosis* [Internet]. UNICEF; 2016. Available from: http://www.unicef.org/media/media_90462.html.

[35] Mutharika A.P. *Malawi Launches Its Health Situation Room*. UNAIDS; 2019.

[36] WebMEd. *The Basics: Cholera*. WebMed; 2019.

[37] Center for Opportunistic. *TB and HIV, Malaria – NICD*. Centre for Opportunistic, Tropical and Hospital Infections; 2015.

[38] USAID. *USAID Announces $ 12 Million for the Prevention of Cervical Cancer in Sub-Saharan Africa*. USAID; 2019.

[39] Piette J.D., Datwani H., Gaudioso S., *et al.* 'Hypertension management using mobile technology and home blood pressure monitoring: Results of a randomized trial in two low/middle-income countries'. *Telemed e-Health* [Internet]. 2012;18(8):613–20. Available from: https://www.liebertpub.com/doi/10.1089/tmj.2011.0271.

[40] WHO. Diabetes [Internet]. WHO; 2019:1–6. Available from: http://www.basel.ru/en/key_facts/.

[41] Novak M.T., Kotanen C.N., Carrara S., Guiseppi-Elie A., and Moussy F.G. 'Diagnostic tools and technologies for infectious and non-communicable diseases in low-and-middle-income countries'. *Health Technol (Berl)*. 2013; 3(4):271–81.

[42] The Nordic Institute of Africa. P*eace and Security Challenges in Southern Africa: Governance Deficits and Lacklustre Regional Conflict Management, Policy Note No 4:2018*; 2018.

[43] Strategic Intelligence. *Tanzania Admits Al-Shabaab Terrorists Attacking Its Citizens and Police*; 2015.

[44] News24. *Mozambique Rebel Leader Dhlakama Accuses Govt Troops of Violating Ceasefire*. News24; 2017.

[45] Coroado H. *Rebels Say Kill 12 Angolan Soldiers in Oil Region Clash*. Reuters; 2016.

[46] Souter D., and Kelly T. *The Role of Information and Communication Technologies in Postconflict Reconstruction*; 2014.

[47] Emergency Telecommunications. *Disruptive Technologies and Their Use in Disaster Risk Reduction and Management*; 2019.

[48] SADC. *Outcome Statement From the Workshop on the Adoption of a Coordinated Regional Transportation Plan for Humanitarian Relief Cargo in SADC*; 2016.

[49] Qadir J., Ali A., ur Rasool R., Zwitter A., Sathiaseelan A., and Crowcroft J. 'Crisis analytics: Big data-driven crisis response'. *J Int Humanit Action* [Internet]. 2016 [cited 2019 Jul 21];1(1):12. Available from: https://doi.org/10.1186/s41018-016-0013-9.

[50] BBC. *Zimbabwe Blocks Facebook, WhatsApp and Twitter Amid Crackdown* [Internet]. BBC News; 2019. Available from: https://www.bbc.com/news/world-africa-46917259.

Chapter 16

Conclusions

Muhammad Nazrul Islam[1]

Information and communication technologies (ICTs) are everywhere and transforming the way we live. The availability of ICTs is also essentially transforming the humanitarian services by better addressing the needs of those affected by humanitarian crises.[*] Therefore, this book strived to focus on areas that include connectivity and communications technologies for humanitarian services; ICTs in disaster mitigation, relief and recovery; humanized and inclusive education; technologies for the women, disabled and aged populations; IoT, big data and blockchain for humanitarian engineering; and the adoption of ICTs in humanitarian sectors.

Nowadays, ICT-based developments and innovations are priority and growing fast to provide the humanitarian assistance. Recent ICT innovations focus on not only the response and recovery but also the prevention and preparedness in the case of humanitarian crisis. As a result, ICT innovations alter the way in which humanitarian assistance is organized and executed. Therefore, future research in the field of humanitarian ICTs should provide more focus on the cyber security, privacy and ethical use of data; how the new ICT solutions could be usable and accessible considering the profiles (technological capabilities) of the end users during the crisis periods; how the ICT systems could be used in an affected area where a limited technological infrastructure is available; explore the contribution of the recent technological innovations like blockchain, big data, AI, IoT and smart cities to provide the humanitarian services; and ensuring the interoperability among the ICT systems (being used during humanitarian crises) as well as between the organizations. Again, though a number of global initiatives facilitate ICT innovations in humanitarian sectors, including the Humanitarian Innovation Fund, UNICEF Innovation Fund, Deloitte Humanitarian Innovation Program, OCHA Humanitarian Research and Innovation Grant, Global Innovation Fund and the Global Alliance for Humanitarian Innovation, the funds allocated to conduct research and development (R&D) in humanitarian sector are inadequate compared to other sectors.*

[1]Department of Computer Science and Engineering, Military Institute of Science and Technology, Dhaka, Bangladesh

[*]Technological innovation for humanitarian aid and assistance; available at: www.europarl.europa.eu/RegData/etudes/STUD/2019/634411/EPRS_STU(2019)634411_EN.pdf.

This book provides new developments, innovations and research outcomes; case studies and lessons learned; and other considerations for the creation and deployment of effective ICTs to provide humanitarian services for the resource-constrained and vulnerable populations in the world in order to improve their lives. In other words, this book highlights the potential of ICT innovations in humanitarian assistance and makes it clear that the use of ICTs for humanitarian services during the humanitarian crises is a worthwhile area for the future potential R&D.

Index